AF368539

Applied Biomechanics: Sport Performance and Injury Prevention

Applied Biomechanics: Sport Performance and Injury Prevention

Editors

Enrique Navarro
Archit Navandar
Santiago Veiga
Alejandro San Juan Ferrer

MDPI • Basel • Beijing • Wuhan • Barcelona • Belgrade • Manchester • Tokyo • Cluj • Tianjin

Editors

Enrique Navarro
Universidad Politécnica de Madrid
Spain

Archit Navandar
Universidad Europea de Madrid
Spain

Santiago Veiga
Universidad Politécnica de Madrid
Spain

Alejandro San Juan Ferrer
Universidad Politécnica de Madrid
Spain

Editorial Office
MDPI
St. Alban-Anlage 66
4052 Basel, Switzerland

This is a reprint of articles from the Special Issue published online in the open access journal *Applied Sciences* (ISSN 2076-3417) (available at: https://www.mdpi.com/journal/applsci/special_issues/biomechanics_sport_injury).

For citation purposes, cite each article independently as indicated on the article page online and as indicated below:

LastName, A.A.; LastName, B.B.; LastName, C.C. Article Title. *Journal Name* **Year**, *Volume Number*, Page Range.

ISBN 978-3-0365-2608-9 (Hbk)
ISBN 978-3-0365-2609-6 (PDF)

Contents

About the Editors

Enrique Navarro was born in Madrid on 30 May 1958. He studied Physics at the Autonomous University of Madrid, and Sports and Exercise Sciences at the Polytechnic University of Madrid. He has been teaching biomechanics at the Faculty of Physical Activity and Sports since 1984. He is currently director of the biomechanics laboratory of the Polytechnic University, with expertise on the use of Automatic Video Motion Capture System (Vicon), dynamometric platforms and surface electromyography. He has extensive experience in software development, having developed a 2D–3D video photogrammetry system. He has published 65 papers (https://www.scopus.com/authid/detail.uri?authorId=24449528700).

Archit Navandar is an Assistant Professor of Sports Biomechanics at the European University of Madrid. His research interests include technique analysis and injury prevention and rehabilitation in different sports at the elite and youth levels. Principally, he has carried out his investigations in soccer, specifically in kicking and hamstring injury prevention and rehabilitation, but has also published a variety of papers in the area of performance analysis applied to soccer, swimming, and basketball. He completed his PhD in sports science from the Polytechnic University of Madrid, Spain in 2016, and prior to that he graduated in Mechanical Engineering at the National Institute of Technology Karnataka (India). He also has a post-graduate degree in sports science from the Polytechnic University of Madrid.

Santiago Veiga was born in Lugo (Spain) in 1982 and is presently living in Madrid (Spain). His academic background includes a degree in Sport Sciences (2005) from the University La Coruña (Spain) (where he was awarded for the Outstanding Graduate of the Year 2005) and a PhD (Hons.) on Sports Performance (2010) for the University Castilla-La Mancha (Spain). Having been a competitive swimmer at the national level until 2002, he was the head coach of Youth Spanish Swimming Team (2017–2021) representing Spain in many international level events. Concurrent with his professional background in swimming coaching, he developed an academic career as a part-time Associate Professor at the Technical University of Madrid from 2011, in the areas of sports biomechanics and skill acquisition. His main research focus has been the biomechanical and performance analysis of competitive swimming, and he has produced more than twenty publications as the main reference author in journals included in JCR.

Alejandro San Juan Ferrer is Associate Professor in the Department of Health and Human Performance at the Polytechnic University of Madrid (Spain), where he lectures on biomechanics and functional recovery. During the last 20 years, he has been working in the field of clinical exercise physiology and biomechanics, devoted to developing new strategies to recover health and quality of life in patients with musculoskeletal injuries or chronic diseases. After he obtained a bachelor's degree in Sport Sciences (2002, European University of Madrid-UEM), he started to investigate the benefits of exercise programs on cancer patients, and this was the topic of his Ph.D. in clinical exercise physiology (2006, UEM). He obtained also an MSs in High Sports Performance (2004, Spanish Olympic Committee), and a bachelor's degree in Physiotherapy (2008, UEM). He was Associate Professor at the Public University of Navarra (2012–2017), where he lectures on physiotherapy in sports. Since 2017, he has been a member of the "Research Group in Biomechanical Analysis", directed by Dr. Navarro. He is working to expand knowledge on evidence-based prevention and recovery of musculoskeletal injuries as well as chronic diseases through exercise programs.

Preface to "Applied Biomechanics: Sport Performance and Injury Prevention"

Dear Colleagues,

It is our pleasure to present this Special Issue entitled "Applied Biomechanics: Sport Performance and Injury Prevention". There are many motivations that encourage us to launch this proposal, which can be summarized in two topics: the social and economic impact of sport, and the role of biomechanics in sport training.

The economic impact of professional sport is obvious, but we must also consider the importance of recreational sport. The number of recreational sports events grows day by day; at present, nonprofessional athletes train regularly, use high-cost equipment, and control their performance using portable sensors. However, while the beneficial effects of sport on health has been widely reported, it is known that the incidence of sport injuries in both professional and amateur athletes is now increasing, producing a concomitant increase of health care costs due mainly to injury rehabilitation treatments.

On the other hand, sports performance should be understood as the search for the best results and the reduction in injury risk. The paradigm of improving performance and preventing injury is currently a fundamental topic in sport sciences. There are many factors that determine human performance, but possibly one of the most important is the mechanical efficacy of movements (i.e., sport technique). In this sense, sports biomechanics as a science that seeks to optimize human movements plays an important role in sports training and injury prevention.

We know that the number of scientific publications on sports sciences is currently very large, and fortunately grows day by day. However, we hope that this Special Issue on sports biomechanics and injury prevention will be received by sport professionals and researchers as a good source of updated information on the biomechanical analysis of motor skills of sports (individual and collective) and its relationships with injuries in both male and female professional/recreational athletes.

Enrique Navarro, Archit Navandar, Santiago Veiga, Alejandro San Juan Ferrer
Editors

Editorial

Applied Biomechanics: Sport Performance and Injury Prevention

Enrique Navarro [1,*], Archit Navandar [2], Santiago Veiga [1] and Alejandro F. San Juan [1]

[1] Department of Health and Human Performance, Faculty of Physical Activity and Sports Sciences-INEF, Universidad Politécnica de Madrid, 28040 Madrid, Spain; santiago.veiga@upm.es (S.V.); alejandro.sanjuan@upm.es (A.F.S.J.)

[2] Faculty of Sports Sciences, Universidad Europea de Madrid, 28670 Madrid, Spain; archit.navandar@universidadeuropea.es

* Correspondence: enrique.navarro@upm.es

Citation: Navarro, E.; Navandar, A.; Veiga, S.; San Juan, A.F. Applied Biomechanics: Sport Performance and Injury Prevention. *Appl. Sci.* **2021**, *11*, 4230. https://doi.org/10.3390/app11094230

Received: 28 April 2021
Accepted: 5 May 2021
Published: 7 May 2021

Publisher's Note: MDPI stays neutral with regard to jurisdictional claims in published maps and institutional affiliations.

This Special Issue had, as its main objective, the compilation of studies on sports performance and its relationship with musculoskeletal injuries. It is a collection of research on eight different sports (soccer, volleyball, swimming, cycling, skiing, golf, athletics, and hockey) considering injuries in general and specific injuries such as hamstring muscle injury, anterior cruciate ligament of the knee, and pain of the pubic symphysis. Additionally, it is noteworthy that most of the studies considered both men and women. Classical biomechanical tools have been used such as 2D and 3D motion analysis, force platforms, and electromyography.

Four studies have focused on soccer, where first the authors addressed differences in the response of muscle activation recorded with surface electromyography between male and female players and its relationship with a hamstring injury [1]. Core stability has also been studied by proposing a test to assess it using professional soccer players [2]. There is an interesting study on a jumping side-volley kicking technique in soccer using a 3D capture system [3]. Finally, we would like to highlight a study on the evaluation of vertical jumps in female soccer players considering fatigue and analyzing the importance of using force platforms instead of contact platforms that only record flight time [4].

Interestingly, we have compiled three studies on volleyball that have focused on the analysis of jumps in relation to anterior cruciate injury when players land on one leg; the authors have used inverse dynamic analysis to conclude that a landing technique based on increased hip, knee, and ankle flexion may decrease the risk of this injury [5]. The authors have also addressed the importance of improving postural balance in female volleyball players as a means of reducing the risk of injury [6]. Finally, we would like to highlight the work on the effect of plyometric training on muscle activation measured with surface electromyography during jumping for the block [7].

The rest of the articles have dealt with different sports. First, we want to present an interesting study [8] about the golf swing technique based on analysis of the plane of motion using a 3D motion capture system; this research was tutored by Dr Young-Hoo Kwon. Notably is the contribution on the effect of the underwater phase in swim starts using 2D photogrammetry [9] with high-level swimmers. The influence of posture on aerodynamic forces in cycling has also been studied to develop a tool to predict forces based on the anthropometry of the subject [10], which may be of great importance for trainers and athletes of this sport. It is always interesting to find information about a sport such as skiing; in this case, the published article carried out a prospective study over four years to find risk factors for musculoskeletal injuries [11]. Again, there is a study that uses 2D movement analysis applied to the 60-meter hurdles in athletics, analyzing runners during an indoor mute championship [12]. Finally, we highlight a study on female rugby players, relating the acute effect of match-play to the strength of the player [13] and research on a

1

group of subjects from different sports on the hamstring/quadriceps torque ratio, pelvic posture, and its relationship with symphysis pubis pain [14].

Funding: This research received no external funding.

Conflicts of Interest: The authors declare no conflict of interest.

References

1. Torres, G.; Armada-Cortés, E.; Rueda, J.; Juan, A.F.S.; Navarro, E. Comparison of Hamstrings and Quadriceps Muscle Activation in Male and Female Professional Soccer Players. *Appl. Sci.* **2021**, *11*, 738. [CrossRef]
2. Etxaleku, S.; Izquierdo, M.; Bikandi, E.; Arroyo, J.G.; Sarriegi, I.; Sesma, I.; Setuain, I. Validation and Application of Two New Core Stability Tests in Professional Football. *Appl. Sci.* **2020**, *10*, 5495. [CrossRef]
3. Zhang, X.; Shan, G.; Liu, F.; Yu, Y. Jumping Side Volley in Soccer—A Biomechanical Preliminary Study on the Flying Kick and Its Coaching Know-How for Practitioners. *Appl. Sci.* **2020**, *10*, 4785. [CrossRef]
4. Armada-Cortés, E.; Barrajón, J.P.; Benítez-Muñoz, J.A.; Navarro, E.; Juan, A.F.S. Can We Rely on Flight Time to Measure Jumping Performance or Neuromuscular Fatigue-Overload in Professional Female Soccer Players? *Appl. Sci.* **2020**, *10*, 4424. [CrossRef]
5. Xu, D.; Jiang, X.; Cen, X.; Baker, J.S.; Gu, Y. Single-Leg Landings Following a Volleyball Spike May Increase the Risk of Anterior Cruciate Ligament Injury More Than Landing on Both-Legs. *Appl. Sci.* **2020**, *11*, 130. [CrossRef]
6. Fuchs, P.X.; Fusco, A.; Cortis, C.; Wagner, H. Effects of Differential Jump Training on Balance Performance in Female Volleyball Players. *Appl. Sci.* **2020**, *10*, 5921. [CrossRef]
7. Wang, M.-H.; Chen, K.-C.; Hung, M.-H.; Chang, C.-Y.; Ho, C.-S.; Chang, C.-H.; Lin, K.-C. Effects of Plyometric Training on Surface Electromyographic Activity and Performance during Blocking Jumps in College Division I Men's Volleyball Athletes. *Appl. Sci.* **2020**, *10*, 4535. [CrossRef]
8. Madrid, M.V.; Avalos, M.A.; Levine, N.A.; Tuttle, N.J.; Becker, K.A.; Kwon, Y.-H. Association between the On-Plane Angular Motions of the Axle-Chain System and Clubhead Speed in Skilled Male Golfers. *Appl. Sci.* **2020**, *10*, 5728. [CrossRef]
9. Stosic, J.; Veiga, S.; Trinidad, A.; Navarro, E. How Should the Transition from Underwater to Surface Swimming Be Performed by Competitive Swimmers? *Appl. Sci.* **2020**, *11*, 122. [CrossRef]
10. Garimella, R.; Peeters, T.; Parrilla, E.; Uriel, J.; Sels, S.; Huysmans, T.; Verwulgen, S. Estimating Cycling Aerodynamic Performance Using Anthropometric Measures. *Appl. Sci.* **2020**, *10*, 8635. [CrossRef]
11. Steidl-Müller, L.; Hildebrandt, C.; Niedermeier, M.; Müller, E.; Romann, M.; Javet, M.; Bruhin, B.; Raschner, C. Biological Maturity Status, Anthropometric Percentiles, and Core Flexion to Extension Strength Ratio as Possible Traumatic and Overuse Injury Risk Factors in Youth Alpine Ski Racers: A Four-Year Prospective Study. *Appl. Sci.* **2020**, *10*, 7623. [CrossRef]
12. González-Frutos, P.; Veiga, S.; Mallo, J.; Navarro, E. Evolution of the Hurdle-Unit Kinematic Parameters in the 60 m Indoor Hurdle Race. *Appl. Sci.* **2020**, *10*, 7807. [CrossRef]
13. Sánchez-Migallón, V.; López-Samanes, A.; Terrón-Manrique, P.; Morencos, E.; Fernández-Ruiz, V.; Navandar, A.; Moreno-Pérez, V. The Acute Effect of Match-Play on Hip Isometric Strength and Flexibility in Female Field Hockey Players. *Appl. Sci.* **2020**, *10*, 4900. [CrossRef]
14. Ludwig, O.; Kelm, J.; Hopp, S. Impact of Quadriceps/Hamstrings Torque Ratio on Three-Dimensional Pelvic Posture and Clinical Pubic Symphysis Pain-Preliminary Results in Healthy Young Male Athletes. *Appl. Sci.* **2020**, *10*, 5215. [CrossRef]

applied sciences

MDPI

Article

Comparison of Hamstrings and Quadriceps Muscle Activation in Male and Female Professional Soccer Players

Gonzalo Torres [†], Estrella Armada-Cortés [†], Javier Rueda, Alejandro F. San Juan [‡] and Enrique Navarro [*,‡]

Health and Human Performance Department, Sport Biomechanics Laboratory, Faculty of Physical Activity and Sports Sciences-INEF, Universidad Politécnica de Madrid, 28040 Madrid, Spain; gonzalo.torresmar@gmail.com (G.T.); cortesarmadaestrella@gmail.com (E.A.-C.); javier.rueda7792@gmail.com (J.R.); alejandro.sanjuan@upm.es (A.F.S.J.)
* Correspondence: enrique.navarro@upm.es
† First two authors contributed equally to this work.
‡ E. Navarro and A. F. San Juan share senior authorship.

Featured Application: Surface electromyography is a valid and useful tool to assess quadriceps and hamstring muscle activation in professional male and female soccer teams. This methodology can be used in hamstring and knee injury prevention and rehabilitation programs.

Abstract: (1) Background: this study aimed to determine if there are differences in quadriceps and hamstring muscle activation in professional male and female soccer players. (2) Methods: muscle activation was recorded by surface electromyography in 27 professional soccer players (19 male and 8 female). The players performed the Bulgarian squat and lunge exercises. Vastus medialis, vastus lateralis, rectus femoris, semitendinosus, and biceps femoris were the muscles analyzed. (3) Results: The statistical analysis of the hamstring:quadriceps ratio showed no significant differences ($p > 0.05$). Significant differences were found in the vastus medialis:vastus lateralis ratio for both the lunge exercise ($t_{20} = 3.35$; $p = 0.001$; d = 1.42) and the Bulgarian squat ($t_{23} = 4.15$; $p < 0.001$; d = 1.76). For the intragroup muscular pattern in the lunge and Bulgarian squat exercises, the female players showed higher activation for the vastus lateralis muscle ($p < 0.001$) than the male players and lower muscle activation in the vastus medialis. No significant differences were found in the rectus femoris, biceps remoris, and semitendinosus muscles ($p > 0.05$). (4) Conclusions: Differences were found in the medial ratio (vastus medialis: vastus lateralis). Moreover, regarding the intramuscular pattern, very consistent patterns have been found. In the quadriceps muscle: VM>VL>RF; in the hamstring muscle: ST>BF. These patterns could be very useful in the recovery process from an injury to return players to their highest performance.

Keywords: electromyography; ratio; prevention; injury

Citation: Torres, G.; Armada-Cortés, E.; Rueda, J.; San Juan, A.F.; Navarro, E. Comparison of Hamstrings and Quadriceps Muscle Activation in Male and Female Professional Soccer Players. *Appl. Sci.* **2021**, *11*, 738. https://doi.org/10.3390/app11020738

Received: 18 December 2020
Accepted: 11 January 2021
Published: 14 January 2021

Publisher's Note: MDPI stays neutral with regard to jurisdictional claims in published maps and institutional affiliations.

1. Introduction

There have been numerous epidemiological studies in soccer for estimating that hamstring strain injuries (HSI) in soccer represent 10–12% of all injuries [1], which indicates five to six injuries per team per season [1]. The cost per injury is estimated to €6355 [2], of which 46% of the players are from relapse [3].

In turn, hamstring breaks have been significantly associated with a low hamstring: quadriceps (H:Q) ratio [4,5]. In soccer, quadriceps injury represents 19% of total injuries and approximately 16% of hamstring injuries [6]. To date, the main risk factors reported for HSI in soccer are previous injuries [7–9], muscle power imbalance [5,10], neuromuscular disorders [11], and fatigue [12–14]. Focusing on soccer, male soccer players get 1.9 times more hamstring injuries than female soccer players, with a 12% absence rate in the season compared to an absence of 6% for female counterparts [15]. Moreover, female soccer players

Appl. Sci. **2021**, *11*, 738. https://doi.org/10.3390/app11020738

have two times more quadriceps injuries than male soccer players, with an 8% absence rate compared to 3% for males [8].

Another important injury in soccer is the anterior cruciate ligament (ACL), which affects 2–9 times more females than males [16,17]. If we focus on knee injury and specifically ACL injury, noncontact ACL injury in athletes has a multifactorial etiology [18]. The factors reported as important are hamstring muscle weakness, excessive quadriceps strength, medial and lateral imbalances, and age [19,20]. In addition, muscular fatigue may increase the risk of the noncontact ACL [18]. In addition, Wojtys and Huston [21] reported that female athletes have a slower response of hamstring activation to anterior stress on the ACL (using anterior tibia translation tests) compared to male athletes in a physical examination. Cowling and Steele [22] reported sex differences in muscle activation strategies of the hamstrings musculature that do not coincide with the findings of [21], who found no significant differences in either segmental alignment or temporal characteristics of the quadriceps muscles shown by males and females on landing. The inter-limb differences in muscle recruitment patterns, muscle strength, and muscle flexibility tend to be greater in females than in males [23].

There are differences between the competition of males and females, and therefore, the incidence of injuries is not the same between the two groups. For males, the most frequent injury is a minor one (65%), while for females the most frequent injury is a moderate one (51%) [12]. Generally speaking, more injuries occur in male's soccer, but it is in female soccer players that the most serious injuries occur [24]. It has been established that the H:Q ratio not only helps to prevent damage to the posterior thigh muscles but also helps to reduce the stress on the ACL of the knee [25]. To help reduce stress on the knee, not only is the H:Q ratio important, but also how each of the quadriceps and the hamstrings muscles are activated, which is called the activation pattern [26].

If attention is paid to the muscle imbalance factor, isokinetic machines have mainly been used to evaluate the relationship between the quadriceps and the hamstrings [27]. These machines measure the maximum total flexor and extensor strength of the knee but do not take into account the participation of each muscle group separately [28]. Another disadvantage of isokinetic machines is that they produce a lot of fatigue in the athlete, which makes it difficult to use them during moments of the season with a lot of intensity of training and official matches [29]. Another tool that has been used to evaluate the muscle activation between the quadriceps and the hamstrings is surface electromyography (sEMG) has also been used as a tool to evaluate muscle activation between the quadriceps and the hamstrings [30–32]. In sEMG studies, basic movements (e.g., forward lunge, Bulgarian squat, and lateral step-ups) are often used to make them easily reproducible [33,34], and these exercises do not cause excessive fatigue, allowing soccer teams to use them at any time during the season. These are simple exercises used to strengthen the hamstring and quadriceps muscles, and their study can provide information on the activation of the different muscles of these muscle groups and see the differences in activation in the phases of exercise [30]. These exercises can be decisive for soccer players, taking into account the muscle group they are targeting, since recording the electrical activity of each of the muscular bellies would help reduce the time in the process of recovery from an injury and also help in its prevention [30].

Exercises with higher H:Q activation ratios may be preferred during early rehabilitation after injury [35]. Previous studies have suggested that the H:Q force ratio should be at least 0.6 to prevent injuries to both the hamstrings and the knee [36]; however, this has been studied using isokinetic machines rather than sEMG. With the use of sEMG, some studies have been carried out with soccer players, most of them males, reporting H:Q ratios from 0.21 to 0.81 [31,32,37] and few studies have compared the ratio of males to females and with a non-soccer specific sample [37,38].

Therefore, the main objective of this study was to determine if there are differences in muscle activation patterns between female and male soccer players for better understanding and application. The following hypotheses were formulated: (1) the H:Q ratio will be different between male and female soccer players; (2) the vastus medialis (VM):vastus lateralis (VL) ratio will be different between male and female soccer players; and (3) there would be a different pattern of intra-hamstrings and intra-quadriceps muscle group coactivation between male and female soccer players.

2. Materials and Methods

2.1. Ethical Considerations

All athletes signed the informed consent form before the study. The study followed the guidelines of the Declaration of Helsinki and was approved by the Ethics Committee of the Universidad Politécnica de Madrid (Spain).

2.2. Participants

Twenty-seven (19 male and 8 female) professional soccer players conducted the study. The male group was made up of players from the Atletico de Madrid youth soccer team. The team was composed of 24 players at the time of data collection, but 5 of them could not perform the test because they were injured and did not have the authorization of the medical staff. The female group was made up of players of the First and Second Spanish Soccer Divisions (Liga Iberdrola and Reto Iberbrola, respectively). All players had medical clearance to conduct the study and were completely healthy and uninjured at the time of data collection. The male soccer team completed the test in the 2015–2016 season. On the other hand, the data collection with female athletes was carried out in the 2018–2019 season. The players who participated in the study were both regular starters and substitutes. However, all players had the same training load, which was 5 days a week plus a regular league match. Both male and female football players had strength training throughout the season, although its frequency varied depending on the time of the season and physical condition to regulate their load control. Inclusion criteria were shown as follows: (1) having medical clearance to conduct the study; (2) not having suffered a musculoskeletal injury one year prior to the date of the protocol (i.e., checked through a previous exclusion questionnaire); (3) presenting neither any cardiovascular, musculoskeletal, and/or neurological disease nor previous ones that could affect participation in the study.

2.3. Study Design

A descriptive study was conducted in the sports biomechanics laboratory, where players had to perform two exercises while quadriceps and hamstring muscle activation was recorded.

Lunge: The starting position was one leg forward and the other leg backward. The knee and the hip were at 90 degrees, both in the front and back legs. The Lafayette Gollehon (Lafayette Instrument Company; Lafayette, IN, USA) goniometer was used in our study to determine degrees. Before starting the exercise, the starting position was established and the distance between each player's feet was measured and marked on the ground. From the starting position, the athletes had to bend the forward leg while keeping their backward leg straight at all times. It is important to note that in our study, we performed the lunge exercise, not the front lunge.

Bulgarian squat: The starting position was one leg forward and the other leg backward placed on a raised surface. As in the lunge exercise, the Lafayette Gollehon goniometer (Lafayette Instrument Company; Lafayette, IN, USA) was used to determine the 90 degrees of the knee and hip. From the initial position, the athletes had to bend the forward leg while keeping their backward leg straight at all times.

2.4. Procedure

The athletes, once they arrived at the laboratory, were measured and weighed only in their underwear. Later, always guided by the club's physical trainer, the players performed a warm-up consisting of a continuous run, followed by joint mobility and core work. The warm-up consisted of a continuous run for 7 min, followed by joint mobility exercises for

3 min and light dynamic stretching. To warm up the central area of the body, the players performed core planks (3 repetitions) for 30 s with 15 s breaks between each plank. After warming up, sEMG sensors were placed on both thighs, on the rectus femoris (RF), vastus lateralis (VL), vastus medialis (VM), semitendinosus (ST), and biceps femoris (BF). The areas where the sensors were going to be placed were shaved and cleaned with alcohol. The physical trainer was responsible for always placing the sensors on the athletes. For the placement of the sensors, the guidelines of the Seniam protocol were followed [39]. sEMG data were recorded using Trigno^{TM} Wireless System (Delsys, Inc., Boston, MA, USA). Data were captured at 1500 Hz.

The players had to repeat the exercise 5 times (Figure 1) with a rhythm of execution of each of the phases (descent, isometric, and ascent phase) of 2 s. The exercises were performed with an external load of 30% of each player's weight.

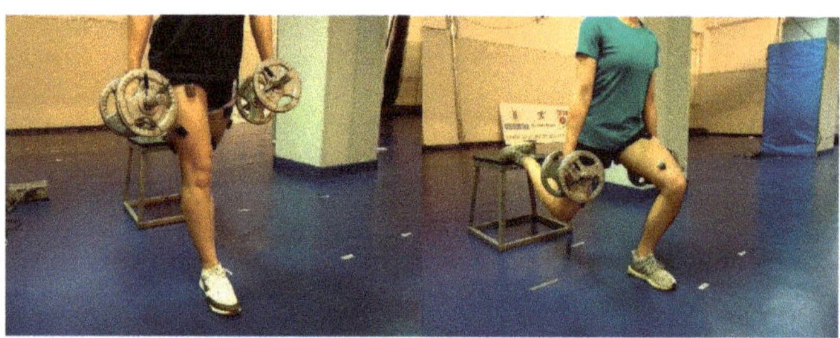

Figure 1. Starting position (**left**) and execution (**right**) of the Bulgarian squat exercise.

2.5. Data Processing

The processing of the data was carried out with EMGWORKS® software (Version 4, Delsys, Inc., Boston, MA, USA). Signal filtering was the first step in data processing using a 2nd-order bandpass Butterworth filter [40] with an attenuation of 40 dB and a cutoff frequency between 20 and 300 Hz [41]. A root mean square (RMS) [42] with a window width of 0.05 s and a window overlap of 0.025 s was later applied to the filtered signal, and the signal offset was removed.

2.6. Statistical Analysis

Dependent variables were calculated following the procedures based on the normalization without the maximum voluntary isometric contraction (MVIC) [32]. A total of seven dependent variables were compared between male and female soccer players in both exercises, H:Q and VM:VL ratios and intragroup muscular ratio. The H:Q ratio was calculated by dividing the mean activity (RMS) of the hamstring muscles by the mean activity of the quadriceps muscles measured in this study. The VM:VL ratio was calculated by dividing the mean activity (RMS) of the VM by the mean activity of the VL. The intragroup muscular pattern expressed the activation of each muscle with respect to the total surface activity of the muscle group. To calculate the intragroup pattern, the activation of each muscle (RMS) was normalized by dividing it by the total activity of the muscle group written as: %RF = RF/RF + VM + VL × 100. The muscle activity selected to calculate the ratio was the mean of the concentric, isometric, and eccentric phases of the 3 central repetitions, following procedures similar to other authors [33].

To compare the differences between the male and female soccer players in H:Q ratios, VM:VL ratios, and intramuscular activations in both exercises, *t*-tests for independent measures were performed for each variable.

The SPSS software 23.0 (Armonk, NY, USA: IBM Corp) was used to perform the statistical analyses. The significance level was set at 0.05, and effect sizes were determined using Cohen's D [43]. Microsoft Excel (Version 2019, Microsoft Corporation, Redmond,

WA, USA) was used to calculate Cohen's D using the means and standard deviation of the samples.

3. Results

The male group was composed of 19 players (age = 19.2 ± 0.5 years, height = 179.7 ± 5.3 cm, weight = 71.0 ± 5.9 kg), and the female group had 8 players (age = 27.3 ± 6.5 years, height = 161.0 ± 0.6 cm, weight = 56.7 ± 4.9 kg). Differences were found between groups in age ($t_{25} = 5.55$; $p < 0.001$; d = 2.33), weight ($t_{25} = 6.02$; $p < 0.001$; d = 2.54), and height ($t_{25} = 9.84$; $p < 0.001$; d = 4.15).

The statistical analysis of the H:Q ratio showed no significant differences (Table 1) between the female and male soccer players in both the lunge and Bulgarian squat exercises ($p > 0.05$).

Table 1. Hamstring: quadriceps (H:Q) ratio (mean $\pm$ SD).

	Female Players		Male Players		Significance	Effect Size (d)
	Mean	SD	Mean	SD		
H:Q ratio in the lunge exercise	0.25	0.18	0.18	0.10	$p > 0.05$	-
H:Q ratio in the Bulgarian squat exercise	0.24	0.16	0.18	0.06	$p > 0.05$	-

Significant differences were found in the VM:VL ratio (Table 2) for both the lunge exercise ($t_{20} = 3.35$; $p = 0.001$; d = 1.42) and the Bulgarian squat ($t_{23} = 4.15$; $p < 0.001$; d = 1.76).

Table 2. VM:VL ratio (mean $\pm$ SD).

	Female Players		Male Players		Significance	Effect Size (d)
	Mean	SD	Mean	SD		
VM:VL ratio in the lunge exercise	1.12	0.36	2.64	1.88	$p = 0.001$	1.42
VM:VL ratio in the Bulgarian squat exercise	1.10	1.88	2.04	0.72	$p < 0.001$	1.76

For the intragroup muscular pattern in the lunge exercise (Figure 2), the female group showed higher activation for the VL muscle ($t_{23} = 4.1$; $p < 0.001$; d = 1.75) than the male group and lower muscle activation in the VM (Table 3) ($t_{23} = -3.8$; $p = 0.001$; d = 1.62) compared to the male group. No significant differences were found in the RF, BF, and ST muscles ($p > 0.05$). Similarly, in the Bulgarian squat exercise (Table 3 and Figure 3), the females also showed higher VL activation ($t_{23} = 3$; $p = 0.006$; d = 1.29) and lower muscle activation in the VM ($t_{23} = -3.9$; $p = 0.001$; d = 1.67). In the same way, no significant differences were found in the RF, BF, and ST muscles ($p > 0.05$).

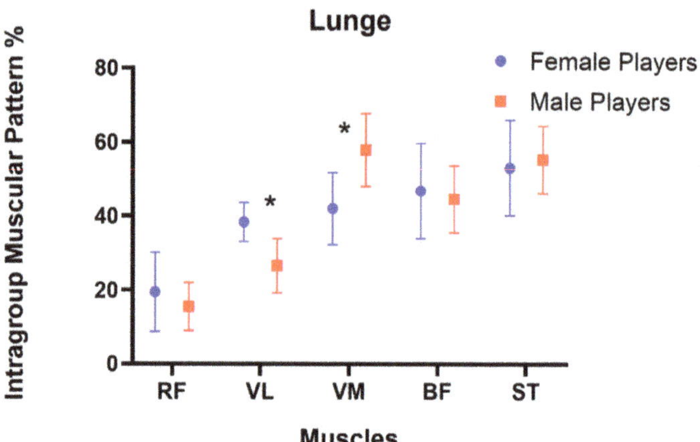

Figure 2. Intragroup muscular pattern (mean ± SD) in percentage. The intragroup muscular pattern expresses the activation of each muscle belly in relation to the total muscle (quadriceps and hamstrings). The asterisk (*) shows that there is a significant difference.

Table 3. Intragroup muscular pattern in percentage.

Exercise	Muscle	Female Players		Male Players		Significance	Effect Size (d)
		Mean (%)	SD (%)	Mean (%)	SD (%)		
	RF	19.52	10.69	15.54	6.47	$p > 0.05$	-
	VL	38.42	5.29	26.53	7.33	$p < 0.001$	1.75
Lunge	VM	42.07	9.75	57.93	9.82	$p = 0.001$	1.62
	BF	46.87	12.92	44.64	9.08	$p > 0.05$	-
	ST	53.13	12.92	55.36	9.08	$p > 0.05$	-
	RF	21.70	10.16	15.90	6.32	$p > 0.05$	-
	VL	36.99	5.57	28.87	6.57	$p = 0.006$	1.29
Bulgarian squat	VM	41.31	8.91	55.23	8.08	$p = 0.001$	1.67
	BF	49.92	1.57	44.64	9.08	$p > 0.05$	-
	ST	50.08	13.57	55.36	9.08	$p > 0.05$	-

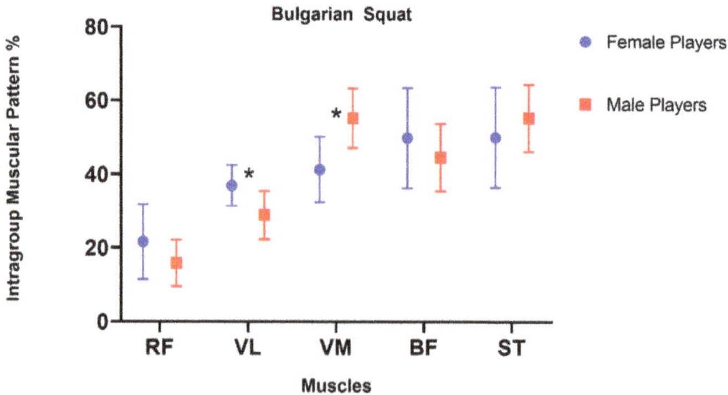

Figure 3. Intragroup muscular pattern (mean ± SD) in percentage. The intragroup muscular pattern expresses the activation of each muscle belly in relation to the total muscle (quadriceps and hamstrings). The asterisk (*) shows that there is a significant difference.

To calculate the intragroup pattern, the activation of each muscle (RMS) was normalized by dividing it by the total activity of the muscle group.

4. Discussion

In the present work, an analysis of the muscular activity in different exercises used in both prevention and rehabilitation was carried out with the aim of finding differences between the male and female professional soccer players. The main objective of this study was to determine if there were differences in muscle activation patterns between female and male soccer players for better understanding and application; however, no significant differences were found. Significant differences in the medial ratio (VM:VL ratio) between the two groups were found for the two exercises. Significant differences were found between both groups for VM and VL activations.

4.1. H:Q Ratio

The results showed no significant differences (Table 1) between the female and male soccer players' H:Q ratios in the lunge and Bulgarian squat exercises. This is perhaps due to the good training and prevention of this muscle in both sexes. Similar studies have been carried out with a team of only males and similar exercises [31]. They carried out the same exercises to professional players and obtained similar results to ours; in the case of those authors, the Bulgarian squat promoted a higher H:Q ratio. Navarro, Chorro, Torres, Navandar, Rueda and Veiga [31] used another type of signal normalization (MVIC) so a possible comparison of results must be done with caution. In our case, the ratios (lunge: female group 0.25; male group 0.18. Bulgarian squat: female group 0.24; male group: 0.18) were slightly lower than those reported by them. Authors, such as McCurdy, et al. [44], set the ideal H:Q ratio in sEMG at 1.67 (unnormalized results). Likewise and more specifically, they concluded that to consider a greater strength of the hamstring with respect to the quadriceps and its decompensation, in females, it should be equal to or greater than 0.71, where our players, neither males nor females, presented such a strength. More recent studies [45] determine the optimum H:Q ratio between 55% and 64% (0.55 and 0.64), and an alteration in the H:Q ratio when there is a difference equal to or greater than 10% between these muscle groups. However, Ruas, Pinto, Haff, Lima, Pinto and Brown [36] reported an H:Q ratio of 0.6 with a range of 0.45–0.59. The results in both studies were too high for our sample. Even so, Ruas, Pinto, Haff, Lima, Pinto and Brown [36] provided insufficient scientific evidence and standardization in this respect, and these ratios mentioned above are not reliable to be used since the methodology in the different studies varied considerably, leading to significantly different results. In addition to the controversy over the ideal H:Q ratio, there is also a controversy over the standardization of electromyographic data [46,47]. The most commonly used signal normalization method is the MVIC [48]. Some studies did not normalize [44] or did so by expressing themselves as a percentage contribution to the total electrical activity of all the muscles tested [30]. Other studies used the highest integrated EMG value among the concentric and eccentric contractions of all exercises, noting that the normalized peak EMG data follow a trend similar to the integrated data [49]. Other authors have used the total activity of each of the muscles (total activity of the quadriceps and the hamstrings) to normalize the signal from each muscular belly [30]. This type of normalization gives us information on how the total workload is distributed on the synergies that occur in the muscle. Another advantage is that this type of normalization can be used at any time during the season. In our case, we considered that not using MVIC has benefits when the sample is elite-professional or when there are injured players.

Studies, such as that of El-Ashker, et al. [50], with isokinetic and sEMG machines, concluded that the functional H:Q ratio was significantly lower in female groups compared to male groups, regardless of the velocity of momentum and the angle of the joint. [51] reported, in the analysis by gender, female soccer players produced lower peak torque H:Q ratios than males involved in the same sport at low speeds. However, at high speeds, there was no difference between the sexes. Different from most of the results presented so far by different authors, Kong and Burns [52] concluded that neither isometric nor isokinetic H:Q ratios differed between males and females, and they pointed out that other characteristics of the subject, such as age and training, may be more relevant.

On the other hand, Gobbi, et al. [53] compared the isokinetic strength between males and females operated on the ACL with two different surgical techniques (i.e., ST and gracilis tendons, or bone-patellar tendon-bone). The authors found that those females who obtained a goosefoot graft (i.e., ST and gracilis tendons) showed worse results in terms of isokinetic forces of knee flexors and extenders than males. No significant differences were found, when the patellar tendon was used as a graft. There are some results presented by various authors that differ from those of the present study; however, it is necessary to point out that the methodology used in all of them is through an isokinetic evaluation.

Electromyographical studies have demonstrated that females may have sex-related differences in the muscle onset time during athletic movement [21,54]. Furthermore, unbalanced medial-to-lateral muscle activations have been associated with increased knee valgus in the frontal plane [54].

4.2. VM:VL Ratio

Studies with female soccer players point out that disproportionate increases in activation of the VL may also result in a low quadriceps medial-to-lateral ratio and an increase in the anterior shear force and the load on the ACL [55]. Considering that the ideal VM:VL ratio is higher than one, our results (Table 2) showed a significantly better VM:VL ratio in male soccer players than in female soccer players. This suggests that the female in our study may be more predisposed to ACL injury.

Deficient motor control, meaning the distribution of electrical activity within the muscles between each of the muscular bellies and also the moment of activation, is problematic, because it affects the load carried by the joints [56]. The activation balance between the medial and lateral parts of the quadriceps is also very important to prevent patellofemoral pain syndrome (PFPS). The stability of the knee is maintained by the dynamic balance between the two muscles [57]. A greater predominance of one of the muscles means that the patella is not aligned and does not make a correct movement. Most problems in the patella arise, when the VM is unable to counteract the activation of the VL. In our case, on the basis of the results, females may be at greater risk of suffering PFPS than males in line with other authors [58]. Although a VM:VL ratio greater than one could be beneficial in reducing joint problems, most studies have reported values slightly below or slightly above one, in both females and males; however, the sample used had a very low sporting level [59,60].

4.3. Intragroup Muscular Pattern

The same intramuscular pattern was repeated in both the lunge and Bulgarian squat. The muscle with the highest weight within the total activation of the quadriceps was the VM, followed by the VL and finally the RF. Similarly, the pattern was repeated in the hamstrings for both exercises, with a higher percentage for ST than for BF. The results are in line with those obtained by other researchers for both the lunge [32,34] and the Bulgarian squat [30,31].

Activation patterns have been used in soccer as a tool in the process of recovery from an injury. One of the ways that were used was to compare the injured leg with the uninjured one, and it was established that when the differences were less than 10–15%, the subject had already recovered from the injury [61,62]. In addition, the activation pattern can be used after an injury as a way to quantify when the player has regained his initial form, so it is important that EMG tests are performed at different times during the season. This can be applied in the case of recurrent pain at the patellar tendon level, where the activation of the vast medialis is the main goal in the early stages of recovery [63].

Other studies have been carried out with female athletes having more analytical exercises, such as quadriceps extension. Those studies give percentages similar to ours (RF: 20–30%; VM: 40–50%; BF: 45–50%; ST: 50–60%), with the only difference being that the VL played a greater role in the total activation when compared to our results (VL: 40–45%) [42]. Other authors, however, point out that it is not correct to establish a relationship between electrical activity and force production in biarticular muscles [64,65]. In short, their results seem to indicate that there are normal activation patterns in the female soccer population. These intramuscular ratios allow both the knee joint and the hamstring muscles to develop their normal activity.

The results obtained for male players are in line with those obtained by other authors [30,34,65,66], with the activation in the order of the VM > VL > RF patterns. In the hamstrings, the ST was more active than the BF. When we compared the data of females and males, we observed that, although they had similar patterns, significant differences were found in the VM and the VL.

Differences in the activation of VM and VL may mainly affect the knee joint, as these muscles are intensively involved in its proper functioning. These differences in magnitude are related to the different compositions of muscle fibers between females and males and their physiological differences [56]. Evidence shows that females have a higher proportion of slow fibers than males [66], which could lead to a change in the activation pattern with a different electromyographic signal distribution than males. It could be expected that the differences between males and females would be greater than those obtained in the study. One of the explanations for these small differences could be that the exercises and loads proposed were of moderate intensity, and in these ranges of effort, no great differences were shown. The greater differences at higher intensity are due, among other things, to the proportion of fast fibers in males [56]. One of the biggest problems when trying to compare the EMG data between males and females is the type of signal normalization. Generally, the MVIC has been used for signal normalization; however, this type of normalization has some problems. The biggest one is its requirement, as it cannot be used with players who are recently injured or at critical times of the season (e.g., near an important match). Another problem is that females sometimes show higher percentages than males in this type of normalization, because the contribution to this specific slow fiber test is greater in females than in males, resulting in data that are difficult to compare between the sexes. That is why many authors are starting to use the same kind of signal normalization [30,32] as we have done in the present study.

The same differences between the male and female groups were observed in the Bulgarian squat exercise (Figure 3), with differences found in VM and VL. Differences in the activation patterns between VM and VL are important primarily to reduce the risk of knee injury. A change in both the activation pattern and the intensity of the quadriceps contraction could increase the stress supported by the ACL [67]. The greater weight in the total activation of the quadriceps by the VM could be good for the stabilization of the knee [67]. Therefore, in both lunge and Bulgarian squat, the patterns found are ideal in a rehabilitation work plan.

The results of our study are in line with other studies [68], although there are differences in the type of normalization and exercises, as most studies use highly analytical exercises, such as isometric tests or quadriceps extension.

In both males and females, the patterns were repeated in the hamstrings, with a greater weight in the total activation by the ST [31,32,69] The increased activation of the ST relative to that of the BF may help to reduce the stress of the latter and reduce its potential for injury, as most hamstring injuries are caused in the BF [70]. The increased activation of the ST may be due to the fact that in both the lunge and Bulgarian squat exercises, the ST is activated to reduce the external rotation of the tibia. In view of the results, both the Bulgarian squat and lunge can be good exercises to implement for both injury prevention and rehabilitation for injured players.

Appl. Sci. **2021**, *11*, 738

5. Conclusions

The sEMG H:Q ratios in the lunge and Bulgarian squat exercises between professional male and female soccer players did not show differences. However, we observed significant differences between sex in the medial ratio (VM:VL ratio). Moreover, regarding the intramuscular pattern, very consistent patterns have been found with differences in VM and VL between males and females. In this sense, the female group showed higher activation in the VL muscle and lower activation in the VM muscle than the male group, without significant differences in the RF, BF, and ST muscles. These patterns could be very useful in the recovery process from an injury to return players to their highest performance.

To generalize our conclusions, future research with a higher sample size, composed of high-level and recreationally trained athletes from different sport modalities and both sexes, is needed.

Author Contributions: Conceptualization, G.T., E.A.-C., A.F.S.J. and E.N.; methodology, G.T., E.A.-C., A.F.S.J. and E.N.; formal analysis, J.R.; investigation, G.T., E.A.-C., J.R., A.F.S.J. and E.N.; resources, G.T., E.A.-C. and E.N.; data curation, G.T., E.A.-C., J.R., A.F.S.J. and E.N.; writing—original draft preparation, G.T., E.A.-C., J.R., A.F.S.J. and E.N.; writing—review and editing, G.T., E.A.-C., J.R., A.F.S.J. and E.N.; visualization, G.T., E.A.-C., J.R., A.F.S.J. and E.N.; supervision, A.F.S.J. and E.N.; project administration, A.F.S.J. and E.N. All authors have read and agreed to the published version of the manuscript.

Funding: This research received no external funding.

Institutional Review Board Statement: The study was conducted according to the guidelines of the Declaration of Helsinki, and approved by the Ethics Committee of Universidad Politécnica de Madrid (Spain).

Informed Consent Statement: Informed consent was obtained from all subjects involved in the study.

Conflicts of Interest: The authors declare no conflict of interest.

References

1. Ekstrand, J.; Waldén, M.; Hägglund, M. Hamstring injuries have increased by 4% annually in men's professional football, since 2001: A 13-year longitudinal analysis of the UEFA Elite Club injury study. *Br. J. Sports Med.* **2016**, *50*, 731–737. [CrossRef]
2. Nouni-Garcia, R.; Asensio-Garcia, M.R.; Orozco-Beltran, D.; Lopez-Pineda, A.; Gil-Guillen, V.F.; Quesada, J.A.; Bernabeu Casas, R.C.; Carratala-Munuera, C. The FIFA 11 programme reduces the costs associated with ankle and hamstring injuries in amateur Spanish football players: A retrospective cohort study. *Eur. J. Sport Sci.* **2019**, *19*, 1150–1156. [CrossRef]
3. Askling, C.; Karlsson, J.; Thorstensson, A. Hamstring injury occurrence in elite soccer players after preseason strength training with eccentric overload. *Scand. J. Med. Sci. Sports* **2003**, *13*, 244–250. [CrossRef]
4. Brockett, C.; Morgan, D.; Proske, U. Predicting hamstring strain injury in elite athletes. *Med. Sci. Sports Exerc.* **2004**, *36*, 379–387. [CrossRef]
5. Croisier, J.L.; Ganteaume, S.; Binet, J.; Genty, M.; Ferret, J.-M. Strength imbalances and prevention of hamstring injury in professional soccer players: A prospective study. *Am. J. Sports Med.* **2008**, *36*, 1469–1475. [CrossRef] [PubMed]
6. Ekstrand, J.; Hagglund, W. Injury incidence and injury patterns in professional football: The UEFA injury study. *Br. J. Sports Med.* **2011**, *45*, 553–558. [CrossRef] [PubMed]
7. Engebretsen, A.H.; Myklebust, G.; Holme, I.; Engebretsen, L.; Bahr, R. Intrinsic risk factors for hamstring injuries among male soccer players: A prospective cohort study. *Am. J. Sports Med.* **2010**, *38*, 1147–1153. [CrossRef] [PubMed]
8. McCall, A.; Carling, C.; Davison, M.; Nedelec, M.; Le Gall, F.; Berthoin, S.; Dupont, G. Injury risk factors, screening tests and preventative strategies: A systematic review of the evidence that underpins the perceptions and practices of 44 football (soccer) teams from various premier leagues. *Br. J. Sports Med.* **2015**, *49*, 583–589. [CrossRef]
9. Navarro, E.; Chorro, D.; Torres, G.; García, C.; Navandar, A.; Veiga, S. A review of risk factors for hamstring injury in soccer: A biomechanical approach. *Eur. J. Hum. Mov.* **2015**, *34*, 52–74.
10. Fousekis, K.; Tsepis, E.; Poulmedis, P.; Athanasopoulos, S.; Vagenas, G. Intrinsic risk factors of non-contact quadriceps and hamstring strains in soccer: A prospective study of 100 professional players. *Br. J. Sports Med.* **2011**, *45*, 709–714. [CrossRef]
11. Cameron, M.; Adams, R.; Maher, C. Motor control and strength as predictors of hamstring injury in elite players of Australian football. *Phys. Ther. Sport* **2003**, *4*, 159–166. [CrossRef]
12. Greco, C.; Silva, W.L.; Camarda, S.R.; Denadai, B.S. Fatigue and rapid hamstring/quadriceps force capacity in professional soccer players. *Clin. Physiol. Funct. Imaging* **2013**, *33*, 18–23. [CrossRef] [PubMed]
13. Greig, M.; Siegler, J.C. Soccer-specific fatigue and eccentric hamstrings muscle strength. *J. Athl. Train.* **2009**, *44*, 180–184. [CrossRef]

14. Small, K.; McNaughton, L.; Greig, M.; Lovell, R. Effect of timing of eccentric hamstring strengthening exercises during soccer training: Implications for muscle fatigability. *J. Strength Condit. Res.* **2009**, *23*, 1077–1083. [CrossRef] [PubMed]

15. Larruskain, J.; Lekue, J.A.; Diaz, N.; Odriozola, A.; Gil, S.M. A comparison of injuries in elite male and female football players: A 5-Season prospective study. *Scand. J. Med. Sci. Sports* **2017**, *28*, 237–245. [CrossRef]

16. Arendt, E.A.; Agel, J.; Dick, R.J. Anterior cruciate ligament injury patterns among collegiate men and women. *J. Athl. Train.* **1999**, *34*, 86.

17. Waldén, M.; Hägglund, M.; Werner, J.; Ekstrand, J. The epidemiology of anterior cruciate ligament injury in football (soccer): A review of the literature from a gender-related perspective. *Knee Surg. Sports Traumatol. Arthrosc.* **2011**, *19*, 3–10. [CrossRef]

18. Alentorn-Geli, E.; Myer, G.D.; Silvers, H.J.; Samitier, G.; Romero, D.; Lázaro-Haro, C.; Cugat, R. Prevention of non-contact anterior cruciate ligament injuries in soccer players. Part 1: Mechanisms of injury and underlying risk factors. *Knee Surg. Sports Traumatol. Arthrosc.* **2009**, *17*, 705–729. [CrossRef]

19. Boling, M.C.; Padua, D.A.; Marshall, S.W.; Guskiewicz, K.; Pyne, S.; Beutler, A. A Prospective Investigation of Biomechanical Risk Factors for Patellofemoral Pain Syndrome:The Joint Undertaking to Monitor and Prevent ACL Injury (JUMP-ACL) Cohort. *Am. J. Sports Med.* **2009**, *37*, 2108–2116. [CrossRef]

20. Kaeding, C.C.; Pedroza, A.D.; Reinke, E.K.; Huston, L.J.; Spindler, K.P. Risk Factors and Predictors of Subsequent ACL Injury in Either Knee After ACL Reconstruction:Prospective Analysis of 2488 Primary ACL Reconstructions From the MOON Cohort. *Am. J. Sports Med.* **2015**, *43*, 1583–1590. [CrossRef]

21. Wojtys, E.M.; Huston, L.J. Neuromuscular performance in normal and anterior cruciate ligament-deficient lower extremities. *Am. J. Sports Med.* **1994**, *22*, 89–104. [CrossRef] [PubMed]

22. Cowling, E.J.; Steele, J.R. Is lower limb muscle synchrony during landing affected by gender? Implications for variations in ACL injury rates. *J. Electromyogr. Kinesiol.* **2001**, *11*, 263–268. [CrossRef]

23. Letafatkar, A.; Rajabi, R.; Tekamejani, E.E.; Minoonejad, H.J.T.k. Effects of perturbation training on knee flexion angle and quadriceps to hamstring cocontraction of female athletes with quadriceps dominance deficit: Pre–post intervention study. *Knee* **2015**, *22*, 230–236. [CrossRef] [PubMed]

24. Cross, K.M.; Gurka, K.K.; Saliba, S.; Conaway, M.; Hertel, J. Comparison of hamstring strain injury rates between male and female intercollegiate soccer athletes. *Am. J. Sports Med.* **2013**, *41*, 742–748. [CrossRef] [PubMed]

25. Markolf, K.L.; O'Neill, G.; Jackson, S.R.; McAllister, D.R. Effects of Applied Quadriceps and Hamstrings Muscle Loads on Forces in the Anterior and Posterior Cruciate Ligaments. *Am. J. Sports Med.* **2004**, *32*, 1144–1149. [CrossRef] [PubMed]

26. Landry, S.C.; McKean, K.A.; Hubley-Kozey, C.L.; Stanish, W.D.; Deluzio, K.J. Neuromuscular and Lower Limb Biomechanical Differences Exist between Male and Female Elite Adolescent Soccer Players during an Unanticipated Side-cut Maneuver. *Am. J. Sports Med.* **2007**, *35*, 1888–1900. [CrossRef]

27. Van Dyk, N.; Bahr, R.; Whiteley, R.; Tol, J.L.; Kumar, B.D.; Hamilton, B.; Farooq, A.; Witvrouw, E. Hamstring and quadriceps isokinetic strength deficits are weak risk factors for hamstring strain injuries: A 4-year cohort study. *Am. J. Sports Med.* **2016**, *44*, 1789–1795. [CrossRef]

28. Evangelidis, P.E.; Pain, M.T.; Folland, J. Angle-specific hamstring-to-quadriceps ratio: A comparison of football players and recreationally active males. *J. Sports Sci.* **2015**, *33*, 309–319. [CrossRef]

29. Kannus, P. Isokinetic evaluation of muscular performance. *Int. J. Sports Med.* **1994**, *15*, S11–S18. [CrossRef]

30. Caterisano, A.; Moss, R.E.; Pellinger, T.K.; Woodruff, K.; Lewis, V.C.; Booth, W.; Khadra, T. The effect of back squat depth on the EMG activity of 4 superficial hip and thigh muscles. *J. Strength Condit. Res.* **2002**, *16*, 428–432.

31. Navarro, E.; Chorro, D.; Torres, G.; Navandar, A.; Rueda, J.; Veiga, S. Electromyographic activity of quadriceps and hamstrings of a professional football team during Bulgarian Squat and Lunge exercises. *J. Hum. Sport Exerc.* **2020**, *1*. [CrossRef]

32. Torres, G.; Chorro, D.; Navandar, A.; Rueda, J.; Fernández, L.; Navarro, E. Assessment of Hamstring: Quadriceps Coactivation without the Use of Maximum Voluntary Isometric Contraction. *Appl. Sci.* **2020**, *10*, 1615. [CrossRef]

33. Begalle, R.L.; DiStefano, L.J.; Blackburn, T.; Padua, D.A. Quadriceps and Hamstrings Coactivation During Common Therapeutic Exercises. *J. Athl. Train. (Allen Press)* **2012**, *47*, 396–405. [CrossRef]

34. Pincivero, D.M.; Aldworth, C.; Dickerson, T.; Petry, C.; Shultz, T. Quadriceps-hamstring EMG activity during functional, closed kinetic chain exercise to fatigue. *Eur. J. Appl. Physiol.* **2000**, *81*, 504–509. [CrossRef] [PubMed]

35. Santana, J. Single-leg training for 2-legged sports: Efficacy of strength development in athletic performance. *Strength Condit. J.* **2001**, *23*, 35. [CrossRef]

36. Ruas, C.V.; Pinto, R.S.; Haff, G.G.; Lima, C.; Pinto, M.D.; Brown, L.E. Alternative Methods of Determining Hamstrings-to-Quadriceps Ratios: A Comprehensive Review. *Sports Med.* **2019**, *5*, 11. [CrossRef]

37. Nimphius, S.; McBride, J.M.; Rice, P.E.; Goodman-Capps, C.L.; Capps, C.R. Comparison of Quadriceps and Hamstring Muscle Activity during an Isometric Squat between Strength-Matched Men and Women. *J. Sports Sci. Med.* **2019**, *18*, 101–108.

38. Youdas, J.W.; Hollman, J.H.; Hitchcock, J.R.; Hoyme, G.J.; Johnsen, J.J. Comparison of hamstring and quadriceps femoris electromyographic activity between men and women during a single-limb squat on both a stable and labile surface. *J. Strength Condit. Res.* **2007**, *21*, 105–111. [CrossRef] [PubMed]

39. Hermens, H.J.; Freriks, B.; Disselhorst-Klug, C.; Rau, G. Development of recommendations for SEMG sensors and sensor placement procedures. *J. Electromyogr. Kinesiol.* **2000**, *10*, 361–374. [CrossRef]

40. Robertson, R.; Dowling, J. Design and responses of Butterworth and critically damped digital filters. *J. Electromyogr. Kinesiol.* **2003**, *13*, 569–573. [CrossRef]
41. De Luca, C.; Gilmore, L.D.; Kuznetsov, M.; Roy, S.H. Filtering the surface EMG signal: Movement artifact and baseline noise contamination. *J. Biomech.* **2010**, *43*, 1573–1579. [CrossRef]
42. Fukuda, T.Y.; Echeimberg, J.O.; Pompeu, J.E.; Lucareli, P.R.G.; Garbelotti, S.; Gimenes, R.O.; Apolinário, A. Root mean square value of the electromyographic signal in the isometric torque of the quadriceps, hamstrings and brachial biceps muscles in female subjects. *J. Appl. Res.* **2010**, *10*, 32–39.
43. Cohen, J. A power primer. *Psychol. Bull.* **1992**, *112*, 155. [CrossRef] [PubMed]
44. McCurdy, K.; O'Kelley, E.; Kutz, M.; Langford, G.; Ernest, J.; Torres, M. Comparison of lower extremity EMG between the 2-leg squat and modified single-leg squat in female athletes. *J. Sport Rehab.* **2010**, *19*, 57–70. [CrossRef] [PubMed]
45. Liporaci, R.F.; Saad, M.C.; Bevilaqua-Grossi, D.; Riberto, M.J.B.O.S.; Medicine, E. Preseason intrinsic risk factors—associated odds estimate the exposure to proximal lower limb injury throughout the season among professional football players. *BMJ Open Sport Exerc. Med.* **2018**, *4*, e000334. [CrossRef] [PubMed]
46. Marras, W.S.; Davis, K.G. A non-MVC EMG normalization technique for the trunk musculature: Part 1. Method development. *J. Electromyogr. Kinesiol.* **2001**, *11*, 1–9. [CrossRef]
47. Suydam, S.M.; Manal, K.; Buchanan, T.S. The advantages of normalizing electromyography to ballistic rather than isometric or isokinetic tasks. *J. Appl. Biomech.* **2017**, *33*, 189–196. [CrossRef]
48. Robertson, G.E.; Caldwell, G.E.; Hamill, J.; Kamen, G.; Whittlesey, S. *Research Methods in Biomechanics*; Human Kinetics: Champaign, IL, USA, 2013.
49. Wright, J.; Delong, T.; Gehlsen, G. Electromyographic Activity of the Hamstrings During Performance of the Leg Curl, Stiff-Leg Deadlift, and Back Squat Movements. *J. Strength Condit. Res.* **1999**, *13*, 168–174.
50. El-Ashker, S.; Carson, B.; Ayala, F.; De Ste Croix, M. Sex-related differences in joint-angle-specific functional hamstring-to-quadriceps strength ratios. *Knee Surg. Sports Traumatol. Arthrosc.* **2017**, *25*, 949–957. [CrossRef]
51. Andrade, M.D.S.; De Lira, C.A.B.; Koffes, F.D.C.; Mascarin, N.C.; Benedito-Silva, A.A.; Da Silva, A.C. Isokinetic hamstrings-to-quadriceps peak torque ratio: The influence of sport modality, gender, and angular velocity. *J. Sports Sci.* **2012**, *30*, 547–553. [CrossRef]
52. Kong, P.W.; Burns, S.F. Bilateral difference in hamstrings to quadriceps ratio in healthy males and females. *Phys. Ther. Sport* **2010**, *11*, 12–17. [CrossRef]
53. Gobbi, A.; Domzalski, M.; Pascual, J. Sports Traumatology, Arthroscopy. Comparison of anterior cruciate ligament reconstruction in male and female athletes using the patellar tendon and hamstring autografts. *Knee Surg. Sports Traumatol. Arthrosc.* **2004**, *12*, 534–539. [CrossRef] [PubMed]
54. Myer, G.D.; Ford, K.R.; Hewett, T.E. The effects of gender on quadriceps muscle activation strategies during a maneuver that mimics a high ACL injury risk position. *J. Electromyogr. Kinesiol.* **2005**, *15*, 181–189. [CrossRef] [PubMed]
55. Monajati, A.; Larumbe-Zabala, E.; Goss-Sampson, M.; Naclerio, F. Surface electromyography analysis of three squat exercises. *J. Hum. Kinet.* **2019**, *67*, 73–83. [CrossRef] [PubMed]
56. Krishnan, C.; Huston, K.; Amendola, A.; Williams, G.N. Quadriceps and hamstrings muscle control in athletic males and females. *J. Orthop. Res.* **2008**, *26*, 800–808. [CrossRef]
57. Jaberzadeh, S.; Yeo, D.; Zoghi, M. The Effect of Altering Knee Position and Squat Depth on VMO: VL EMG Ratio During Squat Exercises. *Physiother. Res. Int.* **2016**, *21*, 164–173. [CrossRef]
58. Panagiotopoulos, E.; Strzelczyk, P.; Herrmann, M.; Scuderi, G.J. Sports Traumatology, Arthroscopy. Cadaveric study on static medial patellar stabilizers: The dynamizing role of the vastus medialis obliquus on medial patellofemoral ligament. *Knee Surg. Sports Traumatol. Arthrosc.* **2006**, *14*, 7–12. [CrossRef]
59. Mostamand, J.; Bader, D.L.; Hudson, Z. Reliability testing of vasti activity measurements in taped and untaped patellofemoral conditions during single leg squatting in healthy subjects: A pilot study. *J. Bodyw. Mov. Ther.* **2013**, *17*, 271–277. [CrossRef]
60. Souza, D.R.; Gross, M.T. Comparison of vastus medialis obliquus: Vastus lateralis muscle integrated electromyographic ratios between healthy subjects and patients with patellofemoral pain. *Phys. Ther.* **1991**, *71*, 310–316. [CrossRef]
61. Araújo, S.R.S.; Medeiros, F.B.; Zaidan, A.D.; Pimenta, E.M.; Abreu, E.A.d.C.; Ferreira, J.C. Comparison of two classification criteria of lateral strength asymmetry of the lower limbs in professional soccer players. *Rev. Brasil. Cineantropometria Desempenho Hum.* **2017**, *19*, 644–651.
62. Menzel, H.-J.; Chagas, M.H.; Szmuchrowski, L.A.; Araujo, S.R.; de Andrade, A.G.; de Jesus-Moraleida, F.R. Analysis of lower limb asymmetries by isokinetic and vertical jump tests in soccer players. *J. Strength Condit. Res.* **2013**, *27*, 1370–1377. [CrossRef] [PubMed]
63. Crossley, K.; Bennell, K.; Green, S.; McConnell, J. A systematic review of physical interventions for patellofemoral pain syndrome. *Clin. J. Sport Med.* **2001**, *11*, 103–110. [CrossRef] [PubMed]
64. Dimitrova, N.; Arabadzhiev, T.; Hogrel, J.-Y.; Dimitrov, G.V. Fatigue analysis of interference EMG signals obtained from biceps brachii during isometric voluntary contraction at various force levels. *J. Electromyogr. Kinesiol.* **2009**, *19*, 252–258. [CrossRef] [PubMed]
65. Fukuda, T.Y.; Alvarez, A.S.; Nassri, L.F.G.; Godoy, C.M.G. Quantitative electromyographic assessment of facial muscles in cross-bite female children. *Rev. Bras. Eng. Biomed.* **2008**, *2008*, 121–129. [CrossRef]

66. Bilodeau, M.; Schindler-Ivens, S.; Williams, D.; Chandran, R.; Sharma, S.S. EMG frequency content changes with increasing force and during fatigue in the quadriceps femoris muscle of men and women. *J. Electromyogr. Kinesiol.* **2003**, *13*, 83–92. [CrossRef]
67. Urabe, Y.; Kobayashi, R.; Sumida, S.; Tanaka, K.; Yoshida, N.; Nishiwaki, G.A.; Tsutsumi, E.; Ochi, M. Electromyographic analysis of the knee during jump landing in male and female athletes. *Knee* **2005**, *12*, 129–134. [CrossRef]
68. Krishnan, C.; Williamns, N. Sex Differences in Quadriceps and Hamstrings EMG–Moment Relationships. *Med. Sci. Sports Exerc.* **2009**, *41*, 1652–1660. [CrossRef]
69. Ninos, J.C.; Irrgang, J.J.; Burdett, R.; Weiss, J.R. Electromyographic analysis of the squat performed in self-selected lower extremity neutral rotation and 30 of lower extremity turn-out from the self-selected neutral position. *J. Orthop. Sports Phys. Ther.* **1997**, *25*, 307–315. [CrossRef]
70. Thelen, D.G.; Chumanov, E.S.; Sherry, M.A.; Heiderscheit, B.C. Neuromusculoskeletal models provide insights into the mechanisms and rehabilitation of hamstring strains. *Exerc. Sport Sci. Rev.* **2006**, *34*, 135–141. [CrossRef]

applied sciences

Article

Single-Leg Landings Following a Volleyball Spike May Increase the Risk of Anterior Cruciate Ligament Injury More Than Landing on Both-Legs

Datao Xu [1], Xinyan Jiang [1], Xuanzhen Cen [1], Julien S. Baker [2,*] and Yaodong Gu [1,*]

[1] Faculty of Sports Science, Ningbo University, Ningbo 315211, China; xudatao3@gmail.com (D.X.); jiangxinyan168@163.com (X.J.); cenxuanzhen@outlook.com (X.C.)

[2] The Department of Sport, Physical Education and Health, Hong Kong Baptist University, Hong Kong 999077, China

* Correspondence: jsbaker@hkbu.edu.hk (J.S.B.); guyaodong@nbu.edu.cn (Y.G.); Tel.: +852-3411-8032 (J.S.B.); +86-574-8760-0456 (Y.G.)

Abstract: Volleyball players often land on a single leg following a spike shot due to a shift in the center of gravity and loss of balance. Landing on a single leg following a spike may increase the probability of non-contact anterior cruciate ligament (ACL) injuries. The purpose of this study was to compare and analyze the kinematics and kinetics differences during the landing phase of volleyball players using a single leg (SL) and double-leg landing (DL) following a spike shot. The data for vertical ground reaction forces (VGRF) and sagittal plane were collected. SPM analysis revealed that SL depicted a smaller knee flexion angle (about 13.8°) and hip flexion angle (about 10.8°) during the whole landing phase, a greater knee and hip power during the 16.83–20.45% ($p = 0.006$) and 13.01–16.26% ($p = 0.008$) landing phase, a greater ankle plantarflexion angle and moment during the 0–41.07% ($p < 0.001$) and 2.76–79.45% ($p < 0.001$) landing phase, a greater VGRF during the 5.87–8.25% ($p = 0.029$), 19.75–24.14% ($p = 0.003$) landing phase when compared to DL. Most of these differences fall within the time range of ACL injury (30–50 milliseconds after landing). To reduce non-contact ACL injuries, a landing strategy of consciously increasing the hip and knee flexion, and plantarflexion of the ankle should be considered by volleyball players.

Keywords: volleyball spike landing; sagittal biomechanics; non-contact ACL injuries; statistical parametric mapping (SPM)

Citation: Xu, D.; Jiang, X.; Cen, X.; Baker, J.S.; Gu, Y. Single-Leg Landings Following a Volleyball Spike May Increase the Risk of Anterior Cruciate Ligament Injury More Than Landing on Both-Legs. *Appl. Sci.* **2021**, *11*, 130. https://dx.doi.org/10.3390/app11010130

Received: 28 October 2020
Accepted: 20 December 2020
Published: 25 December 2020

Publisher's Note: MDPI stays neutral with regard to jurisdictional claims in published maps and institutional affiliations.

1. Introduction

As a competitive sport, volleyball has a high rate of musculoskeletal injury [1]. The special movements of volleyball, such as jumping, landing, blocking, and spiking, need to be combined with fast movements, that put very high demands on the musculoskeletal system of volleyball players. This can cause considerable damage to the musculoskeletal system of players [2,3]. High-level volleyball players are prone to suffering lower limb injury following overuse caused by continuous blocking and attacking [3–5]. This is especially true for the knee joint which is easily injured in the energy transmission process because it comprises of a joint capsule with multiple joints. Previous studies have shown that 63 percent of musculoskeletal injuries in volleyball occur during the basic jump and landing tasks of blocking and spiking. Most of the injuries are located at the knee joint and most injuries are anterior cruciate ligament (ACL) injuries [5]. Anterior cruciate ligament (ACL) injury is a common injury in sports, and 70–90% of ACL injuries are caused by non-contacts [6–8].

From a mechanical point of view, ACL injury is caused by excessive ACL tension. The non-contact injury of ACL is generally caused by the large force or moment generated by the player himself on the knee joint, so that the force of ACL exceeds its ultimate load capacity. The kinematics and kinetics of the sagittal plane are critical to ACL injury during landing tasks, with the increase of the anterior tibial shear force (mainly from the sagittal plane), the load on the ACL (the risk of ACL injury) will increase [9–11]. Previous studies have also shown that the sagittal hip and knee joints play a major role in energy dissipation and buffering during landing, and smaller knee and hip flexion will increase the impact pressure on the knee (the anterior tibial shear force), leading to an increased risk of ACL injury [12–14]. At the same time, the knee and hip flexion angle during the landing phase were associated with ground reaction forces during jump-landing tasks. Increasing the knee and hip flexion angle might reduce impact force during landing tasks [15]. Ryan et al. found that the joint angle and joint moment were strongly correlated with the ACL strain force. When the knee and hip flexion angles decrease and the knee and hip extension moment goes up, the ACL strain force is significantly increased [9,16]. In the sagittal plane, the lower limb joint has a wide range of motion (ROM) and can provide enough shock absorption without excessive compression of the fragile soft tissue structures such as ligaments [17]. For the total energy dissipation of the lower limb joint during a jump-landing task, previous studies have shown that the sagittal plane is about 10 (landing with double-leg) and 20 (landing with single-leg) times larger than the frontal plane [17]. Therefore, the sagittal plane plays a major role in reducing lower limb injuries.

Different landing modes will have a greater influence on the load of the lower limb joints and ligaments [18]. Unreasonable landing patterns include smaller knee and hip flexion angles in the sagittal plane, and greater vertical ground reaction forces (VGRF), which significantly increase the risk of ACL and lower limb injuries [12,13,19]. Volleyball players often land with a single leg after a spike due to a shift in the center of gravity and loss of balance [20]. This will increase the risk of lower limb injury when landing on a single leg, because the impact and energy transfer can only be absorbed by the muscle tissue of the individual leg in contact with the ground. Compared with landing using both legs, previous studies have shown that landing with a single leg will increase knee and hip flexion and extension moments, peak VGRF and power, which result will increase tension and stress of the knee ligament, thus increasing the risk of ACL injury [21–23]. However, there is a lack of research on the biomechanical differences caused by volleyball players landing on a single-leg after a spike, and it is also not clear that landing following a volleyball spike would make a greater difference than a typical landing.

The injury risk of volleyball players while jumping and landing is determined by the spatial and temporal characteristics of the sport. This has been observed especially during landing after blocking and spiking and increases the probability of ACL injury of volleyball players [24]. In landing tasks, previous studies have shown that ACL injury is believed to occur 30–50 ms after initial ground contact [21]. In the biomechanical analysis of landing tasks that may cause damage, the timing of post-landing damage needs consideration. However, current kinematic and kinetic analyses only partially focus on the peak point data and assume that the peak point time in line with the critical time at the injury occurs. Jeffrey and Michael et al. compared and analyzed the peak joint angle and moment (knee and hip flexion, knee and hip moment) between single-leg and double-leg landing tasks [25–28]. Most of them focus on analyzing peak data. Wang and Rachael et al. also compared the peak VGRF and peak knee proximal tibia shear force during different landing tasks [19,29,30].

This analysis method is widely accepted in research, but the comparative peak misses the important time fluctuations that may happen in the whole landing phase. At present, great progress has been made in the study of biomechanics spatial and temporal variability, and a relatively complete model framework has been presented [11,31,32]. Statistical parametric mapping (SPM) is a method, which can test the statistical differences of continuous data such as kinematics and kinetics over the whole period of motion to calculate accurately the significance threshold [33,34]. Therefore, based on the one-dimensional character of joint motion changing with time, combining traditional discrete analysis with statistical parametric mapping of one-dimensional (SPM 1d) was used to conduct statistical analysis on the sagittal biomechanical data of the landing task of volleyball players. This can reveal ACL injury mechanics of volleyball players during landing tasks after a spike and provide injury prevention strategies.

Therefore, the main aim of our study was to investigate the VGRF and kinematics and kinetics of the lower extremity sagittal plane movement in the whole landing stage of volleyball players following a spike landing with single-leg landing (SL) and double-leg landing (DL). This will be achieved using traditional discrete kinematic and kinetics analysis and statistical parametric mapping of one-dimensional (SPM 1d) analysis. We hypothesized that significant biomechanical differences may be observed in the sagittal plane between SL and DL after a spike. Compared with DL, the SL could show smaller knee and hip flexion angles and angular velocities as well as greater moment, power, VGRF and loading rates. Meanwhile, most of these differences may be observed within the 30–50 milliseconds time frame of initial contact with the ground. Our hypothesized results may provide some clues to reveal the injury mechanism of ACL from the temporal kinematic and kinetic differences during landing phase.

2. Materials and Methods

2.1. Participants

The sample size was calculated using the power package RStudio 3.6.1. A power analysis for Paired-samples *T*-test analysis of variance was performed for an effect size of 0.8 (significance level: 0.05; power: 0.8; alternative: two sided). The power analysis indicated that thirteen subjects (the minimal detectable difference threshold values ranged from 11.2 to 14.8) should be recruited for the study, but we only recruited twelve subjects who were eligible for this study. Therefore, twelve healthy semi-professional male volleyball players were recruited to participate in this experiment (age: 23.3 ± 2.2 years, height: 1.87 ± 0.04 m, weight: 80.2 ± 6.3 kg). All were members of the volleyball team from the local university. They were also National Division 2 athletes or above. Subject exclusion criteria included: (1) a history of foot diseases and deformity or lower limbs surgery; (2) lower extremity injuries or medical problems that might affect athletic performance in the six months prior to the trial. All participates had the objectives, demands, experimental procedures and other details of this study explained to them in detail, and all participates provided written informed consent prior to the experiment. The study protocol was approved by the Human Ethics Committee of Ningbo University (Approval Number: RAGH20200121). All experimentation was performed at the same day, from 8:00 to 21:00 (about 13 h).

2.2. Instrumentation

Before the formal test, the height and weight of all subjects were recorded with a calibrated weighing scale and stadiometer. A Vicon system with 8 cameras (Oxford Metric Ltd., Oxford, UK) was used to collect kinematic data during the test, and the data were sampled at a frequency of 200 Hz. An in-ground force plate (AMTI, Watertown, MA, USA) was used to obtain kinetic data, and the sampling frequency was set at 1000 Hz. Vicon Nexus software was used to synchronously capture force plate data and kinematics data. As shown in Figure 1, a total of thirty-six reflective markers with diameter of 12.5 mm were placed on the lower extremities and pelvis to track the motion trajectory according to the experimental requirements and details from a previous study [35]. The reflective markers were located in the bilateral lower limbs and pelvis, which included anterior and posterior (left and right) superior iliac spine, medial and lateral (left and right) condyle, medial and lateral (left and right) malleolus, first and fifth (left and right) metatarsal heads, distal interphalangeal joint (left and right) of the second toe, and six clusters were placed on the middle and lateral (left and right) heel, shank and thigh. The data set of initial coordinates were obtained via collected static experimental data on the subjects in an anatomically neutral position. All negative sagittal data represent flexion or plantarflexion, and all positive data represent extension or dorsiflexion. The data for joint motions were calculated using the Euler equation [36].

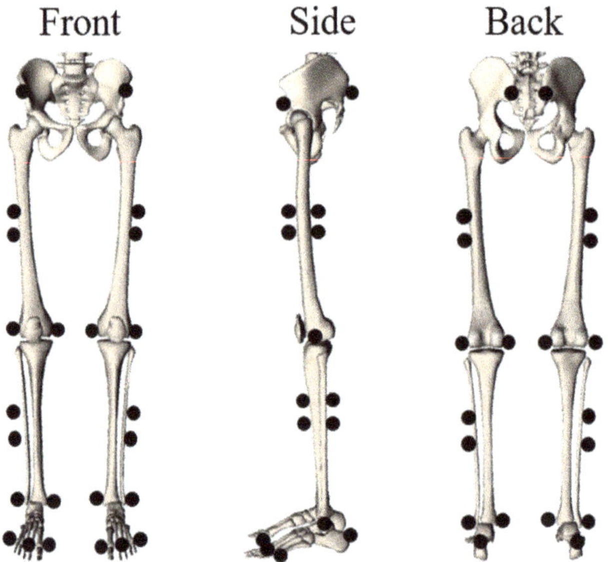

Figure 1. Illustration of lower limb and pelvis reflective markers placement.

2.3. Experimental Procedure

Figure 2 outlines the experimental process. The whole experimental environment was surrounded by eight infrared cameras and the force plate was located between the middle line and the attack line as shown in the figure. Before the formal experiment, participants warmed up for 10 min in a manner of their choice. Following the warm-up, the participants were familiarized with the experimental environment and experimental process and spike landing practice preceded the formal experiment. To avoid the error caused by the differences of participants' flight-time and flight-distance in the air, all participants were asked to take off from force plate and also land on the force plate. All participants were asked to find the most appropriate way (including the most appropriate way to take off, the most appropriate distance between the volleyball and participants) to straight spike during the spike landing practice.

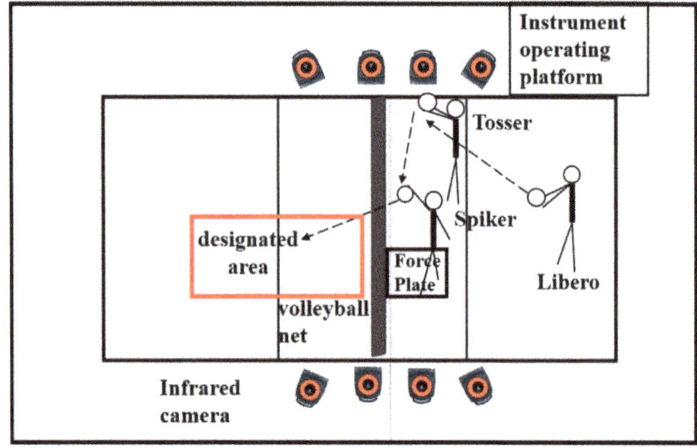

Figure 2. Illustration of the spike landing process.

The average spike height of each subject during the match was recorded, and the height was then marked as the subject's spike height in the experiment. In the spike landing task, the libero held the ball and passed it to the tosser, and the tosser then passed the ball directly in front of the participant (the height of the ball is the height recorded). The participant took off in their own individual way and then spiked the ball with maximal effort on the top of the force plate. After spiking the ball, the subject then landed on the force plate. All participants were asked to use the straight spike and spike the ball into a designated area (Figure 2, red frame area).

The SL is outlined in Figure 3, the participants were asked to land on the force plate with their supporting leg (the supporting leg defined as a leg that is often used for support after landing [37]). The other leg did not contact with the ground to simulate landing with single leg as in an official match. For the DL task (Figure 3), the participants were asked to use the supporting leg to land on the force plate, and the other leg to be placed outside the force plate simultaneously. A successful trial of landing tasks was defined as the subjects not falling after landing, the complete spike sequence agreed with the rules of volleyball, and the ball landed in the designated area. Between each spike landing, participants rested for 30 s to reduce the error due to any acute fatigue caused by successive jumping and landing tasks. Participants were asked to wear professional volleyball shoes during the experiment, and they were randomly (by simple randomization: the subjects were randomly assigned directly) assigned to SL or DL tasks, a total of ten successful experiments were carried out (SL: five times, DL: five times).

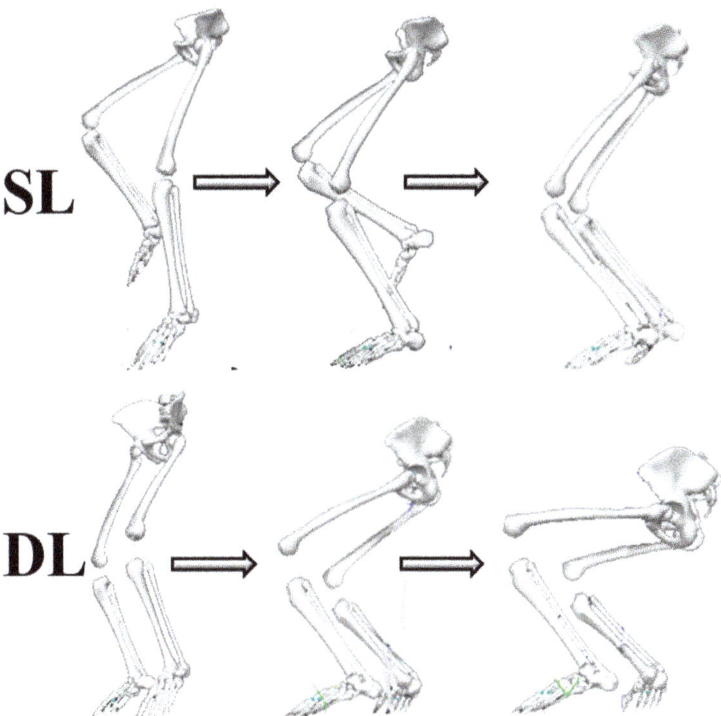

Figure 3. The process of single leg landing (SL) and double-leg landing (DL) following a spike.

2.4. Data Collection and Processing

The force platform had a value of vertical ground reaction force (VGRF) surpassing 10 N, and we defined this as the initial point [11]. Data capture started 2 s prior to the initial contact point and ended 3 s after landing, a total of 5 s of data was collected. The filter frequency selection was based on previous studies by Winter [38]. The dynamic residual analysis was executed in the subsets to confirm the most suitable SNR (signal to noise ratio). Finally, the data were filtered using Butterworth low-pass filters (order: fourth, phase lag: zero, kinematics: 10 Hz, kinetics: 20 Hz). The C3D format file was generated using Vicon Nexus software, and then imported into the C3D file into Visual 3-D software (C-Motion Inc., Germantown, MD, USA) for static modeling and further processing.

The joint angle, joint moment, joint angular velocity and joint power were calculated using Visual 3-D software by inverse kinetics algorithm. Joint power was defined as the product of the joint moment and joint angular velocity. The range of joint motion (ROM) was calculated from the difference between the initial contact joint angle (minimum flexion or plantarflexion) and the joint angle at maximum flexion (dorsiflexion). The jump height was calculated by the change of the center of gravity, and the calculation process also conducted using Visual 3-D software. The flight-time was calculated by the change of time point (the starting point: taking off from the force plate; the end point: landing on the force plate), which was conducted using the Vicon system.

The loading rate of VGRF was defined as the growth rate of VGRF over time. Then, we imported all data into MATLAB (Version: R2019a, The MathWorks, Natick, MA, USA) to process by written script. A custom MATLAB script was used to intercept the time point between the initial contact and the peak knee joint flexion during the landing process.

2.5. Statistical Analysis

The landing phase was defined as the maximum knee flexion after initial contact with the ground. Prior to statistical analysis, all data were subjected to the Shapiro-Wilk normality test. If the data showed nonconformity, the Wilcoxon matched-pairs signed-rank test was executed for data analysis.

For traditional analysis (discrete variable), the MATLAB script was written to extract and calculate the ROM of joint angle, peak data of angle, moment, power, angular velocity, and VGRF. Paired-samples T-test was employed to test differences of ROM of joint angle, peak data of angle, moment, power, angular velocity and VGRF. All traditional analyses were conducted using SPSS 24.0 (SPSSs Inc., Chicago, IL, USA), and the statistical significance was set as 0.05. Cohen's d effect sizes (ES) were split into large (ES > 0.8), medium (0.2 < ES < 0.5), and small (ES < 0.2) benchmark sizes [39].

For SPM 1d analysis, each task generated a separate curve prior to analysis. Then, all data were extracted to expand into 101 data points time series curves (using a custom MATLAB script). The 101 data points represent the landing phase (0–100%). The open-source script of SPM 1d (Paired-samples *T*-test, significance threshold was 0.05) was used in the statistical analysis [34].

3. Results

For the time of landing phase, there were no differences ($p = 0.31$) between SL (193.6 ± 29.8 ms) and DL (197.1 ± 33.4 ms) landing task. For the jump height, there were no differences ($p = 0.22$) between SL (0.59 ± 0.08 m) and DL (0.62 ± 0.10 m) landing task. For the flight-time in the air, there were no differences ($p = 0.09$) between SL (0.66 ± 0.09 s) and DL (0.71 ± 0.12 s) landing task.

For the results of ankle joint. SPM analysis revealed that SL depicted a significantly greater plantarflexion angle than DL during the 0–41.07% (SL: $-19.8{\sim}24.6°$, DL: $-8.6{\sim}29.7°$, $p < 0.001$) landing phase, a significantly greater plantarflexion moment than DL during the 2.76–79.45% (SL: $-1.01{\sim}{-}1.41$ Nm/kg, DL: $-0.45{\sim}{-}0.52$ Nm/kg, $p < 0.001$) landing phase, a significantly greater joint power than DL during the 2.48–9.66% (SL: $-5.20{\sim}{-}31.37$ W/kg, DL: $-2.12{\sim}{-}10.25$ W/kg, $p < 0.001$) and 12.92–26.95% (SL: $-19.71{\sim}{-}2.89$ W/kg, DL: $-6.14{\sim}{-}0.65$ W/kg, $p < 0.001$) landing phase, a significantly greater dorsiflexion angular velocity than DL during the 0–23.54% (SL: $756.3{\sim}102.4°$/s, DL: $509.9{\sim}18.2°$/s, $p < 0.001$) and 44.73–47.63% (SL: $16.2{\sim}12.0°$/s, DL: $17.4{\sim}21.1°$/s, $p = 0.016$) landing phase (Figure 4).

SL (Table 1) depicted greater peak plantarflexion angle ($p = 0.001$), ROM ($p = 0.023$), peak moment ($p = 0.001$), peak power ($p = 0.001$), and peak angular velocity ($p = 0.001$).

For the results of the knee joint. SPM analysis revealed that SL depicted a significantly smaller flexion angle than DL during the 0–100% (SL: $-7.0 \sim -72.2°$, DL: $-20.7 \sim -97.6°$, $p < 0.001$) landing phase, a significantly smaller flexion angular velocity than DL during the 0–32.69% (SL: $-293.9 \sim -157.9°$/s, DL: $-407.3 \sim -275.9°$/s, $p < 0.001$) and 44.36–55.40% (SL: $-107.1 \sim -43.9°$/s, DL: $-195.8 \sim -133.6°$/s, $p = 0.014$) landing phase, a significantly greater extension moment than DL during the 3.86–4.27% ($p = 0.048$) and 11.73–12.32% ($p = 0.045$) and 17.74–21.82% (3.03~4.16 Nm/kg, DL: 2.73~2.49 Nm/kg, $p < 0.001$) landing phase, a significantly greater joint power than DL during the 16.83–20.45% (SL: $-25.9 \sim -29.3$ W/kg, DL: $-16.4 \sim -23.8$ W/kg, $p = 0.006$) landing phase (Figure 4). SL (Table 1) depicted a smaller peak flexion angle ($p = 0.001$) and peak angular velocity ($p = 0.001$), greater peak moment ($p = 0.005$), and peak power ($p = 0.002$).

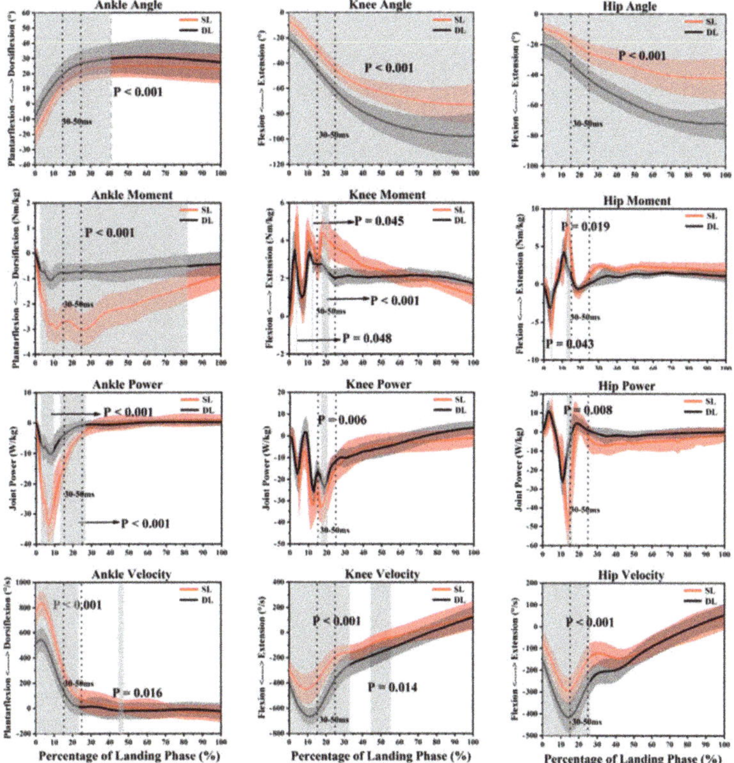

Figure 4. The Statistical Parametric Mapping (SPM) results between single-leg landing (SL) and double-leg landings (DL) tasks following a spike, depicting the mean angle moment power and angular velocity and standard deviation of ankle knee and hip sagittal plane. Two dashed lines are the 30 and 50 ms time frame of initial contact. Grey shaded areas indicate that there are significant differences ($p < 0.05$) between SL and DL during the spike landing phase.

Table 1. Means ± standard deviations for the peak angle moment power and angular velocity of ankle knee and hip sagittal plane, peak VGRF and loading rate between SL and DL.

		SL	DL	p
Ankle	Peak Ankle Plantarflexion Angle (°)	−19.88 ± 8.80 * (initial contact)	−8.58 ± 8.28 * (initial contact)	0.001
	Peak Ankle Dorsiflexion Angle (°)	25.93 ± 7.94 (193.6 ± 29.8 ms)	34.31 ± 14.49 (197.1 ± 33.4 ms)	0.058
	Ankle ROM (°)	45.81 ± 6.38 *	42.89 ± 7.98 *	0.023
	Peak Ankle Moment (Nm/kg)	−3.26 ± 0.55 * (24.2 ± 3.2 ms)	−1.12 ± 0.37 * (18.5 ± 2.3 ms)	0.001
	Peak Ankle Power (W/kg)	−37.97 ± 5.44 * (16.7 ± 2.8 ms)	−11.20 ± 4.15 * (18.7 ± 3.3 ms)	0.001
	Peak Ankle Angular Velocity (°/s)	844.91 ± 95.22 * (8.6 ± 1.7 ms)	559.49 ± 96.93 * (8.4 ± 1.3 ms)	0.001
Knee	Peak Knee Flexion Angle (°)	−74.26 ± 15.18 * (193.6 ± 29.8 ms)	−99.83 ± 17.58 * (197.1 ± 33.4 ms)	0.001
	Peak Knee Extension Angle (°)	−6.96 ± 6.39 * (initial contact)	−20.73 ± 6.84 * (initial contact)	0.003
	Knee ROM (°)	67.30 ± 10.99	79.10 ± 17.33	0.056
	Peak Knee Moment (Nm/kg)	5.52 ± 1.79 * (9.7 ± 3.5 ms)	3.88 ± 1.50 * (8.5 ± 1.9 ms)	0.005
	Peak Knee Power (W/kg)	−42.77 ± 5.77 * (26.3 ± 4.1 ms)	−29.44 ± 5.57 (27.8 ± 4.4 ms)	0.002
	Peak Knee Angular Velocity (°/s)	−464.22 ± 80.49 * (20.6 ± 3.6 ms)	−671.56 ± 77.83 * (24.3 ± 2.8 ms)	0.001
Hip	Peak Hip Flexion Angle (°)	−43.01 ± 12.82 * (193.6 ± 29.8 ms)	−74.31 ± 10.87 * (197.1 ± 33.4 ms)	0.001
	Peak Hip Extension Angle (°)	−9.21 ± 4.09 (initial contact)	−20.04 ± 8.61 (initial contact)	0.021
	Hip ROM (°)	33.80 ± 9.52 *	54.27 ± 13.04 *	0.001
	Peak Hip Moment (Nm/kg)	7.98 ± 2.93 * (28.2 ± 5.7 ms)	4.35 ± 1.99 * (24.3 ± 4.5 ms)	0.007
	Peak Hip Power (W/kg)	−57.90 ± 10.48 * (30.1 ± 6.2 ms)	−30.49 ± 6.91 * (24.7 ± 4.8 ms)	0.001
	Peak Hip Angular Velocity (°/s)	−302.56 ± 68.56 * (30.2 ± 3.7 ms)	−426.14 ± 75.13 * (29.7 ± 4.3 ms)	0.001
VGRF	Peak VGRF (BW)	5.08 ± 0.68 * (48.3 ± 6.6 ms)	3.31 ± 0.55 * (49.8 ± 7.8 ms)	0.001
	Loading Rate (BW/s)	187.23 ± 21.11 *	118.49 ± 14.78 *	0.001

Note: The time corresponding to the occurrence of the peak data point is in parentheses, and the unit is milliseconds (ms). * refer to significance with $p < 0.05$.

For the results of the hip joint, SPM analysis revealed that SL depicted a significantly smaller flexion angle than DL during the 0–100% (SL: −9.2~−41.8°, DL: −20.0~−71.7°, $p < 0.001$) landing phase, a significantly smaller flexion angular velocity than DL during the 0–26.34% (SL: −69.2~−150.1°/s, DL: −157.2~−267.7°/s, $p < 0.001$) landing phase, a significantly greater joint moment than DL during the 3.80–4.60% ($p = 0.043$) and 12.45–14.49% (3.14~7.98 Nm/kg, DL: 4.17~2.37 Nm/kg, $p = 0.019$) landing phase, a significantly greater joint power than DL during the 13.01–16.26% (SL: −29.2~−30.1 W/kg, DL: −21.9~−4.2 W/kg, $p = 0.008$) landing phase (Figure 4). SL (Table 1) that SL depicted a smaller peak flexion angle ($p = 0.001$) and ROM ($p = 0.001$) and peak angular velocity ($p = 0.001$), greater peak moment ($p = 0.007$) and peak power ($p = 0.001$).

For the results of VGRF. SPM analysis revealed that SL depicted a significantly greater VGRF than DL during the 5.87–8.25% (SL: 1.62~1.43 BW, DL: 1.43~1.09 BW, $p = 0.029$), 19.75–24.14% (SL: 3.43~4.71 BW, DL: 2.28~2.94 BW, $p = 0.003$) and 47.09–52.46% (SL: 1.79~1.93 BW, DL: 1.15 ~1.18 BW, $p = 0.004$) landing phase (Figure 5). Table 1 shows that SL depicted greater peak VGRF ($p = 0.001$) and loading rate of VGRF ($p = 0.001$) than DL.

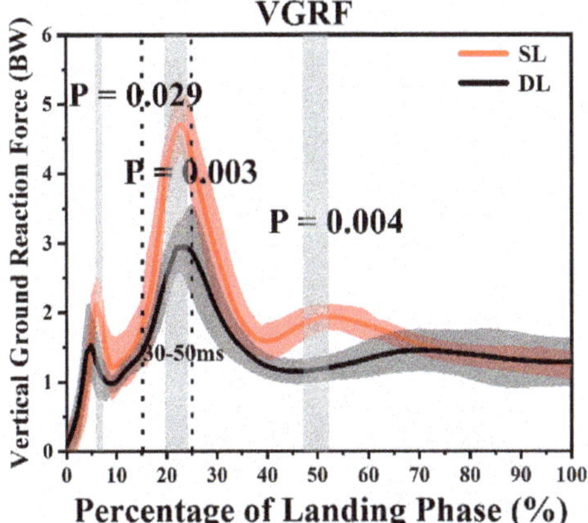

Figure 5. The Statistical Parametric Mapping (SPM) results between single-leg landing (SL) and double-leg landings (DL) tasks following a spike, depicting the mean vertical ground reaction force (VGRF) and standard deviation. Two dashed lines are the 30 and 50 ms time frame of initial contact. Grey shaded areas indicate that there are significant differences ($p < 0.05$) between SL and DL during the spike landing phase.

4. Discussion

We hypothesized that compared with DL, the SL will show smaller knee and hip flexion angles and angular velocities as well as greater moment, power, VGRF, and loading rates. Meanwhile, the SL will show greater ankle plantarflexion angle and dorsiflexion angular velocity, and most of these differences were within the 30–50 milliseconds time frame of initial contact with the ground. Our results agree in part with the proposed hypothesis. SPM demonstrated that most differences occur in the first third of the landing phase. Our results are consistent with ACL injury mechanism in landing tasks, and SL may increase the probability of ACL injury risk.

The wide ROM of the knee and hip sagittal plane are key to lower limb cushioning during landing [17]. Smaller knee flexion during landing can result in lower limb shock absorption, leaving the ACL at a higher risk of injury [40–43]. Our results revealed that SL depicted a smaller knee flexion angle (about 13.8°) and hip flexion angle (about 10.8°) than DL during the landing phase, and SL also has resulted in a flexion angle of less than 45° within the 30–50 ms time frame of initial contact. Markolf and Drahanich et al. both found that the quadriceps force will augment ACL loading, when the knee joint flexion angle is from 0° to 45° [18,44]. Therefore, our current evidence pointed out that the quadriceps force of SL seems to increase the load on the ACL. Meanwhile, SL appears to be a hard landing with the smaller hip flexion angle, and the results of hard landing patterns are followed by an increase in knee and hip moments. Within the 30–50 ms time frame of the initial landing phase, we found that the knee moment of SL is about 2 Nm/kg greater than DL during the 17.74–21.82% (about 35–43 ms after initial ground contact) landing phase. Consistent with previous studies, smaller flexion angles are accompanied by larger moments, and this increases the risk of ACL injury [45]. At the same time, SL depicted a greater negative knee joint power and hip joint power during the landing phase. Previous studies have shown that the case of the negative knee and hip joint power suggests that eccentric work is done by the extensor muscles of the knee and hip joint to consume the impact of energy [12]. With the increase of negative joint power, this means that the eccentricity to the joint extensor increases. The strain on the ACL will increase with the eccentricity to the joint extensor (quadriceps femoris and anterior tibia) then the risk of ACL injury will increase [13,45].

The knee joint plays a significant role in absorbing shock during contact between the body and the ground [12]. During the landing task, the increased VGRF will also increase the contractile force of the quadriceps femoris, thus increasing the load on the ACL [13]. Previous studies have shown that higher VGRF and insufficient VGRF dissipation during the landing phase may also be a risk factor for ACL injury [12,46]. Our results revealed that the SL depicted a significantly greater VGRF than DL during the 19.75–24.14% (38–47 ms after initial ground contact) landing phase, and SL also depicted greater peak VGRF ($p = 0.001$) and loading rate of VGRF ($p = 0.001$) than DL. Cerulli and Lamontagne found that the maximum ACL tension occurred at the moment of the peak VGRF [44,47]. A higher VGRF loading rate indicates a greater impact in a shorter period, thus increasing the risk of ACL injury with a high probability [3,48,49]. Yu et al. found that the angular velocity of the knee and hip flexion was negatively correlated with the peak VGRF during landing tasks, the active knee and hip flexion may be the key factor to reduce the impact force [13]. Consistent with our results, SL depicted a smaller knee and hip flexion angular velocity during the 0–32.69% and 0–26.34% landing phases respectively. It has been shown that increasing the angular velocity of hip and knee flexion during landing can be reduce VGRF, especially in the early landing phase. The difference in VGRF and angular velocity during the landing phase also occurred within the 30–50 ms time frame. This also seems to be associated with ACL injury mechanisms.

During the landing task, the lower limb undergoes a load pattern of distal-to-proximal in which the foot and ankle joint are loaded primarily, and then the load is dissipated to the knee and hip joints. At the moment of initial contact during the landing phase, a greater ankle plantarflexion angle will increase the time to reach the peak VGRF, and a larger plantarflexion landing method can use plantar flexor muscles for attenuation buffering according to previous studies [50,51]. It is also thought to increase lower extremity shock absorption and reduce non-contact ACL injuries [50]. Our results revealed that SL depicted a greater plantarflexion angle (within 83 ms after initial ground contact), power (about 5–54 ms after initial ground contact) during the landing phase. It is worth noting that the time at which these differences occur is also within the framework of the time at which the ACL is thought to occur. The ankle joint in the SL shows higher energy dissipation due to a larger plantarflexion landing method, thus the load impact on the knee joint is relatively reduced. Although the initial contact plantarflexion angle of the ankle joint affects the ROM and peak angular velocity of the knee joint, the knee joint always maintains a constant energy dissipation contribution rate [52,53]. Previous studies have shown that the total energy dissipation of the lower extremities is greater with soft landing (larger knee and hip flexion, ankle plantarflexion) techniques than with hard landing techniques, increasing the total energy dissipation of the lower limb during landing will reduce the potential injury risk of passive joint structure [12,14,54]. Therefore, a landing strategy of consciously increasing the knee and hip flexion, plantarflexion of ankle should be considered by volleyball players in order to reduce non-contact ACL injuries.

There are some limitations that should be considered in the present study. Firstly, we did not use electromyography (EMG) to compare and analyze the muscle activity during the spiking landing task. The muscle activity of the knee extensor also has great significance to this study, and EMG data should be considered for subsequent studies. Secondly, the subjects in our experiment did not include female volleyball players. Previous studies have shown differences between male and female athletes during landing [41,53]. Therefore, female subjects should be considered in future studies. Finally, although we attempted in the experiment to simulate real volleyball match conditions, it is unavoidable that some parameters (e.g., the distance between the ball and participant's hand, the spiked volleyball speed in the air) cannot be quantified. The sample size of our study included twelve semi-professional male volleyball players. This may have contributed to the limitations, so increasing the sample size in the future studies could further verify the results presented here.

5. Conclusions

In summary, this study presents the kinematic and kinetic differences of the lower limb sagittal plane of motion throughout a volleyball players spike landing phase of single and double leg landing. Appendix A exhibits the STROBE Checklist for reporting this study (Appendix A: Table A1). The results of our study are partially consistent with previous studies about typical landings. Following a spike, landing with a single leg shows a smaller flexion angle and angular velocity of the knee and hip joints, a greater joint moment, joint power, VGRF and loading rate. Most of these differences may increase the risk of ACL injury and fall within the time range of ACL injury. SL depicted a smaller knee flexion angle (about 13.8°) and hip flexion angle (about 10.8°), and SL also showed a flexion angle of less than 45° within the 30–50 milliseconds time frame of initial contact with the ground. Therefore, a landing strategy of consciously increasing knee flexion and hip flexion, and plantarflexion of the ankle, should be considered by volleyball players in order to reduce non-contact ACL injuries.

Author Contributions: Conceptualization, D.X. and Y.G.; methodology, D.X., X.J. and Y.G.; software, D.X. and Y.G.; validation, D.X., X.J. and X.C.; investigation, Y.G., X.J. and J.S.B.; writing—original draft preparation, D.X., X.J. and X.C.; writing—review and editing, J.S.B. and Y.G.; All authors have read and agreed to the published version of the manuscript.

Funding: This study was funded by the by Key Project of the National Social Science Foundation Informed Consent Statement: Please add "Informed consent was obtained from all subjects involved in the study." OR "Patient consent was waived due to REASON (please provide a detailed justification)." OR "Not applicable" for studies not involving humans.of China (19ZDA352), National Natural Science Foundation of China (No. 81772423), NSFC-RSE Joint Project (81911530253), and K.C. Wong Magna Fund in Ningbo University.

Data Availability Statement: Data available in a publicly accessible repository.

Conflicts of Interest: The authors declare no conflict of interest.

Appendix A

Table A1. An adapted STROBE Checklist for reporting this study.

Section/Topic	Item Number	Recommendation
Title and abstract	1	Provide in the abstract an informative and balanced summary of "The Risk of Anterior Cruciate Ligament Injury (ACL) Following Different Volleyball Spike Landing Ways"
Introduction		
Background/rationale	2	Explain the scientific background and rationale about ACL injury following different volleyball spike landing ways for the investigation being reported
Objectives	3	State specific objectives, including any prespecified hypotheses
Methods		
Participants	4	Report numbers of individuals of study (e.g., numbers potentially eligible, examined for eligibility, confirmed eligible, included in the study)
Instrumentation	5	Describe the hardware and software equipment used to complete the study (e.g., Vicon motion capture system, AMTI force plate, MATLAB R2019a)

Table A1. *Cont.*

Section/Topic	Item Number	Recommendation
Study design	6	Present key elements of this experimental procedure
Data collection and processing	7	Describe the data collection and processing process of this study
Statistical analysis methods	8	Describe all statistical analysis method
Results		
Main results	9	Present main results of this study
Visual data	10	Present figures and tables of key data
Discussion		
Key results	11	Summarize key results with reference to study objectives
Limitations	12	Discuss limitations of the study, taking into account sources of potential bias (e.g., lack of female subjects, small sample size)
Conclusion	13	Provide a general interpretation of results and discuss a landing strategy to avoid ACL injury
Other information		
Author contributions	14	Describe author contributions of this study
Funding	15	Present the funding of this study

References

1. Briner, W.; Benjamin, H.J. Managing acute and overuse disorders. *Phys. Sportsmed.* **1999**, *27*, 48. [CrossRef] [PubMed]
2. Bere, T.; Kruczynski, J.; Veintimilla, N.; Hamu, Y.; Bahr, R. Injury risk is low among world-class volleyball players: 4-year data from the FIVB Injury Surveillance System. *Br. J. Sports Med.* **2015**, *49*, 1132–1137. [CrossRef] [PubMed]
3. Kabacinski, J.; Murawa, M.; Dworak, L.B.; Maczynski, J. Differences in ground reaction forces during landing between volleyball spikes. *Trends Sport Sci.* **2017**, *24*, 87–92.
4. Salci, Y.; Kentel, B.B.; Heycan, C.; Akin, S.; Korkusuz, F. Comparison of landing maneuvers between male and female college volleyball players. *Clin. Biomech.* **2004**, *19*, 622–628. [CrossRef] [PubMed]
5. Ferretti, A.; Papandrea, P.; Conteduca, F.; Mariani, P.P. Knee ligament injuries in volleyball players. *Am. J. Sports Med.* **1992**, *20*, 203–207. [CrossRef]
6. Griffin, L.Y.; Agel, J.; Albohm, M.J.; Arendt, E.A.; Dick, R.W.; Garrett, W.E.; Garrick, J.G.; Hewett, T.E.; Huston, L.; Ireland, M.L. Noncontact anterior cruciate ligament injuries: Risk factors and prevention strategies. *J. Am. Acad. Orthop. Surg.* **2000**, *8*, 141–150. [CrossRef] [PubMed]
7. Caraffa, A.; Cerulli, G.; Projetti, M.; Aisa, G.; Rizzo, A. Prevention of anterior cruciate ligament injuries in soccer. *Knee Surg. Sports Traumatol. Arthrosc.* **1996**, *4*, 19–21. [CrossRef]
8. McNair, P.; Marshall, R.; Matheson, J. Important features associated with acute anterior cruciate ligament injury. *N. Z. Med.* **1990**, *103*, 537–539. [CrossRef]
9. Bakker, R.; Tomescu, S.; Brenneman, E.; Hangalur, G.; Laing, A.; Chandrashekar, N. Effect of sagittal plane mechanics on ACL strain during jump landing. *J. Orthop. Res.* **2016**, *34*, 1636–1644. [CrossRef]
10. Ali, N.; Robertson, D.G.; Rouhi, G. Sagittal plane body kinematics and kinetics during single-leg landing from increasing vertical heights and horizontal distances: Implications for risk of non-contact ACL injury. *Knee* **2014**, *21*, 38–46. [CrossRef]
11. Xu, D.; Cen, X.; Wang, M.; Rong, M.; István, B.; Baker, J.S.; Gu, Y. Temporal Kinematic Differences between Forward and Backward Jump-Landing. *Int. J. Environ. Res. Public Health* **2020**, *17*, 6669. [CrossRef] [PubMed]
12. Devita, P.; Skelly, W.A. Effect of landing stiffness on joint kinetics and energetics in the lower extremity. *Med. Sci. Sports Exerc.* **1992**, *24*, 108–115. [CrossRef]
13. Yu, B.; Lin, C.F.; Garrett, W.E. Lower extremity biomechanics during the landing of a stop-jump task. *Clin. Biomech.* **2006**, *21*, 297–305. [CrossRef] [PubMed]
14. Zhang, S.N.; Bates, B.T.; Dufek, J.S. Contributions of lower extremity joints to energy dissipation during landings. *Med. Sports Exerc.* **2000**, *32*, 812–819. [CrossRef] [PubMed]
15. Aizawa, J.; Ohji, S.; Koga, H.; Masuda, T.; Yagishita, K. Correlations between sagittal plane kinematics and landing impact force during single-leg lateral jump-landings. *J. Phys. Ther. Sci.* **2016**, *28*, 2316–2321. [CrossRef]
16. Zhou, H.; Ugbolue, U.C. Is There a Relationship Between Strike Pattern and Injury During Running: A Review. *Phys. Act. Health* **2019**, *3*, 127–134. [CrossRef]
17. Yeow, C.H.; Lee, P.V.; Goh, J.C. An investigation of lower extremity energy dissipation strategies during single-leg and double-leg landing based on sagittal and frontal plane biomechanics. *Hum. Mov. Sci.* **2011**, *30*, 624–635. [CrossRef]
18. Jaarsma, E.; Smith, B. The Development and Evaluation of a Participant Led Physical Activity Intervention for People with Disabilities Who Intend to Become More Active. *Phys. Act. Health* **2019**, *3*, 89–107. [CrossRef]

19. Taylor, J.B.; Ford, K.R.; Nguyen, A.-D.; Shultz, S.J. Biomechanical comparison of single-and double-leg jump landings in the sagittal and frontal plane. *Orth. J. Sport Med.* **2016**, *4*, 2325967116655158. [CrossRef]
20. Tillman, M.D.; Hass, C.J.; Brunt, D.; Bennett, G.R. Jumping and landing techniques in elite women's volleyball. *J. Sports Sci. Med.* **2004**, *3*, 30–36.
21. Krosshaug, T.; Slauterbeck, J.R.; Engebretsen, L.; Bahr, R. Biomechanical analysis of anterior cruciate ligament injury mechanisms: Three-dimensional motion reconstruction from video sequences. *Scand. J. Med. Sci. Sports* **2007**, *17*, 508–519. [CrossRef] [PubMed]
22. Markolf, K.L.; O'Neill, G.; Jackson, S.R.; McAllister, D.R. Effects of applied quadriceps and hamstrings muscle loads on forces in the anterior and posterior cruciate ligaments. *Am. J. Sports Med.* **2004**, *32*, 1144–1149. [CrossRef] [PubMed]
23. Hewett, T.E.; Myer, G.D.; Ford, K.R.; Heidt, R.S., Jr.; Colosimo, A.J.; McLean, S.G.; Bogert, A.J.; Paterno, M.V.; Succop, P. Biomechanical measures of neuromuscular control and valgus loading of the knee predict anterior cruciate ligament injury risk in female athletes: A prospective study. *Am. J. Sports Med.* **2005**, *33*, 492–501. [CrossRef] [PubMed]
24. Zwerver, J.; Gouttebarge, V.; Verhagen, E.; Maas, M.; Kilic, O. Incidence, aetiology and prevention of musculoskeletal injuries in volleyball: A systematic review of the literature. *Eur. J. Sport Sci.* **2017**, *17*, 1–29. [CrossRef]
25. Wang, L.; Peng, H. Biomechanical comparisons of single-and double-legged drop jumps with changes in drop height. *Int. J. Sports Med.* **2014**, *35*, 522–527. [CrossRef]
26. Donohue, M.R.; Ellis, S.M.; Heinbaugh, E.M.; Stephenson, M.L.; Zhu, Q.; Dai, B. Differences and correlations in knee and hip mechanics during single-leg landing, single-leg squat, double-leg landing, and double-leg squat tasks. *Res. Sports Med.* **2015**, *23*, 394–411. [CrossRef]
27. Gu, Y.; Ren, X.; Li, J.; Lake, M.; Zhang, Q.; Zeng, Y. Computer simulation of stress distribution in the metatarsals at different inversion landing angles using the finite element method. *Int. Orthop.* **2010**, *34*, 669–676. [CrossRef]
28. Lim, B.; An, K.O.; Cho, E.; Lim, S.T.; Cho, J. Differences in anterior cruciate ligament injury risk factors between female dancers and female soccer players during single-and double-leg landing. *Sci. Sports* **2020**. [CrossRef]
29. Wang, L.I. The lower extremity biomechanics of single-and double-leg stop-jump tasks. *J. Sports Sci. Med.* **2011**, *10*, 151–156. [PubMed]
30. Jiang, C. The Effect of Basketball Shoe Collar on Ankle Stability: A Systematic Review and Meta-Analysis. *Phys. Act. Health* **2020**, *4*, 11–18. [CrossRef]
31. Hébert-Losier, K.; Pini, A.; Vantini, S.; Strandberg, J.; Abramowicz, K.; Schelin, L.; Häger, C.K. One-leg hop kinematics 20 years following anterior cruciate ligament rupture: Data revisited using functional data analysis. *Clin. Biomech.* **2015**, *30*, 1153–1161. [CrossRef] [PubMed]
32. Smeets, A.; Malfait, B.; Dingenen, B.; Robinson, M.A.; Vanrenterghem, J.; Peers, K.; Nijs, S.; Vereecken, S.; Staes, F.; Verschueren, S. Is knee neuromuscular activity related to anterior cruciate ligament injury risk? A pilot study. *Knee* **2019**, *26*, 40–51. [CrossRef] [PubMed]
33. Pataky, T.C.; Robinson, M.A.; Vanrenterghem, J. Vector field statistical analysis of kinematic and force trajectories. *J. Biomech.* **2013**, *46*, 2394–2401. [CrossRef]
34. Pataky, T.C. Generalized n-dimensional biomechanical field analysis using statistical parametric mapping. *J. Biomech.* **2010**, *43*, 1976–1982. [CrossRef]
35. Cappozzo, A.; Cappello, A.; Croce, U.D.; Pensalfini, F. Surface-marker cluster design criteria for 3-D bone movement reconstruction. *Bio-Med. Mater. Eng.* **1997**, *44*, 1165–1174. [CrossRef]
36. Bosco, C.; Luhtanen, P.; Komi, P.V. A Simple Method for Measurement of Mechanical Power in Jumping. *Eur. J. Appl. Physiol.* **1983**, *50*, 273–282. [CrossRef] [PubMed]
37. McNitt-Gray, J.; Hester, D.; Mathiyakom, W.; Munkasy, B. Mechanical demand and multijoint control during landing depend on orientation of the body segments relative to the reaction force. *J. Biomech.* **2001**, *34*, 1471–1482. [CrossRef]
38. Winter, D.A. *Biomechanics and Motor Control of Human Movement*, 2nd ed.; John Wiley & Sons, Inc.: New York, NY, USA, 1990; pp. 154–219. [CrossRef]
39. Cohen, J. *Statistical Power Analysis for the Behavioral Sciences*, 2nd ed.; Lawrence Erlbaum Associates: Hillsdale, NJ, USA, 1990; pp. 19–32.
40. Boden, B.P.; Dean, G.S.; Feagin, J.A.; Garrett, W.E. Mechanisms of anterior cruciate ligament injury. *Orthopedics* **2000**, *23*, 573–578. [CrossRef]
41. Lephart, S.M.; Ferris, C.M.; Riemann, B.L.; Myers, J.B.; Fu, F.H. Gender differences in strength and lower extremity kinematics during landing. *Clin. Orthop. Relat. Res.* **2002**, *401*, 162–169. [CrossRef]
42. Olsen, O.E.; Myklebust, G.; Engebretsen, L.; Bahr, R. Injury mechanisms for anterior cruciate ligament injuries in team handball: A systematic video analysis. *Am. J. Sports Med.* **2004**, *32*, 1002–1012. [CrossRef]
43. Yeow, C.H.; Cheong, C.H.; Ng, K.S.; Lee, P.V.S.; Goh, J.C. Anterior cruciate ligament failure and cartilage damage during knee joint compression: A preliminary study based on the porcine model. *Am. J. Sports Med.* **2008**, *36*, 934–942. [CrossRef] [PubMed]
44. Draganich, L.F.; Vahey, J.W. An in vitro study of anterior cruciate ligament strain induced by quadriceps and hamstrings forces. *J. Biomech.* **1990**, *8*, 57–63. [CrossRef]
45. Nunley, R.M.; Wright, D.; Renner, J.B.; Yu, B.; Garrett, W.E., Jr. Gender comparision of patella tendon tibial shaft angle with weight bearing. *Res. Sports Med.* **2003**, *11*, 173–185. [CrossRef]
46. Boden, B.P.; Torg, J.S.; Knowles, S.B.; Hewett, T.E. Video analysis of anterior cruciate ligament injury: Abnormalities in hip and ankle kinematics. *Am. J. Sports Med.* **2009**, *37*, 252–259. [CrossRef] [PubMed]
47. Cerulli, G.; Benoit, D.L.; Lamontagne, M.; Caraffa, A.; Liti, A. In vivo anterior cruciate ligament strain behaviour during a rapid deceleration movement: Case report. *Knee Surg. Sports Traumatol. Arthrosc.* **2003**, *11*, 307–311. [CrossRef] [PubMed]

48. Shu, Y.; Sun, D.; Hu, Q.L.; Zhang, Y.; Li, J.S.; Gu, Y.D. Lower Limb Kinetics and Kinematics during Two Different Jumping Methods. *J. Biomim. Biomater. Biomed. Eng.* **2015**, *22*, 29–35. [CrossRef]
49. Bates, N.A.; Ford, K.R.; Myer, G.D.; Hewett, T.E. Impact differences in ground reaction force and center of mass between the first and second landing phases of a drop vertical jump and their implications for injury risk assessment. *J. Biomech.* **2013**, *46*, 1237–1241. [CrossRef]
50. Shimokochi, Y.; Ambegaonkar, J.P. Changing sagittal plane body position during single-leg landings influences the risk of non-contact anterior cruciate ligament injury. *Knee. Surg. Sports Traumatol. Arthrosc.* **2013**, *21*, 888–897. [CrossRef]
51. Zhang, B.; Lu, Q. A Current Review of Foot Disorder and Plantar Pressure Alternation in the Elderly. *Phys. Act. Health* **2020**, *4*, 95–106. [CrossRef]
52. Yu, P.; Gong, Z.; Meng, Y.; Baker, J.S.; István, B.; Gu, Y. The Acute Influence of Running-Induced Fatigue on the Performance and Biomechanics of a Countermovement Jump. *Appl. Sci.* **2020**, *10*, 4319. [CrossRef]
53. Decker, M.J.; Torry, M.R.; Wyland, D.J.; Sterett, W.I.; Steadman, J.R. Gender differences in lower extremity kinematics, kinetics and energy absorption during landing. *Clin. Biomech.* **2003**, *18*, 662–669. [CrossRef]
54. Norcross, M.F.; Blackburn, J.T.; Goerger, B.M.; Padua, D.A. The association between lower extremity energy absorption and biomechanical factors related to anterior cruciate ligament injury. *Clin. Biomech.* **2010**, *25*, 1031–1036. [CrossRef] [PubMed]

Article

How Should the Transition from Underwater to Surface Swimming Be Performed by Competitive Swimmers?

Jelena Stosic [1], Santiago Veiga [1,*], Alfonso Trinidad [2] and Enrique Navarro [1]

[1] Health and Human Performance Department, Universidad Politécnica de Madrid, 28040 Madrid, Spain; jstosic@gmail.com (J.S.); enrique.navarro@upm.es (E.N.)

[2] Faculty of Education and Humanities, Francisco de Vitoria University, 28223 Pozuelo de Alarcón-Madrid, Spain; alfonso.trinidad@ufv.es

* Correspondence: santiago.veiga@upm.es

Abstract: Despite the increasing importance of the underwater segment of start and turns in competition and its positive influence on the subsequent surface swimming, there is no evidence on how the transition from underwater to surface swimming should be performed. Therefore, the aim of the present study was to examine the role of segmental, kinematic and coordinative parameters on the swimming velocity during the pre-transition and transition phases. A total of 30 national male swimmers performed 4×25 m (one each stroke) from a push start at maximum velocity while recorded from a lateral view by two sequential cameras (50 Hz), and their kinematic and coordinative swimming parameters were calculated by means of two-dimensional direct linear transformation (DLT) algorithms. Unlike pre-transition, backward regression analysis of transition significantly predicted swimming velocity in all strokes except breaststroke (R^2 ranging from 0.263 in front crawl to 0.364 in butterfly). The inter-limb coordination was a predictor in butterfly stroke ($p = 0.006$), whereas the body depth and inclination were predictors in the alternate strokes (front crawl ($p = 0.05$) and backstroke ($p = 0.04$)). These results suggest that the body position and coordinative swimming parameters (apart from kicking or stroking rate and length) have an important influence on the transition performance, which depends on the swimming strokes.

Keywords: kinematics; inter-limb coordination; performance; direct linear transformation

Citation: Stosic, J.; Veiga, S.; Trinidad, A.; Navarro, E. How Should the Transition from Underwater to Surface Swimming Be Performed by Competitive Swimmers?. *Appl. Sci.* **2021**, *11*, 122. https://doi.org/10.3390/app11010122

Received: 30 November 2020
Accepted: 21 December 2020
Published: 24 December 2020

Publisher's Note: MDPI stays neutral with regard to jurisdictional claims in published maps and institutional affiliations.

1. Introduction

In the last few years, the importance of the underwater segment of the start and turns in swimming competitions has increased both quantitatively and qualitatively. The contribution of underwater swimming to the total race distances has considerably increased over the last 20 years [1] and elite swimmers now spend between 15 and 25% of the 100 m race distances underwater [2]. Additionally, swimmers achieving the fastest underwater velocities (especially on 100 m events) or the longest underwater distances (on 200 m events) have a critical advantage on their performances during World Championships [3]. The lower drag resistance experienced by swimmers underwater [4] as well as the improvement on the undulatory techniques [5], allow swimming competitors to achieve the fastest race velocities and to prioritize these segments over the mid-pool swimming techniques.

The benefits of the underwater swimming for competitive swimmers also include a transfer of momentum to the subsequent surface swimming during starts and turns [2]. Elite swimmers exhibited 5–10% faster swimming velocities after the start emersion compared to mid-pool swimming, with small increases of both stroke rate (SR) and length (SL) [2]. After the turn, those swimmers achieving the longest underwater distances (backstroke and butterfly) also obtained 5–6% faster velocities at emersion compared to mid-pool [2]. This positive impact of underwater on surface swimming can be achieved if a correct transition from underwater (the so-called breakout) is performed. Swimmers must restart their arm propulsion at this point and they no longer maintain the hydrodynamic position they adopt during underwater kicking [6]. While doing so, they must optimize the stroke timing in order to finish the arm pull (and begin their aerial recovery) when the head reaches the water surface. Additionally, in this phase, swimmers change location (from fully submerged to emerged at water surface) that could have implications on active drag [7], associated with an increase in wave drag [8]. The simultaneous (butterfly and breaststroke) or alternate (front crawl and backstroke) arm stroke techniques, the supine (backstroke) or prone (front crawl, butterfly and breaststroke) body position and the beginning of the flutter kick versus the undulatory kick [5] are other constraints that swimmers must handle in this phase.

In relation to the beginning of arm propulsion, previous studies have quantified the inter-limb coordination by measuring the time gap between the propulsive actions of arms and/or legs. Chollet et al. [9] found positive correlations between shorter inter-limb time gaps and the free-swimming velocity, although dependent on the swimmer skill [10], stroke [11], gender [12] and the use of added resistance [13]. For the swimmers' body positions, both body inclination and body depth have been examined as the different locations of the centre of mass and centre of flotation cause an active rotational torque, which influences the body projected frontal area [14]. An increased body angle has been negatively correlated with swimming underwater velocity [15] and surface velocity [16] due to increased projected area and, therefore, frontal drag [17,18]. On the other hand, body depth has been correlated positively with underwater velocity [19] due to decreased swimming resistance [20,21]. Finally, for the underwater kicking, Arellano and colleagues [22,23] indicated that the underwater undulatory swimming (UUS) velocity is dependent on kicking kinematics (i.e., kicking rate, amplitude, and length (distance per kick)), the kicking rate being the most important parameter. When competitive swimmers begin the first UUS cycle underwater, they are located at relatively greater depth (-0.92 m from water surface) [24] and display greater kicking rate and amplitude compared to the end of underwater swimming [25].

Although these kinematic parameters have been examined in the mid-pool swimming or in the underwater undulatory kicking phases, no information about their role in the transition from underwater to surface swimming has been explored as yet. For example, it is unknown whether great depth could be beneficial during transition given that it would require greater body angles (i.e., higher active drag) to reach water surface in a relatively restricted distance. Therefore, the aim of the present study was to examine the role of the segmental, kinematic and coordinative parameters on the swimming velocity during the pre-transition and transition phases of the push start. Considering the positive influence of underwater on the subsequent surface swimming, it could be expected that swimmers would decrease the time gaps between the propulsive arm actions, would minimize body inclinations and would maximize UUS velocity for faster transitioning velocity.

2. Materials and Methods

2.1. Participants

A total of thirty national level male swimmers (16.80 ± 1.44 years, 1.73 ± 0.05 m and 64.56 ± 6.78 kg) with a personal best time in the 100 m front crawl, backstroke, butterfly or breaststroke events within the 85% of world record (86.03 ± 2.34%), volunteered to participate in this study. Inclusion criteria for the experiment comprised a continuous attendance to the training sessions (9 in water sessions per week) as well as the absence of a major injury in the last three months before the experiment. The participants (>18 years) or their legal guardians (<18 years) signed a written consent document to participate in the study that was approved by the local Ethics Committee (number 45/2018) and in accordance with the Declaration of Helsinki [26].

2.2. Procedure

Swimming tests took place from 09:00 h until 14:00 h in a 50 × 25 m pool heated to 28° C and with 65% humidity. Participants maintained their usual training routine in the previous days before data collection and were asked to avoid physical activities (that were not related to swimming practice) at least 24 h before every session. Additionally, the participants were asked to avoid heavy food consumption and energy drinks at least 3–4 h before each test session. In a longitudinal lane of a pool, swimmers performed four repetitions (one of each stroke in a random order) of 25 m at their maximal velocity from a push start and with a 3 min rest between them. The push start consisted of swimmers pushing off the wall in-water from an initial position where one arm was in contact with the starting wall and the other arm was in-water. Swimmers were instructed to start with their feet at approximately one meter below the water surface and to push off the wall at the level of the feet to minimize the effect of wave drag [27]. Prior to the trials, black joint markers of round shape and made of waterproof adhesive tape (diameter 25 mm) were attached to the wrist, shoulder, hip, knee and ankle joint centers. Then, swimmers conducted a standardized warm-up which consisted of 10 min of warm-up activities on dry land and 20 min ($\approx$1.0 km) in the water.

2.3. Data Acquisition and Processing

Swimmers were recorded with two cameras placed behind the underwater windows (JVC GY-DV500E, 50 Hz, 1/1000 s) and with their optical axis perpendicular to the sagittal plane of the swimmers. Both cameras were located 12 m at the side of the swimmers' lane, the first camera being at 7.5 m and the second camera at 15 m from the starting wall. Swimming recordings were manually digitized and temporally coded by an experienced observer using biomechanical model of four points (shoulder, hip, knee and ankle). Two-dimensional direct linear transformation (2D-DLT) algorithms [28] were employed to transform the two dimensional screen coordinates (in pixels) to real meter coordinates. A rectangular structure of PVC (2 m high and 5 m wide) with known coordinates located in the swimmer's plane of movement was used for the camera's calibration prior to recordings in each camera view. The origin of the reference system was placed on the starting wall of the pool at the water surface level [29], with the horizontal and vertical axes aligned towards the starting direction and the above water direction, respectively.

The transitioning from underwater to surface swimming was divided in two phases: (1) pre-transition—the last underwater kicking action (downward kick plus upward kick movements, [30]) before the beginning of arm propulsion and (2) transition (breakout)—the first complete arm stroke cycle after pre-transition (Figure 1). The pre-transition was defined from the swimmers' feet in the highest position (or lowest position in backstroke), whereas the transition phase was defined from the swimmers' hands separation to the first or second water entry, respectively, for the butterfly or alternate strokes. In breaststroke, both pre-transition and transition phases were defined from the maximum flexed knee position on the first and subsequent breaststroke kick after the underwater arm pullout.

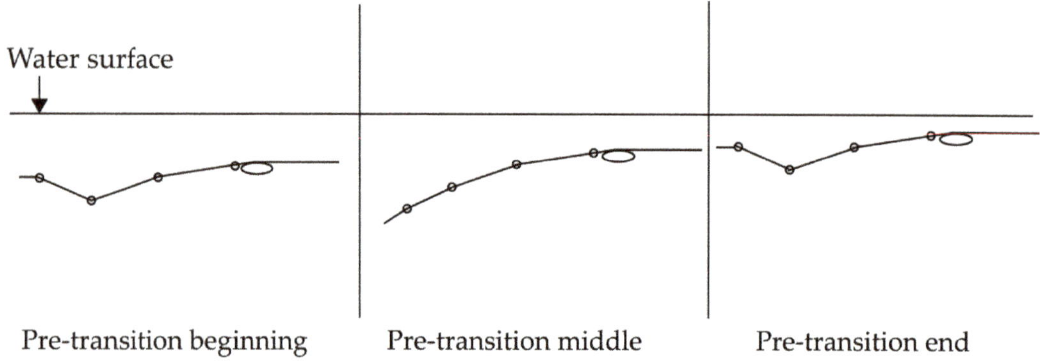

Pre-transition beginning   Pre-transition middle   Pre-transition end

a

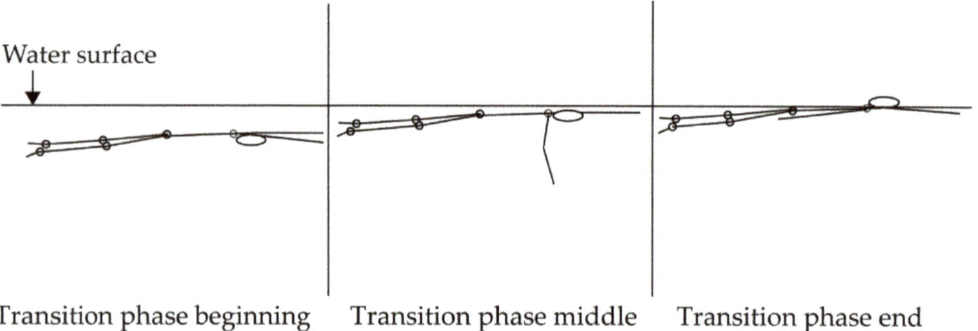

Transition phase beginning   Transition phase middle   Transition phase end

b

Figure 1. Schematic illustration of the pre-transition (**a**) and transition (**b**) phases from underwater to surface swimming.

2.4. Variables Definition

For each phase, the following kinematic parameters were calculated. In relation to the body position: (i) trunk inclination as the angle (°) between shoulder-hip line and horizontal, (ii) body inclination as the angle (°) between shoulder-knee and horizontal and (iii) body depth as vertical distance from water surface to hip position at the beginning of pre-transition or transition phases. Both inclination parameters were obtained at the beginning (hand separation from the hydrodynamic kicking position) and in the middle (hand pulling on the shoulder vertical) of the arm pull and were denoted as inclination 1 and inclination 2, respectively. For the kicking parameters: (i) kick amplitude was the vertical displacement of the ankle from the highest to the lowest position, (ii) kick length was the horizontal displacement of the hip in an entire kick (ascending and descending) cycle, and (iii) kick rate, as the inverse of time employed in an entire kick. For the stroking parameters: (i) SR was considered as the inverse of the time needed to complete one full stroke cycle, (ii) SL was the horizontal displacement of the hip during one stroke cycle and (iii) swimming velocity the multiplication of SL and SR [31].

For the coordinative parameters, the selected events whose lag times were calculated were adapted from previous work on the four competitive strokes [11]. Those lag times were then divided by the duration of the swimming cycle and multiplied by 360 degrees in order to get discrete relative phase (DRP) angles [32]. In front crawl and backstroke, two lag times (expressed as discrete relative phase angles) were calculated: DRP1, the end of the propulsive action (arm exit in front crawl and arm extension in backstroke) of the first pulling arm and the beginning of the propulsive action (arm pull) of the second pulling arm; DRP2, the end of the propulsive action of the second pulling arm and the beginning of the propulsive action of the third pulling arm (during the second swimming cycle). In butterfly, four discrete relative phase angles: DRP1, the hand separation from the streamlined position and the feet in the highest position during kick phase; DRP2, the beginning of the arms' pull and the feet in the lowest position in the kick cycle; DRP3, the beginning of the arms' push and the feet in the highest point of the kick period; DRP4, the arms' exit from water and the feet in the lowest position of the kick cycle. Finally, in breaststroke four discrete relative phase angles were obtained: DRP1, the end of leg propulsion (leg extension) and the beginning of the arm propulsion (arms' backward movement); DRP2, the feet together after leg extension and the beginning of arm propulsion (arms' backward movement); DRP3, the beginning of arm recovery (arms moving forward) and the beginning of leg recovery (legs moving forward); DRP4, the end of arm recovery (arms fully extended) and the end of leg recovery (legs fully flexed).

2.5. Data Analysis

After checking the normal distribution of the data by Kolmogorov–Smirnov test, parametric statistics was applied. One-way ANOVA (stroke) was applied in order to examine differences within the pre-transition and transition phase regarding kinematics (velocities, lengths and frequencies) and segmental variables (body and trunk inclinations, body depths). When main effects were identified, Bonferroni adjustments with effect sizes (ES, as partial eta-squared values) were applied to interpret meaningful effects. Multiple regression analysis (backward model) was used to identify the influence of body position and coordinative parameters on the average swimming velocity. Stroking or kicking parameters (length and rate) were not input variables for backward regression analysis due to their direct influence on swimming or kicking velocity, respectively. In addition, correlation coefficients (Pearson product moment correlation) were identified between each independent variable and dependent variable (velocity). The threshold values of the correlation coefficient that represented small, moderate, large, very large and nearly perfect correlations were 0.1, 0.3, 0.5, 0.7 and 0.9, respectively, according to recommendations in the literature [33]. Finally, an analysis of variance on the stroke was performed for all kinematic and coordinative parameters. All the analysis was performed using Statistical Package for Social Sciences (IBM SPSS for Windows, Version 20.0.; IBM Corp., Armonk, NY, USA), arranging data by stroke (front crawl, backstroke, butterfly and breaststroke). Data were expressed as mean ± standard deviation with 95% confidence intervals, unless otherwise indicated.

3. Results

Table 1 shows the kinematic parameters of competitive swimmers during pre-transition and transition phases of the push start. In the pre-transition phase, there were no differences on the kicking parameters of the front crawl, backstroke and butterfly strokes. However, there was stroke effect on the body depth ($F = 18.18$, $p = 0.000$, ES = 0.31), trunk inclination 1 ($F = 13.06$, $p = 0.000$, ES = 0.24), trunk inclination 2 ($F = 5.82$, $p = 0.001$, ES = 0.13) and body inclination 1 ($F = 12.71$, $p = 0.000$, ES = 0.24). In the transition phase, differences were observed between strokes on the SL ($F = 6.03$, $p = 0.001$, ES = 0.13) and SR ($F = 54.09$, $p = 0.000$, ES = 0.57) values, as well as on the majority of segmental parameters: body depth ($F = 39.47$, $p = 0.000$), trunk inclination 2 ($F = 17.97$, $p = 0.000$, ES = 0.31), body inclination 1 ($F = 51.77$, $p = 0.000$, ES = 0.57) and body inclination 2 ($F = 8.22$, $p = 0.000$, ES = 0.17).

Table 1. Segmental kinematics—mean (standard deviation)—of competitive swimmers when transiting from underwater to surface swimming in all four strokes.

	Front Crawl	Backstroke	Butterfly	Breaststroke
Pre-transition				
Kick length (m/cycle)	0.77 (0.12)	0.72 (0.10)	0.77 (0.12)	3.66 (0.50)
Kick rate (cycles/s)	2.15 (0.31)	2.14 (0.30)	2.14 (0.35)	0.39 (0.07)
Kick amplitude (m)	0.29 (0.06)	0.27 (0.18)	0.31 (0.06)	-
Body depth (m)	−0.39 (0.13)	−0.62 (0.22) [a]	−0.36 (0.12) [a]	−0.53 (0.16) [ac]
Trunk inclination 1 (°)	16.20 (5.26)	8.25 (6.13) [a]	16.13 (5.35) [a]	10.56 (7.30) [ac]
Trunk inclination 2 (°)	4.58 (3.07)	9.01 (5.22) [a]	4.91 (3.57) [a]	7.42 (6.38)
Body inclination 1 (°)	14.31 (5.07)	6.17 (4.55) [a]	13.14 (4.87) [a]	9.25 (8.19)
Body inclination 2 (°)	6.44 (3.96)	4.42 (3.70)	6.32 (4.50)	6.63 (5.39)
Transition				
Body depth (m)	−0.20 (0.11)	−0.52 (0.17) [a]	−0.20 (0.11) [a]	−0.34 (0.14) [abc]
Trunk inclination 1 (°)	9.00 (4.77)	9.02 (7.98)	9.00 (5.07)	9.80 (5.77)
Trunk inclination 2 (°)	7.45 (3.67)	12.78 (6.07) [a]	14.28 (6.76) [b]	17.50 (6.39) [bc]
Body inclination 1 (°)	11.66 (6.46)	6.37 (4.49) [a]	12.62 (5.56) [a]	25.26 (6.83) [abc]
Body inclination 2 (°)	9.66 (4.27)	12.98 (5.99)	11.15 (4.75)	15.42 (5.76) [ac]

Note: a, b and c superscripts denote statistical differences ($p < 0.05$) from first, second or third left column.

Figure 2 shows relationships of the pre-transition and transition parameters with the average velocity of each phase. During pre-transition, average velocity was small to moderately correlated with body depth in front crawl ($r = 0.262$) and butterfly ($r = 0.283$) strokes, whereas no correlations ($p > 0.05$) were observed for the body inclination. For the kicking variables, kick rate and kick length (but not amplitude) were moderate to largely correlated with faster velocities in all strokes, except the backstroke kicking rate. During transition, body inclination and body depth were negatively correlated with velocity in backstroke whereas no relationships were observed for the remaining strokes. In this phase, some of the coordinative parameters were moderately related to transition velocity, indicating shorter time gaps related to faster velocities except for DRP2 in front crawl and DRP3 in breaststroke.

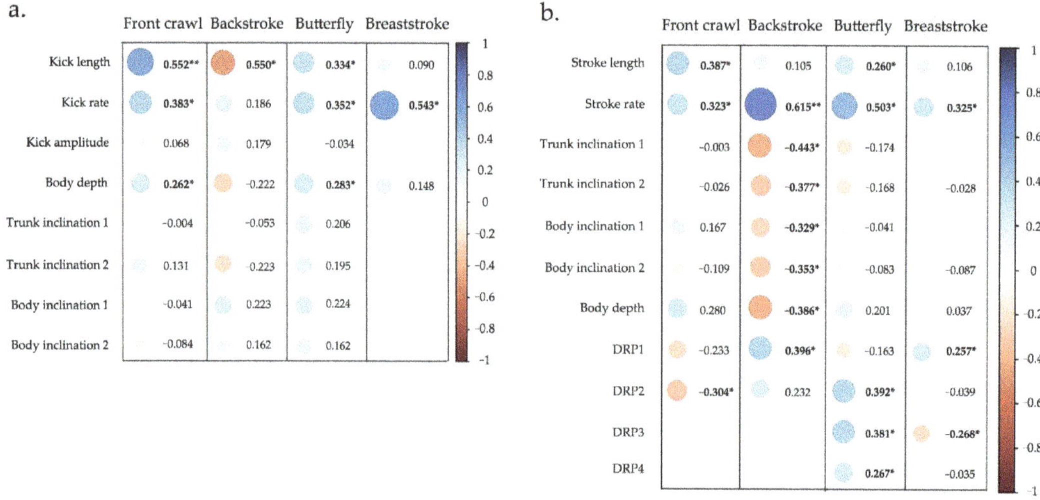

Figure 2. Relationships between the kinematic, segmental and coordination variables with average velocity of the competitive swimmers in the pre-transition (**a**) and transition (**b**) of the push start. Note: * denotes correlations at a $p < 0.05$ level.

Backward regression analysis of pre-transition phase performance revealed that the body position parameters did not statistically predict ($p > 0.05$) the average velocity in this phase. In transition (Table 2), the regression model significantly predicted swimming velocity in all strokes except breaststroke (with R^2 ranging from 0.263 in front crawl to 0.364 in butterfly). All inclination and depth parameters were predictors of velocity in front crawl ($p = 0.044$) and backstroke ($p = 0.039$) whereas all coordinative parameters were predictors in butterfly ($p = 0.006$).

Table 2. Backward regression analysis of the transition phases in all four strokes.

Dependent Variables	Backward Regression Model			
	Front Crawl	Backstroke	Butterfly	Breaststroke
Beta standardized coefficients				
Body depth	0.32	−0.40	1.80	0.28
Trunk inclination 1	−0.36	0.32		
Body inclination 1	0.38	−0.30		
DRP1		0.33	−0.52	0.82
DRP2	−0.35		0.62	−0.49
y-intercept (constant)	1.67 ± 0.08	1.92 ± 0.12	1.69 ± 0.08	1.25 ± 0.07
R^2	0.263	0.321	0.364	0.230
R^2 adjusted	0.154	0.213	0.293	0.125
Standard error of estimate	0.161	0.178	0.152	0.111
F	2.407	2. 959	5.141	2.191
p	0.044 *	0.039 *	0.006 **	0.118
Number of observations	30	30	30	30

Note: beta standardized coefficients presented for independent variables where $p < 0.05$.

4. Discussion

The aim of the current study was to examine the factors relevant for swimming performance during transition from underwater to surface swimming by applying correlation and regression analysis. Previously, it had been indicated that swimming velocities of elite swimmers after emersion from underwater were faster than during mid-pool swimming [2]. However, no information about how swimmers should perform this transition phase had been revealed. Our results indicate that body position and coordinative swimming parameters (besides SR and SL) have an influence on the transition performance, which magnitude depends on the swimming strokes.

4.1. Pre-Transition Phase

Before swimmers emerged from underwater, during the last underwater kicking cycle (pre-transition phase), the body position parameters (depth and inclination) did not statistically predict the forward velocity. At this point, only the kicking parameters (kicking length and rate) were correlated with velocity (Figure 2) and this highlighted the importance of kicking propulsion rather than technical position of the body underwater [34]. Values of kicking kinematics during pre-transition were similar to those obtained in USS by Arellano et al. [22,23], Hochstein et al. [35] and Alves et al. [36] although the magnitude of correlations with forward velocity were in line with previous research [23] both for the kicking rate ($r = 0.519$ compared to $r = 0.383$ in the present research) and kicking length ($r = 0.630$ compared to $r = 0.552$). The kicking amplitude, on the other hand, did not present statistical relationships with pre-transition velocity.

The body position parameters on the pre-transition phase were between $6°$ and $16°$ for the body inclination (respect to horizontal) and from −0.36 m (butterfly) to −0.62 m (backstroke) for the body depth (Table 1). These inclinations were in line with previous data from Arellano et al. [22] who reported body angles of $17°$ degrees during USS. However, the pre-transition body depth was obviously lower than the optimum (from −0.74 to −1.03 m) for UUS [24]. These data, interestingly, suggest that, when transiting for underwater to surface swimming, swimmers diminish depth but maintain body inclinations with a correct alignment almost parallel to water surface, not different from the UUS [15]. Otherwise, there would be an increase in frontal drag [7] with a concomitant velocity reduction [37]. As previously mentioned, the body position parameters presented small or no correlations with pre-transition velocity. However, the direction of relationships for the body depth depended on strokes (Figure 2). In backstroke, where body depth was greatest, negative correlation between depth and forward velocity were detected. On the other hand, in the ventral techniques (where body depth was lower) greater depths were related to faster velocities. This highlighted the stroke-dependent strategies when transiting from underwater to surface swimming.

4.2. Transition Phase

During transition, the swimmers' SR in all strokes showed greater correlations (medium to large) with average velocity than the SL (small to medium) (Figure 2). This was in line with the increased SR values of elite swimmers after start and turn emersion compared to mid-pool swimming [2], due to a shorter relative duration of the propulsive stroke phases. In a situation of an increased overall drag due to wave drag emergence [8], un-fatigued swimmers [38], probably prioritized to increase the SR to overcome resistance. Compared to mid-pool swimming, the stroking parameters during transition exhibited lower relationships with forward velocity (moderate to large) than previously reported (for instance, r values were from 0.87 to 0.92 for the SR [39–41]). This highlights the specific characteristic of the transition phase, where other kinematic factors (depth, inclination …) besides stroking parameters seem to have an important role for building velocity. Indeed, the regression models proposed in the present research, based on the body position and inter-limb coordination of transition, explained up to the 30% of the average velocity for the butterfly stroke, which represents a meaningful finding for swimmers and coaches.

When beginning the transition phase, swimmers in the present research were located between 0.20 and 0.52 m below the water surface and, interestingly, the body inclinations averaged 9° degrees regardless of stroke. These body inclinations were similar (10.68°) to those reported by Kjendlie et al. [16] in surface swimming. Probably, maximum velocity at which swimming trials were performed in the present research and the swimmer's skill level helped them to counteract the active rotational torque and maintaining a close body alignment to horizontal [14,42]. This could be the reason why the correlations between body inclinations and average velocity were low, especially in alternate strokes where no undulations were present. For the body depth, values close to 0.4 m below the water surface were in line to the minimum suggested by Lyttle et al. [20] to avoid wave drag. Interestingly, at this point, backstroke swimmers presented the deepest body positions compared to the ventral strokes that could be influenced by the more incongruent body side (compared to prone position) approaching water surface that would cause higher friction and form drag [43]. However, additionally, different arm pull position in backstroke compared to the ventral strokes (hand trajectory below or lateral to the body position) could explain depth differences.

For the coordinative parameters, a shorter time gap between the arm propulsion of both backstroke arms was related to faster transition velocities, which is in line with positive correlations between inter-limb coordination and free swimming velocity [9,40]. However, surprisingly, the second discrete relative phase in front crawl was negatively correlated with transition velocity (more time gap, faster velocity). This could be related to the role of breathing on the inter-limb coordination [44] after the second arm pull, as swimmers are usually encouraged to avoid breathing on the first arm propulsion after underwater. In the simultaneous strokes, shorter time gaps between the arm and leg butterfly propulsion were correlated with faster velocities during transition, indicating a preferable propulsion continuity to overcome drag forces. In breaststroke, on the contrary, longer times gaps between the arm and leg propulsion (DRP1) were related to faster velocities. Although motor continuity is generally recommended [45], breaststroke swimmers in transition from underwater must recover arms from the underwater pullout (hands close to thighs) instead of a regular arm pull. Therefore, the time gap between the propulsive phases of arms and legs could be expected to be longer. This represents an interesting point linked to the correct coordination of the underwater pullout [46].

According to regression analysis of transition-phase, the body position and inter-limb coordinative parameters of competitive swimmers statistically predicted an important amount (between 15 and 29% depending on the stroke, Table 2) of the variance in the forward velocity. Statistical models included the body depth in all four strokes and the body or trunk inclination in the alternate strokes (front crawl and backstroke) for predicting transition performance, highlighting the importance of an appropriate body positioning when approaching the water surface. This should be performed while minimizing body depth and maintaining trunk inclination in backstroke but minimizing inclination in front crawl and maintaining depth with water surface. Obviously, on the simultaneous techniques (butterfly and breaststroke) where body undulations occur [47,48], the effect of trunk/body position for velocity was probably hindered. For the inter-limb propulsion coordination, in butterfly the time gaps related to the beginning of arm propulsion (DRP1 and 2) were included in the regression model. This data probably indicates that swimmers had to primarily adapt their arm to leg movements at the beginning of the arms propulsion (not in DRP3 and 4) when the body position was still underwater. For the alternate strokes, shorter times gaps at the end of the first arm pull in backstroke and longer time gaps at the end of the second arm pull in front crawl were included in the regression model. These results highlight the technical demands of swimmers who successfully adapted to the changing constraints of the transition phase, i.e., increased drag forces (underwater to surface swimming), modified body position (while ascending to water surface) and arm-to-leg propulsion (compared to primarily leg propulsion in underwater).

4.3. Study Limitations and Future Research

The present results provide interesting insights on how swimmers could organize their movements when transiting from underwater to surface swimming. The underwater kicking execution seems to be the priority for swimmers before transiting to surface swimming. However, during the first arm stroke cycle, swimmers should carefully control their body inclination and depth as well as the time gap between propulsive arm or leg movements. At this point, the stroking parameters are still the most correlated variables with average velocity, but some other kinematic parameters can represent meaningful improvements for competitive swimmers. It should be also acknowledged that further research is still needed for kinematics parameters during transition as all measurements in the present study were discrete values. Continuous analysis of swimmers' inclination and depth when transiting from underwater would provide greater information relating to their body positioning. This is especially relevant for alternate strokes, where inclinations and body depth were part of the regression model.

5. Conclusions

The body position and inter-limb coordination of competitive swimmers when transiting from underwater to surface swimming represented important factors on the swimming velocity, explaining from 15 to 30% on the variance during the first arm stroke cycle. Differences were observed from simultaneous to alternate strokes as, in the simultaneous strokes, the arm-to-leg coordination at the beginning of arm propulsion was the predictor variable of the swimming velocity, whereas for the front crawl and backstroke (alternate) the body depth and inclination seemed to be the key factors. In the pre-transition phase at the end of the underwater kicking phase, no influence of the body position was observed as the kicking parameters (length and rate) were the predictor variables for the average velocity. Swimmers should carefully control their body inclination and depth as well as the inter-limb coordination on the first arm stroke cycle after underwater swimming.

Author Contributions: Conceptualization, S.V.; data curation, J.S. and A.T.; methodology, S.V., J.S. and E.N.; software, E.N.; validation, S.V., A.T. and E.N.; writing—original draft, J.S. and S.V.; writing—review and editing, S.V. All authors have read and agreed to the published version of the manuscript.

Funding: This research received no external funding.

Institutional Review Board Statement: The study was conducted according to the guidelines of the Declaration of Helsinki and approved by the Institutional Review Board (or Ethics Committee) of Universidad Politécnica de Madrid (45/2018).

Informed Consent Statement: Informed consent was obtained from all subjects involved in the study.

Data Availability Statement: The data presented in this study are available on request from the corresponding author.

Acknowledgments: Authors would like to thank Xiao Qiu for her assistance with manuscript figures.

Conflicts of Interest: The authors declare no conflict of interest.

References

1. Veiga, S.; Cala, A.; Mallo, J.; Navarro, E. A new procedure for race analysis in swimming based on individual distance measurements. *J. Sports Sci.* **2013**, *31*, 159–165. [CrossRef] [PubMed]
2. Veiga, S.; Roig, A. Effect of the starting and turning performances on the subsequent swimming parameters of elite swimmers. *Sport. Biomech.* **2017**, *16*, 34–44. [CrossRef] [PubMed]
3. Veiga, S.; Roig, A.; Gómez-Ruano, M.A. Do faster swimmers spend longer underwater than slower swimmers at World Championships? *Eur. J. Sport Sci.* **2016**, *16*, 919–926. [CrossRef] [PubMed]
4. Nicolas, G.; Bideau, B. A kinematic and dynamic comparison of surface and underwater displacement in high level monofin swimming. *Hum. Mov. Sci.* **2009**, *28*, 480–493. [CrossRef]
5. Takeda, T.; Sakai, S.; Takagi, H. Underwater flutter kicking causes deceleration in start and turn segments of front crawl. *Sport. Biomech.* **2020**. [CrossRef]
6. Naemi, R.; Easson, W.J.; Sanders, R.H. Hydrodynamic glide efficiency in swimming. *J. Sci. Med. Sport.* **2010**, *13*, 444–451. [CrossRef]
7. Kolmogorov, S.V.; Duplisheva, O.A. Active drag, useful mechanical power output and hydrodynamic force coefficient in different genders and performance levels. *J. Biomech.* **1992**, *25*, 311–318. [CrossRef]
8. Vennell, R.; Pease, D.; Wilson, B. Wave drag on human swimmers. *J. Biomech.* **2006**, *39*, 664–671. [CrossRef]
9. Chollet, D.; Chalies, S.; Chatard, J.C. A new index of coordination for the crawl: Description and usefulness. *Int. J. Sports Med.* **2000**, *21*, 54–59. [CrossRef]
10. Schnitzler, C.; Seifert, L.; Alberty, M.; Chollet, D. Hip velocity and arm coordination in front crawl swimming. *Int. J. Sports Med.* **2010**, *31*, 875–881. [CrossRef]
11. Chollet, D.; Seifert, L. Inter-limb coordination in the four competitive strokes. In *World Book of Swimming: From Science to Performance*; Nova Science Publishers, Inc.: Hauppauge, NY, USA, 2011; pp. 153–172. ISBN 9781616682026.
12. Seifert, L.; Boulesteix, L.; Chollet, D. Effect of gender on the adaptation of arm coordination in front crawl. *Int. J. Sports Med.* **2004**, *25*, 217–223. [CrossRef] [PubMed]
13. Schnitzler, C.; Brazier, T.; Button, C.; Seifert, L.; Chollet, D. Effect of velocity and added resistance on selected coordination and force parameters in front crawl. *J. Strength Cond. Res.* **2011**, *25*, 2681–2690. [CrossRef] [PubMed]
14. Kjendlie, P.L.; Stallman, R.K.; Stray-Gundersen, J. Passive and active floating torque during swimming. *Eur. J. Appl. Physiol.* **2004**, *93*, 75–81. [CrossRef] [PubMed]
15. Houel, N.; Elipot, M.; André, F.; Hellard, P. Influence of angles of attack, frequency and kick amplitude on swimmer's horizontal velocity during underwater phase of a grab start. *J. Appl. Biomech.* **2013**, *29*, 49–54. [CrossRef] [PubMed]
16. Kjendlie, P.L.; Ingjer, F.; Stallman, R.K.; Stray-Gundersen, J. Factors affecting swimming economy in children and adults. *Eur. J. Appl. Physiol.* **2004**, *93*, 65–74. [CrossRef] [PubMed]
17. Wang, K.F.; Wang, L.Z.; Yan, W.X.; Li, D.J.; Xiong, S. A new device for estimating active drag in swimming at maximal velocity. *J. Sports Sci.* **2007**, *25*, 375–379.
18. Kolmogorov, S.V.; Rumyantseva, O.A.; Gordon, B.J.; Cappaert, J.M. Hydrodynamic characteristics of competitive swimmers of different genders and performance levels. *J. Appl. Biomech.* **1997**, *13*, 88–97. [CrossRef]
19. Ruschel, C.; Araujo, L.G.; Pereira, S.M.; Roesler, H. Kinematical analysis of the swimming start: Block, flight and underwater phases. In Proceedings of the XXV ISBS Symposium, Ouro Preto, Brazil, 23–25 August 2007; pp. 385–388.
20. Lyttle, A.D.; Blanksby, B.A.; Elliot, B.C.; Lloyd, D.G. The effect of depth and velocity on drag during the streamlined guide. *J. Swim. Res.* **1998**, *13*, 15–22.
21. Machado, L.; Ribeiro, J.; Costa, L.; Silva, A.J.; Rouboa, A.I.; Mantripragada, N.; Marinho, D.A.; Fernandes, R.; Vilas-Boas, J.P. The effect of depth on the drag force during underwater gliding: A CFD approach. *Int. Soc. Biomech. Sport.* **2010**, *4*–5.

22. Arellano, R.; Pardillo, S.; Gavilán, A. Underwater undulatory swimming: Kinematic characteristics, vortex generation and application during the start, turn and swimming strokes. In Proceedings of the XX International Symposium on Biomechanics in Sports, Cáceres, Spain, 1–5 July 2002.
23. Arellano, R.; Pardillo, S.; Gavilán, A. Usefulness of the strouhal number in evaluating human underwater undulatory swimming. In Proceedings of the IX Symposium Mondial Biomécanique et Médecine de la Natation, Saint-Etienne, France, 21–23 June 2002; pp. 33–38.
24. Tor, E.; Pease, D.L.; Ball, K.A. Comparing three underwater trajectories of the swimming start. *J. Sci. Med. Sport.* **2015**, *18*, 725–729. [CrossRef]
25. De Jesus, K.; de Jesus, K.; Machado, L.; Fernandes, R.J.; Vilas-Boas, J.P. Linear kinematics of the underwater undulatory swimming phase performed after two backstroke starting techniques. In Proceedings of the 30th International Conference on Biomechanics in Sports, Melbourne, Australia, 2–6 July 2012; pp. 371–374.
26. World Medical Association. World Medical Association Declaration of Helsinki. Ethical principles for medical research involving human subjects. *Bull. World Health Organ.* **2001**, *79*, 373–374.
27. Lyttle, A.; Blanksby, B. A look at gliding and underwater kicking in the swim turn. In Proceedings of the 18th International Symposium on Biomechanics in Sports, Hong Kong, China, 25–30 June 2000.
28. Abdel-Aziz, Y.I.; Karara, H.M. Direct linear transformation from comparator coordinates into space coordinates in close range photogrammetry. In Proceedings of the Symposium on Close Range Photogrammetry, Urbana, IL, USA, 26–29 January 1971; pp. 1–18.
29. Takeda, T.; Takagi, H.; Tsubakimoto, S. Effect of inclination and position of new swimming starting block's back plate on track-start performance. *Sport. Biomech.* **2012**, *11*, 370–381. [CrossRef] [PubMed]
30. Atkison, R.R.; Dickey, J.P.; Dragunas, A.; Nolte, V. Importance of sagittal kick symmetry for underwater dolphin kick performance. *Hum. Mov. Sci.* **2014**, *33*, 298–311. [CrossRef] [PubMed]
31. Barbosa, T.M.; de Jesus, K.; Abraldes, J.A.; Ribeiro, J.; Figueiredo, P.; Vilas-Boas, J.P.; Fernandes, R.J. Effects of protocol step length on biomechanical measures in swimming. *Int. J. Sports Physiol. Perform.* **2015**, *10*, 211–218. [CrossRef] [PubMed]
32. Wheat, J.; Glazier, P. Techniques for Measuring Coordination and Coordination Variability. In *Variability in the Movement System: A Multi-Disciplinary Perspective*; Davids, K., Bennett, S.J., Newell, K., Eds.; Human Kinetics: Champaign, IL, USA, 2005.
33. Hopkins, W.G.; Marshall, S.W.; Batterham, A.M.; Hanin, J. Progressive statistics for studies in sports medicine and exercise science. *Med. Sci. Sports Exerc.* **2009**, *41*, 3–12. [CrossRef]
34. Nakashima, M. Simulation analysis of the effect of trunk undulation on swimming performance in underwater dolphin kick of human. *J. Biomech. Sci. Eng.* **2009**, *4*, 94–104. [CrossRef]
35. Hochstein, S.; Blickhan, R. Body movement distribution with respect to swimmer's glide position in human underwater undulatory swimming. *Hum. Mov. Sci.* **2014**, *38*, 305–318. [CrossRef]
36. Alves, F.; Lopes, P.; Veloso, A.; Martins-Silva, A. Influence of body position on dolphin kick kinematics. In Proceedings of the 24th International Symposium on Biomechanics in Sports, Salzburg, Austria, 14–18 July 2006; pp. 3–6.
37. Toussaint, H.M.; de Groot, G.; Savelberg, H.H.; Vervoorn, K.; Hollander, A.P.; van Ingen Schenau, G.J. Active drag related to velocity in male and female swimmers. *J. Biomech.* **1988**, *21*, 435–438. [CrossRef]
38. Suito, H.; Ikegami, Y.; Nunome, H.; Sano, S.; Shinkai, H.; Tsujimoto, N. The effect of fatigue on the underwater arm stroke motion in the 100 M front crawl. *J. Appl. Biomech.* **2008**, *24*, 316–324. [CrossRef]
39. Wakayoshi, K.; D'Acquisto, L.J.; Cappaert, J.M.; Troup, J.P. Relationship between oxygen uptake, stroke rate and swimming velocity in competitive swimming. *Int. J. Sports Med.* **1995**, *16*, 19–23. [CrossRef]
40. Seifert, L.; Chollet, D.; Bardy, B. Effect of swimming velocity on arm coordination in the front crawl: A dynamic analysis. *J. Sports Sci.* **2004**, *22*, 651–660. [CrossRef] [PubMed]
41. Barbosa, T.M.; Keskinen, K.L.; Fernandes, R.; Colaço, P.; Carmo, C.; Vilas-Boas, J.P. Relationships between energetic, stroke determinants, and velocity in butterfly. *Int. J. Sports Med.* **2005**, *26*, 841–846. [CrossRef] [PubMed]
42. Strzala, M.; Krezalek, P. The body angle of attack in front crawl performance in young swimmers. *Hum. Mov.* **2010**, *11*, 23–28. [CrossRef]
43. Anderson, E.J.; McGillis, W.R.; Grosenbaugh, M.A. The boundary layer of swimming fish. *J. Exp. Biol.* **2001**, *204*, 81–102. [PubMed]
44. Seifert, L.; Chollet, D.; Allard, P. Arm coordination symmetry and breathing effect in front crawl. *Hum. Mov. Sci.* **2005**, *24*, 234–256. [CrossRef] [PubMed]
45. Seifert, L.; Leblanc, H.; Chollet, D.; Delignières, D. Inter-limb coordination in swimming: Effect of speed and skill level. *Hum. Mov. Sci.* **2010**, *29*, 103–113. [CrossRef]
46. Seifert, L.; Vantorre, J.; Chollet, D. Biomechanical analysis of the breaststroke start. *Int. J. Sports Med.* **2007**, *28*, 970–976. [CrossRef]
47. Sanders, R.H.; Cappaert, J.M.; Devlin, R.K. Wave characteristics of butterfly swimming. *J. Biomech.* **1995**, *28*, 9–16. [CrossRef]
48. Sanders, R.H.; Cappaert, J.M.; Pease, D.L. Wave characteristics of olympic breaststroke swimmers. *J. Appl. Biomech.* **1998**, *14*, 40–51. [CrossRef]

 *applied sciences*

Article

Estimating Cycling Aerodynamic Performance Using Anthropometric Measures

Raman Garimella [1,2,*,†], Thomas Peeters [1,†], Eduardo Parrilla [3], Jordi Uriel [3], Seppe Sels [4], Toon Huysmans [5] and Stijn Verwulgen [1,*]

1   Department of Product Development, Faculty of Design Sciences, University of Antwerp, 2000 Antwerp, Belgium; Thomas.Peeters2@uantwerpen.be

2   NA, Voxdale bv, 2110 Wijnegem, Belgium

3   Instituto de Biomecánica, Universitat Politecnica de Valencia, 46022 Valencia, Spain; eduardo.parrilla@ibv.org (E.P.); jordi.uriel@ibv.org (J.U.)

4   Op3Mech, Faculty of Applied Engineering, University of Antwerp, 2020 Antwerp, Belgium; seppe.sels@uantwerpen.be

5   Section on Applied Ergonomics and Design, Faculty of Industrial Design Engineering, Delft University of Technology, 2628 CE Delft, The Netherlands; t.huysmans@tudelft.nl

*   Correspondence: raman.garimella@uantwerpen.be (R.G.); stijn.verwulgen@uantwerpen.be (S.V.)

†   These authors are co-first authors.

Received: 11 November 2020; Accepted: 1 December 2020; Published: 2 December 2020

Abstract: Aerodynamic drag force and projected frontal area (A) are commonly used indicators of aerodynamic cycling efficiency. This study investigated the accuracy of estimating these quantities using easy-to-acquire anthropometric and pose measures. In the first part, computational fluid dynamics (CFD) drag force calculations and A (m^2) values from photogrammetry methods were compared using predicted 3D cycling models for 10 male amateur cyclists. The shape of the 3D models was predicted using anthropometric measures. Subsequently, the models were reposed from a standing to a cycling pose using joint angle data from an optical motion capture (mocap) system. In the second part, a linear regression analysis was performed to predict A using 26 anthropometric measures combined with joint angle data from two sources (optical and inertial mocap, separately). Drag calculations were strongly correlated with benchmark projected frontal area (coefficient of determination $R^2 = 0.72$). A can accurately be predicted using anthropometric data and joint angles from optical mocap (root mean square error (RMSE) = 0.037 m^2) or inertial mocap (RMSE = 0.032 m^2). This study showed that aerodynamic efficiency can be predicted using anthropometric and joint angle data from commercially available, inexpensive posture tracking methods. The practical relevance for cyclists is to quantify and train posture during cycling for improving aerodynamic efficiency and hence performance.

Keywords: 3D shape modeling; aerodynamics; computational fluid dynamics; cycling; projected frontal area; inertial sensors; joint biomechanics; motion capture system

---

1. Introduction

In road cycling, aerodynamic drag force (or 'drag') contributes to 70–90% of the resistance to the cyclist on level ground [1]. Improving cycling performance is a priority for elite and amateur cyclists. A wind tunnel is the gold standard for measuring aerodynamic force, drag area, and studying air flow behavior. However, a wind tunnel is not easily accessible or affordable even for elite athletes. Several alternative methods for measuring aerodynamic performance in athletes in controlled environments are described in the literature including photogrammetry [2]; power meters [3]; and air pressure and speed sensors [4,5]. The approach of computational fluid dynamics (CFD) has been used on models of cyclists and bicycles under various conditions to determine the optimum cyclist pose and selection of equipment and accessories [6–10] as well as the influence of aerodynamics during professional cycling races [11–13].

To use CFD in the field of cycling aerodynamics, two-dimensional (2D) or three-dimensional (3D) models of the cyclist and equipment are required. A 3D scanning device provides accurate models, but state-of-the-art scanning equipment is difficult to access for most athletes. Indirect methods to predict the 3D shape of a given human using select anthropometric data have been described in the literature [14–17]. If the 3D model of a cyclist is available, directly or indirectly, but not in a desirable pose configuration, it is possible to re-pose the model using animation techniques [18,19]. In this study, we used 3D models obtained from an algorithm [16,17] and re-posed them to various pose configurations to investigate if the cyclist aerodynamic drag could be predicted by a combination of anthropometric data and CFD analyses.

Aerodynamic drag force F (N) on an object moving through a fluid is given by

$$F = 0.5 \cdot C_D \cdot \rho \cdot v^2 \cdot A, \tag{1}$$

where C_D is the drag coefficient, a dimensionless quantity that is a function of Reynold's number, Mach number, the form drag and skin friction of the object (C_D is generally found from experiment); ρ is the air density at a given pressure and temperature (kg/m^3); v is the velocity of air relative to the object (m/s); and A is the projected frontal area of object (m^2).

Projected frontal area of the cyclist is the biggest factor influencing drag during cycling (up to 90%) [8] where the cyclist accounts for up to 70% and the rest is accounted for by the bicycle [20]. The importance of the projected frontal area as a key metric for cycling performance has been stressed in the literature, with several methods proposed for measurement and estimation including (digital) photogrammetry, planimetry, wind tunnel tests, and field tests [21–25].

The projected frontal area of a cyclist depends upon the shape and pose of the cyclist. The shape of a human can be estimated from anthropometric measures or features as described in [26]. The pose of a human can be uniquely described by a set of joint angles. In this study, we combined the shape and pose information to ultimately predict the projected frontal area of a cyclist. We did this by first predicting the 3D shape of 10 participants in a standardized (standing) pose using basic body measures. 3D shapes of the participants in the cycling pose will be produced using joint angle information using motion capture (mocap). From these final shapes, the projected frontal area can be calculated.

Anthropometric measures in this study were recorded manually by the researchers. Joint angles were recorded using two mocap techniques: optical and inertial. Optical mocap systems are considered the gold standard for human mocap [27], but it is typically restricted to indoor settings. Bike-fitting and aerodynamic analysis usually take place in indoor environments, where several outdoor conditions are neglected [28]. Analyzing cycling movements in realistic outdoor circumstances would be an added value. Therefore, mocap using inertial measurement units (IMUs), which have been shown as reliable for human mocap, can be used for this application [27]. The advantage of inertial mocap is that full body joint angles in multiple degrees of freedom can be provided continuously and analyzed directly in real-time or afterward.

The aim of this study was to investigate low-cost and easy-to-deploy methods to predict the drag and projected frontal area of cyclists using rudimentary anthropometric and joint angle data. Using information that is easy to acquire even by untrained personnel, we provide a proof of principle that aerodynamic analyses and pose-training can be done at home, indoors, potentially outdoors, and for various pose configurations at scale.

2. Materials and Methods

2.1. Participants

Ten male amateur cyclists were recruited in the study (n = 10, age = 32.5 ± 6.7 years, body height = 176.1 ± 5.8 cm and body mass = 74.6 ± 15.1 kg). Ethical approval and consent were obtained prior to the measurements (17/21/261, Ethics Committee, University of Antwerp, Antwerp, Belgium). Table 1 lists the anthropometric parameters collected from tape measures by one researcher corresponding with the ISAK guidelines as well as ISO 8559 guidelines.

Table 1. List of participant data that was recorded from measurements and optical and inertial motion capture.

Anthropometric Data (cm)		2D Joint Angles Optical Mocap (°)	3D Joint Angles Inertial Mocap (°)
Body Height	Shoulder Breadth (acromion)	Left Knee	Left Knee
Chest Circumference	Hip Breadth (standing)	Left Elbow	Left Shoulder
Under-Bust Circumference	Chest Circumference (scye)	Left Hip	Left Elbow
Waist Circumference (minimum)	Hip Circumference	Back	Right Knee
Waist Circumference (trousers)	Arm Circumference (scye)	Head	Right Shoulder
Neck Circumference (shirt)	Shoulder Breadth (bideltoid)	Left Shoulder	Right Elbow
Neck Circumference (tight, hull)	Hip Breadth (sitting)		Neck (2D)
Lower Arm Circumference (mid, Hull)	Upper Arm Length		Left Hip
Biceps Circumference Hull	Lower Arm Length		Right Hip
Spine-Shoulder Length	Sternum to Femur Length		Pelvis
Arm Length	Upper Leg Length		Left Wrist
Back Length (shirt)	Lower Leg Length		Right Wrist
Torso Length (shirt)	Body Mass (kg)		Chest
Age (years)	Gender (m/f)		

2.2. Optical Mocap

The camera of a smartphone (Lenovo Group Limited, Hong Kong, China) was used to record 2D angles of the participant. The camera, which could record pictures as well as videos (at 30 Hz), was positioned on the left-hand side of the participant at a fixed distance, with the lens parallel to the sagittal plane of the participant. The images and videos were analyzed for joint angles using an open-source image-processing algorithm (Detectron, Facebook AI Research, Facebook Inc., Cambridge, MA, USA) to identify the human in frame, the major joints of the skeleton, and the lines between the joints. The joint centers were defined as (Figure 1a): ankle joint as the *lateral malleolus*, knee joint as the patella, hip joint as the *greater trochanter*, shoulder joint as the *acromion*, elbow joint as the *lateral epicondyle*, and wrist joint as the *lunate bone*. Based on the joint coordinates, flexion/extension angles were determined (Table 1). Joint angles of the right-hand side of the participants were provided by the algorithm, but were excluded from analyses to avoid unreliable data induced by parallax errors.

The projected frontal area of the participants was captured using an infrared depth-sensing camera (Intel® RealSense™ Depth Camera D415, Intel Corporation, Mountain View, CA, USA) with a sampling frequency of 3 Hz placed in front of the cyclist (Figure 2). Furthermore, an average of ten iterations was calculated to eliminate the influence of noise and different pedal position [29].

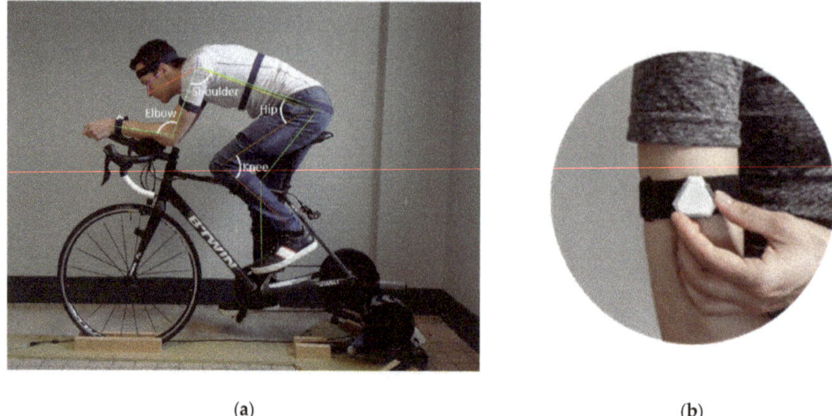

(a)   (b)

Figure 1. (**a**) A participant in the time trial position on the stationary bike. The angles obtained from the optical mocap are illustrated. The wearable inertial sensors on the participant can also be seen. (**b**) A single sensor unit.

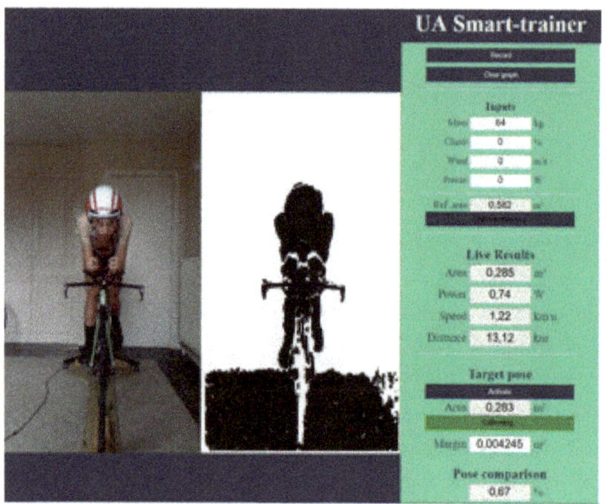

Figure 2. Screenshot of projected frontal area camera capture interface. A cyclist is seen in RGB (red green blue) picture format and their projected area is shown as a silhouette. A calibration step is required to eliminate the floor from the projected frontal area calculation. The area in m^2 is also present in the interface and is recorded at 3 Hz.

2.3. Inertial Mocap

The inertial sensors comprised of a set of 11 wearable units (Figure 1b) (Notch Interfaces Inc., Brooklyn, NY, USA) strapped to various body segments of the participant (Figure 3). The units consisted of accelerometers, gyroscopes, and magnetometers. The accuracy of the IMUs was 2° for yaw, pitch, and roll rotations, whereas the accuracy of the whole system was optimized by considering several factors such as a correct steady pose, a tight fit of the sensors during the measurements, and avoiding magnetic interference from the environment. They were calibrated to continuously obtain 3D joint angle measurements of knees, hips, shoulders, elbows, wrists, neck, pelvis, and chest (full list in Table 1). Hip and pelvis joint angles were excluded from the analysis since the hip strap moves during cycling due to unavoidable contact with the upper legs and this induced errors in recording angles.

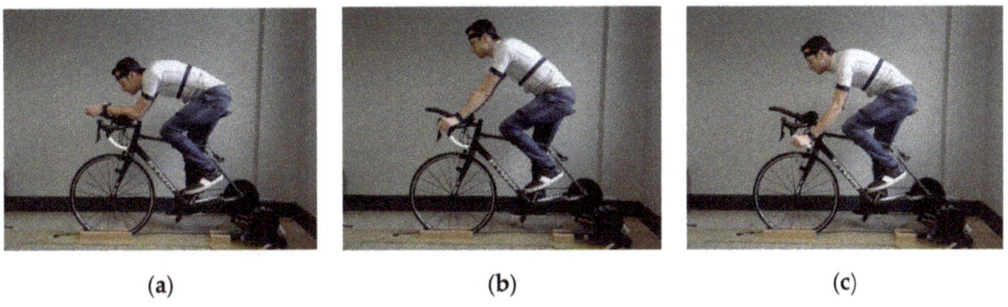

(a) **(b)** **(c)**

Figure 3. A participant photographed in three poses as instructed in the Static protocol (**a**) *TT*, (**b**) *Hoods*, and (**c**) *Drops*.

2.4. Protocol

Participants were asked to follow a series of instructions on a road bike (Btwin, Decathlon S.A., Lille, France) fixed on a stationary mount. All participants were provided with tight-fitting clothing. The experiment consisted of:

1. Static protocol, where the instruction was to maintain three poses (Figure 3) for at least five seconds each, with the right leg extended to the bottom of the pedal stroke:

    - *Time Trial (TT)*: race pose with arms resting horizontally on the clip-on handlebars;
    - *Hoods*: relaxed pose with hands on the hoods of the handlebar; and
    - *Drops*: race pose with hands on the dropped handlebars.

    For each pose, a picture with the sagittal camera and a frontal camera were simultaneously recorded.

2. Dynamic protocol, which lasted roughly two minutes per participant. The cyclists were instructed to pedal at a cadence of roughly 1 Hz and position their hands on the handlebar hoods and:

    - bend their back from the highest possible to lowest possible inclination and back over 30 s;
    - pronate knees from closest to farthest from top tube and back over 30 s;
    - extend neck from lowest possible to highest possible angle and back over 30 s; and
    - proceed to perform 30 s of comfortable cycling.

    Along with a video recording at a rate of 30 frames per second from the sagittal camera, the projected area (3 Hz) and joint angles from IMUs were registered continuously (10 Hz).

2.5. Procedure

The measurements from the static protocol were used to investigate the relation between CFD drag force and benchmark projected frontal area. Predicted 3D models of the participants in a standing pose were obtained using anthropometric measures. These models were re-posed to the cycling pose using the angles from the optical mocap and were used in CFD analysis to calculate drag force.

Second, the data from the dynamic protocol were used to perform a regression analysis with the aim to predict the projected frontal area based on anthropometric data and joint angles. Figure 4 shows an overview of the methods.

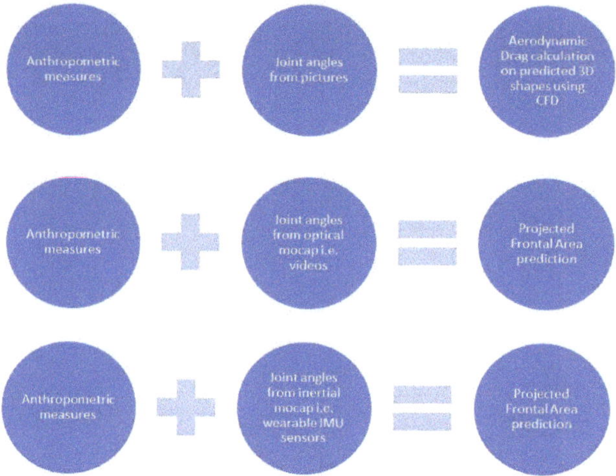

Figure 4. Schematic diagram of the methods evaluated in this study.

2.5.1. 3D Models

The models were generated using a statistical shape model using the methodology described in the literature [17]. The method utilizes a partial least square regression between body measurements and principal component analysis (PCA) components. Using this method, given a set of measurements (in this study: age, gender, body height, body mass, chest circumference, hip circumference maximum, and arm length), we can obtain PCA components that can be used to predict a 3D shape of a human body. The database and body measurements are described in [16]. Each model is a combination of 100 (PCA) components with a corresponding weight (or 'score') attached to each component. The model provides the best-fit human shape for the given inputs. The physical meaning of individual components is not readily apparent, but the relevant modes will be discussed in the Results section.

The 3D models were obtained in a standardized standing 'A-pose' (Figure 5). Since all models in the statistical shape model were registered to match certain body landmarks (i.e., joints), it was easy to rig a skeleton in the 3D models (Figure 6a). The skeleton was based on the models proposed by the International Society of Biomechanics and was optimized using large datasets of 3D human scans in different poses.

Using joint angles from the optical mocap, the standing models were re-posed to *TT*, *Hoods*, and *Drops* poses (Figure 6b,c) using animation software (Blender, Blender Foundation, Amsterdam, The Netherlands). The skeleton in the optical mocap method had fewer bones compared to the skeleton of the predicted 3D models. The angles from the optical mocap were accordingly adjusted to fit the 3D model skeleton (e.g., the back was modeled as one straight bone as a simplification). We assumed that the upper body of the 3D models were symmetric, as instructed to the participants. Hence, the left-hand side upper body angles were sufficient for re-posing these models to the unique pose of every participant. The pictures were used as a reference when required.

Projected frontal area of the 3D models was calculated using drawing software (PTC Creo, PTC Inc., Boston, MA, USA). This data will be compared to the benchmark projected frontal area obtained from the depth sensing camera.

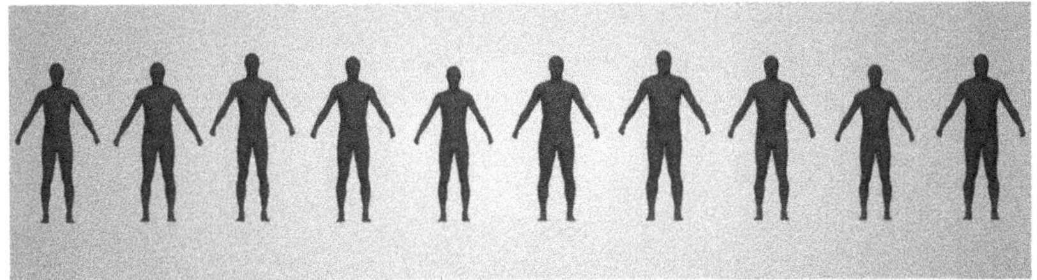

Figure 5. A front-view of the standing 3D models generated of the 10 participants.

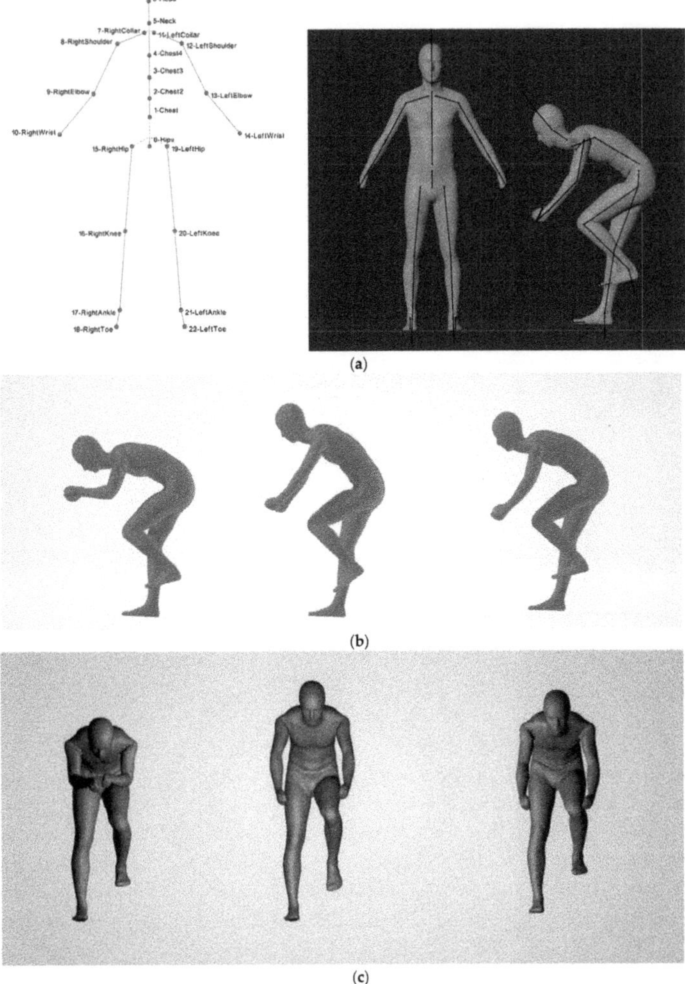

Figure 6. (**a**) (Left) The body joints used to define the rig/skeleton are listed. (Right) The rig in the standing model of one participant and the rig adapted to the *Drops* pose of the same participant. (**b**) 3D shapes of the participant in *TT, Hoods,* and *Drops* poses. (**c**) Front view of the 3D models of one participant in the same poses as above.

2.5.2. Computational Fluid Dynamics (CFD) Analysis

CFD calculations on the 3D models were conducted (Star CCM+, Siemens Industry Software Inc., Plano, TX, USA) in a domain that consisted of a wind tunnel of 6.5 m × 2.5 m × 2.5 m (L × B × H), modeled after a wind tunnel facility [30]. Cyclist models were positioned at 2 m from the inlet of the test section, with the models 'facing' the opposite direction of the wind. The surface of the objects was assumed to be uniform and smooth. The bicycle was not included in the CFD domain. We adopted parameters based on previous literature [1], which are summarized in Table 2.

Table 2. List of parameters for computational fluid dynamics (CFD) analysis.

Parameter	Value
Air Viscosity	1.81×10^{-5} Pa·s
Inlet Velocity/ Air Speed U_∞ (free stream)	16.7 m/s = 60 km/h
Outlet Pressure	0 Pa (static)
Temperature	295.3 K
Relative Humidity	39.9 %
Atmospheric Pressure	100,480 Pa
Saturation Vapor Pressure	269,225.2 Pa
Air Density	1.181 kg/m^3
Turbulence Model	Shear Stress Transport (SST) k-ω
Meshers	Polyhedral Surface Remesher Prism Layer Mesher
Base Cell Size	0.15 m
Prism Layers	5
Reynolds number Re	1.33×10^6
L (dimension of object)	<1 m
C, f (skin friction coefficient)	3.46×10^{-3}
Wall Shear Stress τ	2.84×10^{-1} m/s
U* (dimensionless velocity)	4.91×10^{-1} m/s
Y (first layer cell height)	0.00281304 m
Y+	90
δ (prism layer height)	0.039 m

The size of individual cells in the domain determines total cell count, accuracy, and computation time. For a given domain, a lower base cell size implies higher mesh resolution and accuracy, but implies longer computational time. To obtain the least base cell size for which reliable drag values can be calculated, a mesh convergence study was conducted. We investigated drag for one cyclist model in the domain over 700 iterations each with base cell sizes of 0.1500 m, 0.1000 m, 0.080 m, 0.060 m, 0.050 m, and 0.0039 m. The results of the mesh convergence study revealed that the average drag for the last 50 iterations for the model in each of these domains was 32.75 N, 33.53 N, 31.96 N, 32.23 N, 33.97 N, and 32.35 N. It was concluded that the drag force obtained from the calculations at a base cell size of 0.15 m (~200,000 cells) can provide reliable results for the sake of obtaining trends in drag area at a significantly lower computational time (an average duration of 90 min per simulation, which is roughly 70% faster than the same for the finest mesh). To obtain a high degree of absolute accuracy of modeling the air flow (e.g., a detailed visualization of the wake of the cyclist), we recommend a higher resolution in the domain. However, the aim of the present study was to estimate drag area and compare trends therein with values reported in the literature.

2.5.3. Regression Analyses

The regression between the drag force of the 3D models and the projected area from the frontal camera is a simple least squares method to fit the data.

For the second part, the projected frontal area and joint angles data were first synchronized using a hand raise at the beginning of the dynamic protocol. At this point, the projected frontal area reached a maximum, which was matched with a maximal back/chest and elbow flexion angle. Furthermore, the projected frontal area had a sample rate of 3 Hz, whereas it was 30 Hz for the optical mocap and 10 Hz for the inertial mocap. To align the data, average values of one-second intervals were used. One subject was excluded due to an error in projected frontal area calculation, resulting in a total of 914 data points in the final regression analyses.

Two linear regression models were generated to predict the projected frontal area based on different input data, consisting of

(1)   anthropometric data (Table 1), 2D joint angles from the optical mocap system (Table 1), and the weights of the first 20 principal components of body shape; and

(2)   anthropometric data and joint angles from the inertial mocap system (Table 1).

The stepwise linear regression method (SPSS Statistics 27, IBM, Armonk, NY, USA) was used to define the optimal model. From the optimal model, the importance of each included variable in the equation was analyzed using standardized beta coefficients.

To determine the accuracy of both regression models, the equation was cross-validated by subsequently leaving the data of one subject out to form a linear regression equation and applying these results to the excluded subject, repeating this procedure for each subject. The cross-validated prediction values were compared with the actual projected frontal area to determine the intraclass correlation coefficient (ICC) [31,32]. The two-way-random ICC-model, type absolute agreement, was calculated for the entire dataset. Furthermore, the root mean square error (RMSE) (m^2) projected frontal area between the prediction and actual values was calculated as well as the relative error of the predicted value compared to the actual projected frontal area.

3. Results

3.1. Drag Force versus Projected Frontal Area

Aerodynamic drag force from the CFD simulations of the 3D models of all 10 participants is shown in Table 3. Using these and the corresponding projected frontal area, drag coefficients C_D can be obtained from Equation (1) as air density and wind velocity are known and are constant (Table 2). These C_D values are also listed in Table 3. Drag force is plotted with projected frontal area in Figure 7. This area was compared with area from the depth camera and the comparison per pose is shown in Figure 8. The regression analysis of drag and corresponding area for all poses (3 poses × 10 participants = 30 data points) yielded a coefficient of determination of R^2 of 0.72.

Table 3. Drag force of the 3D models from the Static protocol. C_D was obtained by using Equation (1) and the area from the projected frontal area camera.

	Participant	1	2	3	4	5	6	7	8	9	10
TT	Drag (N)	28.79	26.81	27.49	26.94	26.84	29.59	31.30	28.71	29.86	35.71
	C_D	0.70	0.58	0.64	0.63	0.69	0.72	0.69	0.63	0.76	0.76
Hoods	Drag (N)	37.43	35.99	37.31	44.23	31.80	39.07	45.03	36.13	35.97	41.67
	C_D	0.74	0.66	0.70	0.81	0.70	0.79	0.78	0.66	0.73	0.75
Drops	Drag (N)	32.29	33.15	32.40	32.56	30.07	33.99	35.08	31.32	25.99	40.01
	C_D	0.66	0.66	0.69	0.65	0.72	0.76	0.69	0.64	0.63	0.76

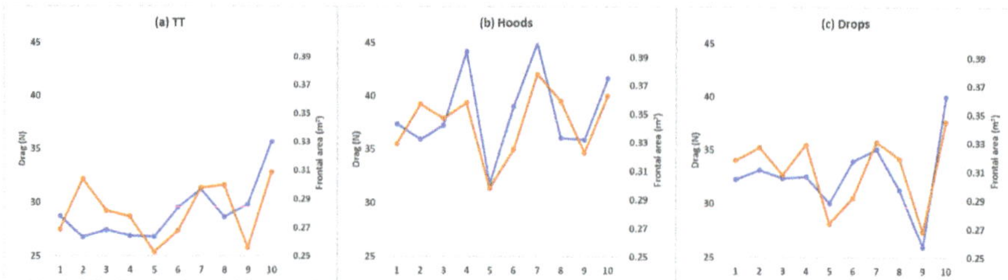

Figure 7. Drag (N, in blue) and area obtained from the projected frontal area camera (m², in orange) for the 10 participants are plotted on primary and secondary axes, respectively. (**a**) *TT*, (**b**) *Hoods*, and (**c**) *Drops* pose.

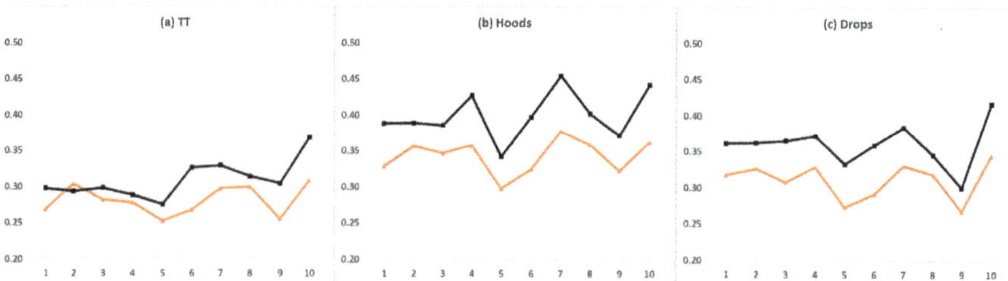

Figure 8. Area obtained from the projected frontal area camera (orange) and 3D models (black) are compared in this figure for (**a**) *TT*, (**b**) *Hoods*, and (**c**) *Drops*. The area is plotted on the y-axis, with units m² against the participant numbers on thee x-axis.

3.2. Projected Frontal Area Prediction Based on Anthropometrics and Joint Angles

The regression model to predict the projected frontal area (m²) using anthropometric data (cm), 2D joint angles (°) from the optical mocap, and the scores associated with the 20 principal components of the 3D body shape is shown in Equation (2). The regression model based on inertial joint angles is shown in Equation (3). The adjusted R^2 value corresponding to Equation (2) is 0.78 ($p < 0.001$) and 0.81 ($p < 0.001$) for Equation (3).

$$
\begin{aligned}
\text{Projected Frontal Area (Optical)} = {} & 0.166678 + 0.000505 \times \text{Back Flexion} + 0.000075 \times \\
& \textit{Score}6 + 0.005565 \times \text{Upper Arm Length} + 0.000326 \times \text{Left Elbow Flexion} + \\
& 0.001023 \times \text{Neck Circumference (tight)} - 0.000188 \times \text{Head Flexion} - 0.00335 \times \text{Hip} \\
& \text{Breadth (sitting)} + 0.000805 \times \text{Left Hip Flexion} + 0.000393 \times \text{Left Knee Flexion} - \\
& 0.002276 \times \text{Lower Arm Length} + 0.000176 \times \text{Left Shoulder Flexion,}
\end{aligned}
\tag{2}
$$

$$
\begin{aligned}
\text{Projected Frontal Area (Inertial)} = {} & -0.322372 + 0.006888 \times \text{Neck Circumference (tight)} \\
& - 0.000998 \times \text{Chest Anterior Tilt} - 0.00027 \times \text{Neck Flexion} + 0.002636 \times \text{Shoulder} \\
& \text{Breadth (Acromion)} + 0.005635 \times \text{Lower Arm Length} + 0.000704 \times \text{Chest Lateral Tilt} \\
& + 0.012827 \times \text{Upper Leg Length} + 0.000426 \times \text{Left Elbow Flexion} - 0.00047 \times \text{Right} \\
& \text{Knee Flexion} - 0.000388 \times \text{Left Knee Flexion} - 0.005606 \times \text{Back Length} + 0.000708 \times \\
& \text{Chest Rotation} - 0.000349 \times \text{Right Shoulder Flexion} + 0.000296 \times \text{Right Elbow} \\
& \text{Supination} + 0.001107 \times \text{Neck Circumference (Shirt)} + 0.000187 \times \text{Left Shoulder} \\
& \text{Internal Rotation} + 0.000072 \times \text{Right Shoulder Internal Rotation,}
\end{aligned}
\tag{3}
$$

Table 4 shows the beta coefficients for each included variable for both the optical and inertial regression model. Table 5 shows the accuracy of the regression models after cross validation in terms of ICC, RMSE, and relative error.

Table 4. Standardized beta coefficients for the included variables in the regression model based on the optical and inertial mocap data.

Optical Regression		Inertial Regression	
Variable	Beta coefficient	Variable	Beta coefficient
Back angle	0.19	Neck circumference (tight)	0.79
Score 6	0.49	Chest anterior tilt	−0.80
Upper arm length	0.31	Neck flexion	−0.31
Left elbow flexion	0.23	shoulder breadth (acromion)	0.18
Neck circumference (tight)	0.12	Lower arm length	0.31
Head flexion	−0.12	Chest lateral tilt	0.44
Hip Breadth (while sitting)	−0.31	Upper leg length	0.66
Left hip flexion	0.36	Left elbow flexion	0.32
Left knee flexion	0.17	Right knee flexion	−0.20
Lower arm length	−0.13	Left knee flexion	−0.30
Left shoulder flexion	0.09	Back length	−0.52
		Chest rotation	0.35
		Right shoulder flexion	−0.27
		Right elbow supination	0.34
		Neck circumference (shirt)	0.13
		Left shoulder internal rotation	0.16
		Right shoulder internal rotation	0.09

Table 5. The accuracy of both regression models in terms of ICC, RMSE, and relative error.

	Optical Regression	Inertial Regression
ICC	0.43 ($p < 0.001$)	0.51 ($p < 0.001$)
RMSE (m^2)	0.037	0.032
Relative error (%)	−0.18 ± 9.98	1.70 ± 8.72

4. Discussion

The results show that the projected frontal area is an indicator of drag force and can be considered the benchmark for practical purposes in the analysis in this study. Furthermore, the linear regression analysis indicates that the projected frontal area can be predicted using anthropometric data combined with joint angle data, providing several practical applications in cycling.

4.1. Drag Force versus Projected Frontal Area

The instructed poses ranged from least aerodynamic to most aerodynamic (i.e., *TT* is more aerodynamically efficient than *Drops*, which is more efficient than *Hoods*) [33]. From Table 3, the drag values follow this order. The C_D values of the participants agree with those found in the literature [21]. Considering that drag force was predicted from the input parameters that contained very simple anthropometric data (i.e., age, gender, height, weight, chest girth, hip circumference maximum, and arm length), the R^2 value of 0.72 is promising. The angles were also obtained from a low-cost, basic smartphone camera using an open source algorithm. When compared to the cost of state-of-the-art wind tunnel measurements or 3D scanners, these results are affordable for the amateur cyclist. The predicted drag force can, for instance, be compared across poses for evaluating the relative ranking of these on-bike poses. In this way, the notion of a 'virtual wind tunnel' that enables aerodynamic bike fitting, pose evaluation, and on-bike posture training can be introduced to amateur cyclists.

The 3D models were not identical replicas of the shapes of the participants. The other anthropometric data (up to 15) of the participants in Table 1 were not always an exact match with those of the 3D models (for instance, forearm length was different for all participants). The models were generated considering the shape data of thousands of human shapes. Hence, individual differences are expected. For future research, we recommend evaluating the methods proposed in this study with 3D shapes obtained from other algorithms to predict human shape in motion [34–37]. For instance, these state-of-the-art algorithms can also accurately model the soft tissue deformation, especially around joints with extreme flexion.

The image-processing algorithm Detectron can capture 2D angles reliably. Hence, we considered the left-hand side angles in our methods and we assumed that the left and right upper body were symmetric. However, there is the chance of parallax error despite the best efforts from researchers and participants to align with the camera. Additionally, the method of modeling the head angle was not the same in the skeleton of the 3D model (Figure 6a). In cycling aerodynamics, the head (and helmet) is one of the most important factors of a cyclist's aerodynamic posture. The sensitivity of the head is suspected to be a contributing factor for the discrepancies in the drag values in the *TT* pose, which did not have as good a correlation with the projected frontal area as the other poses. Whether the discrepancies in the drag of the models in the TT pose are due to the soft tissue deformations or due to the sensitivity of the various body segments in the extreme position are to be investigated.

The area obtained from the projection of 3D models did not include the bicycle, but the projected frontal area camera recorded total projected area of the bike and cyclist combination. The bicycle alone was found to be 0.15 m^2. Hence, the area from 3D models was expected to be lower than the camera values by around 0.15 m^2. However, the area from 3D models was consistently higher. However, the error seemed to be systematic, as observed in Figure 8. This leads us to believe that this could be due to an offset error in the camera calibration.

Area is a strong indicator of drag and can be considered the benchmark for practical purposes in the analysis in this study. For future research, we recommend comparison of CFD drag with gold standard wind tunnel drag to validate the methods described in this study. As mentioned in the Materials and Methods, the CFD predictions can become closer to ground truth drag with a more detailed mesh, longer domain, and accurate modeling of surface roughness (bike, clothes, helmets, accessories, etc.).

4.2. Projected Frontal Area Prediction Based on Anthropometrics and Joint Angles

Furthermore, this study investigated the opportunities to predict the projected frontal area as an indicator of aerodynamic drag using several body dimensions and joint angles. Previous research showed the correlation between projected frontal area and several joint angles [38], which indicates the opportunities to link aerodynamic efficiency and mocap. The optical regression method uses 2D joint angles from pictures, added with scores of the parametric 3D body shapes and anthropometric measurements. The inertial regression method provides the same anthropometric data, added with more extensive 3D joint angles data from IMUs.

The optical model uses 11 variables to predict the projected frontal area, where the inertial mocap model includes 17 variables. Regarding anthropometric data, the tight neck circumference and lower arm length are included in both regression models, whereas the upper arm length (optical model) and the neck circumference (inertial model) have the biggest influence on the projected frontal area. The change in body shape corresponding with principal component (or 'mode') #6 has the biggest influence on predicting projected frontal area (beta coefficient 0.49), where other components were not included in the regression model. From Figure 9, mode 6 appears to be linked with the upper body girth and the width of the upper legs.

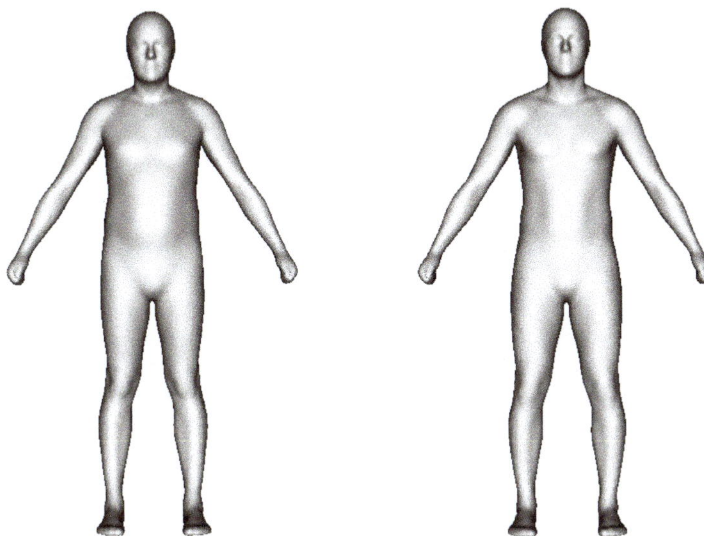

Figure 9. The weight associated with mode #6 of the statistical shape model had the highest influence in Equation (2). From visual inspection, mode #6 appears to be linked with the upper body girth and the width of the upper legs. This figure shows the +3σ (left) and −3σ (right) of mode #6.

Regarding joint angles, previous studies indicated that torso, shoulder, head, and elbow angle have the most considerable influence on aerodynamics [39,40], which were all included in our models. The hip flexion angle had the highest beta coefficient for the optical model. However, the hip angle was excluded in the inertial mocap analysis due to moving of the strap during cycling, which causes unreliable data. This issue can be solved by using double-sided tape to attach the inertial sensor directly on the hip-joint level or using the sensor at the rear rather than front of the participant. It could be an improvement to use a similar method for the legs as well, since the straps should be really tightened to prevent from coming loose, which can be impeding or annoying for the participant during cycling.

Including the back/chest angle, elbow flexion angle, and shoulder flexion angle in the regression models is obvious since these angles determine the position of the upper body. The higher these angles, the more upright the position and the higher the projected frontal area. In both regression models, joint angles are more used to predict the aerodynamics compared to anthropometric data, since they also consider the movement during cycling, where anthropometric data can only be used to predict the general influence on the projected frontal area independent of the cycling pose.

The inertial regression model includes 3D joint angles, but the added value in predicting the projected frontal area is limited, since only the lateral tilt and rotation of the chest and rotation of the elbow and shoulder are included. To improve the accuracy of both models, analyzing regression models using relative values compared to the pose with the minimal projected frontal area as a reference pose was considered, but did not result in improved accuracy. Furthermore, logarithmic and exponential models can be used in future studies to optimize accuracy. Finally, combined effects of anthropometric data can be worth considering (e.g., multiplication of body height and weight can be an approach of the shape of the body).

The accuracy of both models is equivalent, where the inertial mocap model has a slightly higher accuracy in general. The results of the cross-validation showed an RMSE of 0.037 m^2 for the optical model and 0.032 m^2 for the inertial mocap model. Each step in the process of predicting the projected frontal area induces possible errors. Anthropometric data are obtained by hand measurements by one researcher, which means that the error in this step is negligible. The accuracy of the pictures and videos from the side camera is dependent on the visibility of joints that are used to calculate 2D angles and can be neglected in this study design since the correctness of joint angles is checked by the researchers. The highest inaccuracy occurred for the IMUs, with a possible error of 1 to 2° for an individual sensor. The calibration, placing of sensors, and magnetic interference can cause even more considerable errors for full body 3D joint angles.

However, the accuracy of the regression models indicates that both models can have advantageous applications. The inertial mocap model can provide an estimation of the projected frontal area without the use of any software or postprocessing method, which means that it can be used for real-time outdoor estimating of the aerodynamic quality of a certain cycling pose.

5. Conclusions

Drag force and projected frontal area are two commonly used metrics in evaluating the aerodynamic efficiency of a cyclist. This study investigated low-cost methods to estimate these two quantities using easy to acquire anthropometric measures, CFD analyses, a smartphone camera in combination with open-source optical mocap, and inertial mocap from commercially available plug-and-play wearable sensors. Drag calculations were strongly correlated with benchmark projected frontal area measurements (R^2 = 0.72). Projected frontal area can accurately be predicted using anthropometric data and joint angles from optical motion capture (RMSE = 0.037 m^2) and inertial measurement units (RMSE = 0.032 m^2). These methods have practical relevance for amateur as well as elite cyclists: crucially, the training toward an optimized aerodynamic posture with the goal to improve performance. An individual cyclist can compare several poses to determine the relative aerodynamic efficiency of a given pose without accessing expensive wind tunnel facilities or expensive 3D scanners. The methods can be implemented in indoor cycling as an addition to indoor training platforms to provide real-life effects of posture and movement changes of the cyclist with the aim to enable real-time analysis of the biomechanical and aerodynamic effects and hence has added value over some current commercial offerings [41,42]. Furthermore, this method can potentially be employed in outdoor cycling settings given the portability of the inertial mocap system, where real-time estimations of the aerodynamic efficiency can be provided during training or for analysis afterward. This can be interesting for professional riders to observe how accurately they can retain their optimal aerodynamic pose for long durations or for amateurs to provide an indication of the effect of different cycling poses.

Author Contributions: Conceptualization, R.G. and T.P.; Methodology, R.G. and T.P.; Software, R.G., T.P., S.S., E.P., and J.U.; Validation, R.G. and T.P.; Formal analysis, R.G., T.P., and S.S.; Investigation, R.G. and T.P.; Resources, R.G. and T.P.; Data curation, R.G. and T.P.; Writing—original draft preparation, R.G. and T.P.; Writing—review and editing, R.G. and T.P.; Visualization, R.G., T.P., S.S., E.P., and J.U.; Supervision, S.V. and T.H.; Project administration, S.V.; Funding acquisition, R.G., T.P., T.H., and S.V. All authors have read and agreed to the published version of the manuscript.

Funding: This work was supported by VLAIO (The Flemish Agency for Innovation and Entrepreneurship) under grant code "Baekeland Mandaat HBC.2016.0602" and the University of Antwerp Stimpro PS ID: 36536. The computational resources and services used in this work were provided by the Flemish Supercomputer Center (VSC), funded by the Research Foundation-Flanders (FWO) and the EWI department (Economics, Science, Innovation) of the Flemish Government.

Conflicts of Interest: The authors declare no conflict of interest. The funders had no role in the design of the study; in the collection, analyses, or interpretation of data; in the writing of the manuscript, or in the decision to publish the results.

References

1. Blocken, B.; van Druenen, T.; Toparlar, Y.; Andrianne, T. Aerodynamic analysis of different cyclist hill descent positions. *J. Wind Eng. Ind. Aerodyn.* **2018**, *181*, 27–45. [CrossRef]

2. Ainegren, M.; Jonsson, P. Drag Area, Frontal Area and Drag Coefficient in Cross-Country Skiing Techniques. *Proceedings* **2018**, *2*, 313. [CrossRef]

3. Martin, J.C.; Milliken, D.L.; Cobb, J.E.; McFadden, K.L.; Coggan, A.R. Validation of a mathematical model for road cycling power. *J. Appl. Biomech.* **1998**, *14*, 276–291. [CrossRef]

4.   Merkes, P.F.J.; Menaspà, P.; Abbiss, C.R. Validity of the Velocomp powerpod compared with the verve cycling infocrank power meter. *Int. J. Sports Physiol. Perform.* **2019**, *14*, 1382–1387. [CrossRef]

5.   Valenzuela, P.L.; Alcalde, Y.; Gil-Cabrera, J.; Talavera, E.; Lucia, A.; Barranco-Gil, D. Validity of a novel device for real-time analysis of cyclists' drag area. *J. Sci. Med. Sport* **2020**, *23*, 421–425. [CrossRef]

6.   Íñiguez-De-La Torre, A.; Íñiguez, J. Aerodynamics of a cycling team in a time trial: Does the cyclist at the front benefit? *Eur. J. Phys.* **2009**, *30*, 1365–1369. [CrossRef]

7.   Fintelman, D.M.; Hemida, H.; Sterling, M.; Li, F.X. CFD simulations of the flow around a cyclist subjected to crosswinds. *J. Wind Eng. Ind. Aerodyn.* **2015**, *144*, 31–41. [CrossRef]

8.   Defraeye, T.; Blocken, B.; Koninckx, E.; Hespel, P.; Carmeliet, J. Computational fluid dynamics analysis of cyclist aerodynamics: Performance of different turbulence-modelling and boundary-layer modelling approaches. *J. Biomech.* **2010**, *43*, 2281–2287. [CrossRef] [PubMed]

9.   Godo, M.; Corson, M.; Legensky, S. An Aerodynamic Study of Bicycle Wheel Performance Using CFD. In Proceedings of the 47th AIAA Aerospace Sciences Meeting Including the New Horizons Forum and Aerospace Exposition, Orlando, FL, USA, 5–8 January 2009. [CrossRef]

10.   Godo, M.; Corson, D.; Legensky, S. A Comparative Aerodynamic Study of Commercial Bicycle Wheels Using CFD. In Proceedings of the 48th AIAA Aerospace Sciences Meeting Including the New Horizons Forum and Aerospace Exposition, Orlando, FL, USA, 4–7 January 2010. [CrossRef]

11.   Blocken, B.; Toparlar, Y. A following car influences cyclist drag: CFD simulations and wind tunnel measurements. *J. Wind Eng. Ind. Aerodyn.* **2015**, *145*, 178–186. [CrossRef]

12.   Blocken, B.; van Druenen, T.; Toparlar, Y.; Malizia, F.; Mannion, P.; Andrianne, T.; Marchal, T.; Maas, G.J.; Diepens, J. Aerodynamic drag in cycling pelotons: New insights by CFD simulation and wind tunnel testing. *J. Wind Eng. Ind. Aerodyn.* **2018**, *179*, 319–337. [CrossRef]

13.   Blocken, B.; Toparlar, Y.; Andrianne, T. Aerodynamic benefit for a cyclist by a following motorcycle. *J. Wind Eng. Ind. Aerodyn.* **2016**, *155*, 1–10. [CrossRef]

14.   Huysmans, T.; Goto, L.; Molenbroek, J.; Goossens, R. DINED Mannequin. *Tijdschr. Voor Hum. Factors* **2020**, *45*, 4–7.

15.   Ballester, A.I.; Piérola, A.; Parrilla, E.; Uriel, J.; Ruescas, A.V.; Perez, C.; Durá, J.V.; Alemany, S. 3D Human Models from 1D, 2D and 3D Inputs: Reliability and Compatibility of Body Measurements. In Proceedings of the 9th International Conference on 3D Body Scanning Technologies, Lugano, Switzerland, 16–17 October 2018. [CrossRef]

16.   Ballester, A.; Parrilla, E.; Vivas, J.A.; Pierola, A.; Uriel, J.; Puigcerver, S.A.; Piqueras, P.; Solve, C.; Rodriguez, M.; Gonzalez, J.C.; et al. Low-cost data-driven 3D reconstruction and its applications. In Proceedings of the 6th International Conference on 3D Body Scanning Technologies, Lugano, Switzerland, 27–28 October 2015. [CrossRef]

17.   Allen, B.; Curless, B.; Popović, Z. Exploring the Space of Human Body Shapes: Data-driven Synthesis under Anthropometric Control. *Sae Int.* **2004**. [CrossRef]

18.   Stave, D.Å. Cyclist Posture Optimisation Using CFD. 2018. Available online: https://brage.bibsys.no/xmlui/handle/11250/2562590?platform=hootsuite (accessed on 27 November 2020).

19.   Garimella, R.; Beyers, K.; Huysmans, T.; Verwulgen, S. Rigging and Re-posing a Human Model from Standing to Cycling Configuration. In *International Conference on Applied Human Factors and Ergonomics*; Springer: Cham, Switzerland; Washington, DC, USA, 2019; pp. 525–532. [CrossRef]

20.   Defraeye, T.; Blocken, B.; Koninckx, E.; Hespel, P.; Carmeliet, J. Aerodynamic study of different cyclist positions: CFD analysis and full-scale wind-tunnel tests. *J. Biomech.* **2010**, *43*, 1262–1268. [CrossRef]

21.   Debraux, P.; Grappe, F.; Manolova, A.V.; Bertucci, W. Aerodynamic drag in cycling: Methods of assessment. *Sport. Biomech.* **2011**, *10*, 197–218. [CrossRef]

22.   Heil, D.P. Body mass scaling of projected frontal area in competitive cyclists. *Eur. J. Appl. Physiol.* **2001**, *85*, 358–366. [CrossRef]

23.   Olds, T.; Olive, S. Methodological considerations in the determination of projected frontal area in cyclists. *J. Sports Sci.* **1999**, *17*, 335–345. [CrossRef]

24.   Debraux, P.; Bertucci, W.; Manolova, A.V.; Rogier, S.; Lodini, A. New method to estimate the cycling frontal area. *Int. J. Sports Med.* **2009**, *30*, 266–272. [CrossRef]

25.   Barelle, C.; Chabroux, V.; Favier, D. Modeling of the time trial cyclist projected frontal area incorporating anthropometric, postural and helmet characteristics. *Sport Eng.* **2010**, *12*, 199–206. [CrossRef]

26. Danckaers, F.; Huysmans, F.; Lacko, D.; Sijbers, J. Evaluation of 3D Body Shape Predictions Based on Features. In Proceedings of the Proceedings of the 6th International Conference on 3D Body Scanning Technologies, Lugano, Switzerland, 27–28 October 2015; pp. 258–265. [CrossRef]

27. Garimella, R.; Peeters, T.; Beyers, K.; Truijen, S.; Huysmans, T.; Verwulgen, S. *Capturing Joint Angles of the Off-Site Human Body*; IEEE: New Delhi, India, 2018; pp. 1244–1247. [CrossRef]

28. Smith, M.F.; Davison, R.C.R.; Balmer, J.; Bird, S.R. Reliability of mean power recorded during indoor and outdoor self-paced 40 km cycling time-trials. *Int. J. Sports Med.* **2001**, *22*, 270–274. [CrossRef]

29. Griffith, M.D.; Crouch, T.; Thompson, M.C.; Burton, D.; Sheridan, J.; Brown, N.A.T. Computational Fluid Dynamics Study of the Effect of Leg Position on Cyclist Aerodynamic Drag. *J. Fluids Eng.* **2014**, *136*, 101105. [CrossRef]

30. Celis, B.; Ubbens, H.H. Design and Construction of an Open-circuit Wind Tunnel with Specific Measurement Equipment for Cycling. *Procedia Eng.* **2016**, *147*, 98–103. [CrossRef]

31. Weir, J. Quantifying test-retest reliability using the intraclass correlation coefficient and the SEM. *J. Strength Cond Res.* **2005**, *19*, 231–240.

32. Hawkins, D.M.; Basak, S.C.; Mills, D. Assessing Model Fit by Cross-Validation. *J. Chem. Inf. Model.* **2003**, *43*, 579–586. [CrossRef]

33. Barry, N.; Burton, D.; Sheridan, J.; Thompson, M.; Brown, N.A.T. Aerodynamic performance and riding posture in road cycling and triathlon. *J. Spors Eng. Technol.* **2015**, *229*, 28–38. [CrossRef]

34. Bogo, F.; Kanazawa, A.; Lassner, C.; Gehler, P.; Romero, J.; Black, M.J. Keep it SMPL: Automatic estimation of 3D human pose and shape from a single image. In *European Conference on Computer Vision*; Springer: Amsterdam, The Netherlands; Cham, Switzerland, 2016; Volume 9909, pp. 561–578. [CrossRef]

35. Black, M.J. Estimating Human Motion: Past, Present, and Future. Talk. October 2018. Available online: https://www.youtube.com/watch?v=5jU9rIqBz7M (accessed on 27 November 2020).

36. Bogo, F.; Romero, J.; Loper, M.; Black, M.J. FAUST: Dataset and evaluation for 3D mesh registration. In Proceedings of the IEEE Conference on Computer Vision and Pattern Recognition, Columbus, OH, USA, 23–28 June 2014; pp. 3794–3801. [CrossRef]

37. Pons-Moll, G.; Romero, J.; Mahmood, N.; Black, M.J. Dyna: A model of dynamic human shape in motion. *ACM Trans. Graph.* **2015**, *34*, 4. [CrossRef]

38. Peeters, T.; Garimella, R.; Francken, E.; Henderieckx, S.; van Nunen, L.; Verwulgen, S. The Correlation between Frontal Area and Joint Angles During Cycling. *Adv. Intell. Syst. Comput.* **2020**, *1206*, 251–258. [CrossRef]

39. Fintelman, D.M.; Sterling, M.; Hemida, H.; Li, F.X. Optimal cycling time trial position models: Aerodynamics versus power output and metabolic energy. *J. Biomech.* **2014**, *47*, 1894–1898. [CrossRef] [PubMed]

40. Smurthwaite, J. How to Be More Aero on Your Road Bike (Video), Cycling Weekly. 2015. Available online: https://www.cyclingweekly.com/videos/fitness/be-faster-by-being-more-aero (accessed on 25 November 2020).

41. Airshaper—Aerodynamics Made Easy. Available online: www.airshaper.com (accessed on 27 November 2020).

42. Bioracer Aero. Available online: https://bioracermotion.com/en/bioracer-aero (accessed on 10 May 2020).

Publisher's Note: MDPI stays neutral with regard to jurisdictional claims in published maps and institutional affiliations.

 *applied sciences*

Article

Evolution of the Hurdle-Unit Kinematic Parameters in the 60 m Indoor Hurdle Race

Pablo González-Frutos [1,*], Santiago Veiga [2], Javier Mallo [2] and Enrique Navarro [2]

[1] Faculty of Health Sciences, Universidad Francisco de Vitoria, 28223 Madrid, Spain
[2] Health and Human Performance Department, Technical University of Madrid, 28040 Madrid, Spain;
    santiago.veiga@upm.es (S.V.); javier.mallo@upm.es (J.M.); enrique.navarro@upm.es (E.N.)
*   Correspondence: p.gfrutos.prof@ufv.es; Tel.: +34-659-83-26-09

Received: 16 October 2020; Accepted: 30 October 2020; Published: 4 November 2020

Featured Application: Coaches and athletes should implement their training programs to have an impact on some of these variables according to the specific demands of each hurdle-unit phase and gender.

Abstract: The aims of this study were to compare the five hurdle-unit split times from the deterministic model with the hurdle-to-hurdle model and with the official time, to compare the step kinematics of each hurdle-unit intervals, and to relate these variables to their respective hurdle-unit split times. The temporal and spatial parameters of the 60 m hurdles race were calculated during the 44th Spanish and 12th IAAF World Indoor Championships (men: n = 59; women: n = 51). The hurdle-unit split times from the deterministic model showed a high correlation (r = 0.99; $p < 0.001$) with the split times of the hurdle-to-hurdle model and faster split times were related to shorter step and flight times in hurdle steps for both genders. At the first hurdle, male athletes tended to increase their flight and contact times while the tendency of female athletes was to decrease their contact and flight times. In addition, at the first hurdle, both genders presented shorter take-off distance, shorter landing distance, and greater step width than in the remaining hurdles of the race. Therefore, coaches should implement training programs that have an impact on these key variables according to the specific demands of each hurdle-unit phase and gender.

Keywords: track and field; kinematics; performance analysis; competition; DLT algorithms

---

1. Introduction

The hurdles events have been analyzed through different approaches such as biomechanics, conditional, teaching and learning, talent identification, psychological, or medical [1,2]. Within biomechanical analyses, two types of studies stand out: temporal and spatial. The ubication of the hurdles serves as a useful reference for the division of the event into the following phases [3]: approach run phase (from the starting line to the first hurdle), hurdle unit phase (race and clearance of the hurdles), and run-in phase (from the last hurdle to the finishing line).

Temporal analyses are commonly focused on the split times between each hurdle clearance (foot landing) [4–7]. The times from the athlete's foot landing after one hurdle to the following hurdle are employed for practical reasons, as it allows an easy visual identification of the splits. Tsionakos et al. [6] showed that the correlation between intermediate times and final performance was decisive from the fifth obstacle onwards in the 110 m obstacle event (r = 0.77 to 0.98), while Panoutsakopoulos et al. [7] found a correlation between split times with the official time in the 60 m indoor event in most of the intervals, both for elite men (r = 0.65 to 0.84) and women (r = 0.55 to 0.87). Similarly, González Frutos et al. [8] found an almost perfect correlation (r = 0.99) between the mean of the hurdle-unit split time and the official time in the 60 m hurdle race. In addition to the split time analysis, it is frequent to refer to the flight time over the hurdle (from take-off to landing) in temporal studies. Panoutsakopoulos et al. [7] found a correlation in not all of the hurdles between the flight time and the official time, while González Frutos et al. [8] found a very large correlation (r = 0.82) between the mean flight time of the five hurdle steps and the official time.

Among the spatial analyses, three-dimensional (3D) studies stand out, even though they are habitually carried out in a single hurdle-unit interval (between the second and fifth hurdle), in a training situation or employing a small sample of subjects due to the costly analysis of the data [9–16]. The organization of the steps that make up the hurdle-unit phase within the studies depends on the selection of the phase of interest: following a deterministic model which comprises the preparatory, hurdle, landing and recovery steps [9,17], only the hurdle and landing steps [10,11,13–15], or from hurdle to hurdle [12]. There is a single precedent of the analysis of all ten hurdles during the 110 m hurdle event [18], who performed a two-dimensional analysis and calculated take-off (2.04 ± 0.07 m) and landing (1.47 ± 0.03 m) distances. Complementarily, González-Frutos et al. [8] analyzed the average of every step of the hurdle-unit phase, in addition to the approach run and run-in phases, finding differences between international and national hurdlers in various parameters during 60 m hurdles event, for both men and women.

Most of these kinematical studies indicate that an efficient hurdle clearance technique is associated with the take-off contact time, take-off and landing distances, take-off angle, and to the hurdle flight time. However, the differences found between some studies may be due to the analysis of different single hurdle intervals, since the kinematic parameters in hurdles and sprint events change as the running speed varies [18,19]. Also, differences could rely on the different criteria when defining the split times of each hurdle-unit [11,13–15] as not many studies consider both the preparatory and recovery steps on it.

Therefore, the aims of the present research were: (1) to compare the five hurdle-unit split times from a deterministic model with the hurdle to hurdle model and with the official time, (2) to compare the step kinematics of each hurdle-unit intervals, and (3) to relate these variables to their respective hurdle-unit split times. Our hypothesis is that both models correlate with each other and with the final result in the race, while there should be differences in the registered values of each hurdle-unit interval. Additionally, we hypothesize that these differences between hurdle-unit phases, and the correlations between variables and their hurdle-unit split times, will be diverse for men and women.

2. Materials and Methods

2.1. Subjects

All the races were filmed during the 60 m hurdle event of the 44th Spanish Indoor Championship and 12th IAAF World Indoor Championship (2008) with the license of both organizing committees. The best performance of each athlete (men, n = 59; women, n = 51) from the heat, semifinal, and final rounds was included in the study. All experimental procedures were carried out in accordance with the Declaration of Helsinki and were approved by the Ethics Committee of the Technical University of Madrid.

2.2. Procedures

The races were filmed with six fixed JVC GY-DV300 video cameras (Japan Victor Company, Yokohama, Japan) located at the main stands and operating at 50 fps (frames per second; shutter speed: 1/1000), similar to that previously used in other studies [5,9,11,12,15]. Four cameras had a predominantly side view: camera one recorded the first 13 m; camera two from 13 to 30 m; camera three from 30 to 47 m; camera four the last 13 m (47 to 60 m) of the race. To avoid visual occlusion of the athletes, cameras five and six were placed to capture the frontal view: camera five recorded the first 30 m (including the referees' starting shot) and camera six the last 30 m. A common event (foot landing or take-off) in the field of view of two different cameras were employed for the camera synchronization.

An experienced observer manually digitized the hurdlers' foot landing and take-off points from the race footage using Photo 23D software (Technical University of Madrid, Spain) [20]. The official lane marks were employed as control points (six in each camera) for calibration purposes and direct lineal transformation algorithms [21] were used to reconstruct the real coordinates (in meters) from the screen coordinates (in pixels). Thirty points of known spatial coordinates (different from those used for calibration purposes) represented by the official lane marks and uniformly distributed in the field of view were used for error estimation. The root mean square error [22] was less than 0.04 m for the step length (x-axis) and less than 0.02 m for the step width (y-axis) on the six cameras, providing a validity of the measurements similar to those of other previous research on race analysis [23]. The step length and step width were calculated by the difference on the x (longitudinal) or y (transverse) axes, respectively, of the most forward point of the foot on the ground during successive contacts on it.

The definition (Figure 1) proposed by McDonald and Dapena [9] was used for the organization of the temporal and spatial parameters of the five hurdle-unit phases of the 60 m hurdles, which integrated the preparatory step, hurdle step (divided into take-off distance and landing distance), landing step, and recovery step. The calculation of the split times used the same sequence, including the time elapsed from the beginning of the preparatory step to the beginning of the next preparatory step. Alternatively, the split times were also calculated from the landing step to the next landing step as done by the temporal studies [4–7], for comparative purposes.

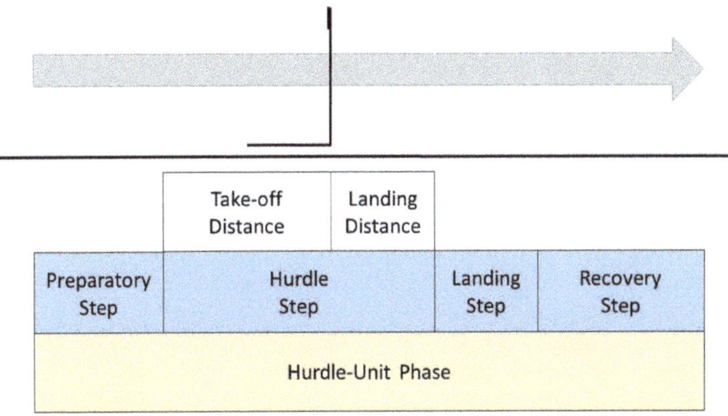

Figure 1. Definitions employed for the hurdle-unit phase analysis according to McDonald and Dapena [9].

2.3. Statistical Analysis

All results are expressed as the mean ± standard deviation (SD). Split times, step times, contact times, flight times, step lengths, and step width of the athletes were compared with a repeated measures analysis of variance according to the hurdle-unit phase (first, second, third, fourth, and fifth hurdle, and mean of the five hurdles) for both genders of athletes (male or female). Planned repeated contrast tests between successive laps were carried out. Post hoc tests were used to determine statistical effects ($p < 0.05$) between factors using Bonferroni corrections and were interpreted using effect sizes ($\eta p2$) with 0.2, 0.5, and 0.8 threshold values for small, medium, and large effects [24]. Cohen d effect sizes between the hurdle-unit phases and mean hurdle-unit phase were calculated to display practical significance. Effect sizes of 0.00 to 0.19, 0.20 to 0.59, 0.60 to 1.19, 1.20 to 1.99, 2.00 to 3.99, and ≥4.00 were interpreted as trivial, small, moderate, large, very large, and nearly perfect, respectively [25]. Pearson correlation coefficients were used to relate all the kinematic race parameters to the hurdle-unit split time or race result, being 0.1, 0.3, 0.5, 0.7, and 0.9, the threshold values that represented small, moderate, large, very large, and nearly perfect correlations [25]. Statistical analyses were performed using the IBM Statistical Package for Social Sciences Statistics, version 22.0 (IBM Inc., Armonk, NY, USA).

3. Results

3.1. Split Times

The mean split times on the hurdle-unit phase (men: 1.10 ± 0.06 s; women: 1.09 ± 0.07 s) measured from preparatory to the next preparatory step showed nearly perfect relationships with the split times measured from landing to the next landing step (r = 0.99; $p < 0.001$), and also with the race results (r = 0.99; $p < 0.001$) both for men and women.

Hurdle-unit split times changed along the 60 m hurdles race (men: $F_{3.718} = 3.28$, $p = 0.014$, $\eta p2 = 0.05$, and women: $F_{3.299} = 6.60$, $p = 0.001$, $\eta p2 = 0.12$) with the greatest differences observed in hurdle one for both gender and hurdle five for men (Table 1). Split times in all the hurdles showed very large to nearly perfect correlations with the race results, with correlation coefficients increasing from the first to the fourth hurdle (Table 1).

Table 1. Evolution of the hurdle-unit split times (s) and relationships (r) with the 60 m hurdle race results during the 44th Spanish Indoor and 12th IAAF World Indoor Championships.

	Hurdle 1	Hurdle 2	Hurdle 3	Hurdle 4	Hurdle 5
Split Time					
Men	1.11 ± 0.06 [*4]	1.10 ± 0.06	1.10 ± 0.08	1.09 ± 0.07	1.11 ± 0.07 [*4]
Women	1.10 ± 0.06 [*34#5]	1.09 ± 0.07	1.09 ± 0.08	1.08 ± 0.09	1.08 ± 0.07
Correlation					
Men	0.906 [#]	0.914 [#]	0.927 [#]	0.955 [#]	0.874 [#]
Women	0.942 [#]	0.970 [#]	0.975 [#]	0.980 [#]	0.944 [#]

Inter-hurdle-unit phase differences and correlations: * $p < 0.05$, # $p < 0.001$. Superscripts numbers indicate the hurdle-unit (1–5) where differences exist.

3.2. Kinematic Differences between Hurdles

In the preparatory step, women showed differences between hurdles in the step ($F_{3.50} = 4.15$, $p = 0.005$, $\eta p2 = 0.08$), contact ($F_{3.77} = 3.23$, $p = 0.015$, $\eta p2 = 0.06$), and flight times ($F_{3.74} = 10.84$, $p = 0.001$, $\eta p2 = 0.18$), whereas men (Table 2) only experienced differences in the flight time ($F_{3.95} = 2.92$, $p = 0.023$, $\eta p2 = 0.05$). Women (Table 3) showed the shortest preparatory step, contact, and flight times on the first hurdle. In the hurdle step, changes were observed through the race in the step (men: $F_{3.53} = 6.35$, $p = 0.001$, $\eta p2 = 0.10$; women: $F_{3.37} = 5.26$, $p = 0.001$, $\eta p2 = 0.10$) and flight (men: $F_{3.33} = 5.92$, $p = 0.001$, $\eta p2 = 0.09$; women: $F_{3.69} = 3.19$, $p = 0.017$, $\eta p2 = 0.06$) times, with the greatest values observed in the first hurdle both for men (Table 2) and women (Table 3). In the landing step, men (Table 2) showed differences along the race in the step ($F_{4.00} = 4.35$, $p = 0.002$, $\eta2 = 0.07$) and contact times ($F_{3.96} = 3.04$, $p = 0.018$, $\eta p2 = 0.05$), while, in the recovery step, changes were observed for women (Table 3) in step time ($F_{3.69} = 6.20$, $p = 0.001$, $\eta p2 = 0.11$), with maximum step times accounting in the first hurdle.

When compared to the mean race values, the greatest differences for women were observed in the first hurdle for contact (moderate effect, Figure 2D) and flight times (large effect, Figure 2F) of the preparatory step, and in the third hurdle for preparatory flight time (moderate effect, Figure 2F). However, the differences regarding the mean race values for men hurdlers did not show a moderate effect size for the step, contact, and flight times (Figure 2A–C,E).

Step length also changed through the 60 m hurdle race in every step of the hurdle-unit phase, except for the landing step in the women's race. The greatest changes were observed in the landing distance during the men's race ($F_{3.82} = 27.91$, $p = 0.001$, $\eta p2 = 0.33$) and in the take-off distance of the women's race ($F_{3.74} = 62.29$, $p = 0.001$, $\eta p2 = 0.56$). Step width also displayed meaningful differences according to the hurdle order, specifically in the hurdle step for men ($F_{2.99} = 11.97$, $p = 0.001$, $\eta p2 = 0.17$) and women ($F_{3.04} = 11.64$, $p = 0.001$, $\eta p2 = 0.19$).

Table 2. Evolution of the spatial-temporal variables of men hurdlers participating in the 44th Spanish Indoor and 12th IAAF World Indoor Championships.

		Hurdle 1	Hurdle 2	Hurdle 3	Hurdle 4	Hurdle 5
Step Time	Preparatory step	0.20 ± 0.00	0.20 ± 0.00	0.20 ± 0.00	0.20 ± 0.00	0.20 ± 0.00
	Hurdle Step	0.51 ± 0.01	0.50 ± 0.01	0.50 ± 0.01	0.49 ± 0.01 [#125]	0.50 ± 0.01
	Landing Step	0.16 ± 0.00	0.15 ± 0.00	0.15 ± 0.00	0.16 ± 0.00 [¥2#3]	0.16 ± 0.00
	Recovery Step	0.25 ± 0.00	0.25 ± 0.00	0.24 ± 0.00	0.24 ± 0.00	0.25 ± 0.00
Contact Time	Preparatory step	0.11 ± 0.00	0.11 ± 0.00	0.11 ± 0.00	0.11 ± 0.00	0.11 ± 0.00
	Hurdle Step	0.12 ± 0.00	0.12 ± 0.00	0.12 ± 0.00	0.12 ± 0.00	0.12 ± 0.00
	Landing Step	0.09 ± 0.00 [¥3]	0.08 ± 0.00	0.08 ± 0.00	0.09 ± 0.00	0.09 ± 0.00
	Recovery Step	0.12 ± 0.00	0.12 ± 0.00	0.12 ± 0.00	0.12 ± 0.00	0.12 ± 0.00
Flight Time	Preparatory step	0.09 ± 0.00	0.09 ± 0.00	0.09 ± 0.00	0.09 ± 0.00	0.09 ± 0.00
	Hurdle Step	0.39 ± 0.01 [#4]	0.38 ± 0.00	0.38 ± 0.01	0.37 ± 0.01	0.38 ± 0.01 [¥3]
	Landing Step	0.07 ± 0.00	0.07 ± 0.00	0.07 ± 0.00	0.08 ± 0.00	0.07 ± 0.00
	Recovery Step	0.12 ± 0.00	0.13 ± 0.00	0.12 ± 0.00	0.12 ± 0.00	0.12 ± 0.00
Step Length	Preparatory step	1.82 ± 0.01 [*24#3]	1.88 ± 0.02	1.91 ± 0.02	1.87 ± 0.02	1.86 ± 0.01 [¥3]
	Hurdle Step	3.76 ± 0.03	3.73 ± 0.02	3.89 ± 0.03 [#1245]	3.71 ± 0.02	3.72 ± 0.02
	Take-off distance	2.01 ± 0.02 [*4#235]	2.11 ± 0.02	2.14 ± 0.02	2.08 ± 0.02 [¥35]	2.12 ± 0.02
	Landing distance	1.74 ± 0.03 [#245]	1.62 ± 0.03	1.75 ± 0.03 [#245]	1.63 ± 0.03	1.60 ± 0.02
	Landing Step	1.42 ± 0.02 [¥5#4]	1.46 ± 0.02	1.45 ± 0.02 [*4]	1.50 ± 0.02	1.48 ± 0.02
	Recovery Step	1.98 ± 0.01 [*4]	2.01 ± 0.02	1.98 ± 0.01 [*4]	2.03 ± 0.01	2.02 ± 0.02
Step Width	Preparatory step	0.13 ± 0.01	0.14 ± 0.01	0.15 ± 0.01 [*4]	0.12 ± 0.01	0.13 ± 0.01
	Hurdle Step	0.16 ± 0.01 [¥23#45]	0.09 ± 0.01	0.10 ± 0.01	0.08 ± 0.01	0.08 ± 0.01
	Landing Step	0.13 ± 0.01	0.12 ± 0.01	0.12 ± 0.01	0.12 ± 0.01	0.16 ± 0.01 [*1¥2#34]
	Recovery Step	0.21 ± 0.01	0.24 ± 0.01	0.23 ± 0.01	0.24 ± 0.01	0.29 ± 0.01 [*2¥34#1]

Inter-hurdle difference: * $p < 0.05$; ¥ $p < 0.01$; # $p < 0.001$. Superscripts numbers indicate the hurdle-unit (1–5) where differences exist.

Table 3. Evolution of the spatial-temporal variables of women hurdlers participating in the 44th Spanish Indoor and 12th IAAF World Indoor Championships.

		Hurdle 1	Hurdle 2	Hurdle 3	Hurdle 4	Hurdle 5
Step Time	Preparatory step	0.20 ± 0.00 [¥3]	0.21 ± 0.00	0.21 ± 0.00	0.21 ± 0.00	0.21 ± 0.00
	Hurdle Step	0.46 ± 0.01	0.46 ± 0.01	0.45 ± 0.01 [*12]	0.45 ± 0.01	0.45 ± 0.01
	Landing Step	0.18 ± 0.00	0.17 ± 0.00	0.17 ± 0.00	0.17 ± 0.00	0.17 ± 0.00
	Recovery Step	0.26 ± 0.00 [*2#5]	0.25 ± 0.00	0.25 ± 0.00	0.25 ± 0.00	0.25 ± 0.00
Contact Time	Preparatory step	0.11 ± 0.00 [*4¥5]	0.11 ± 0.00	0.11 ± 0.00	0.11 ± 0.00	0.11 ± 0.00
	Hurdle Step	0.11 ± 0.00	0.11 ± 0.00	0.11 ± 0.00	0.11 ± 0.00	0.11 ± 0.00
	Landing Step	0.09 ± 0.00	0.09 ± 0.00	0.09 ± 0.00	0.09 ± 0.00	0.09 ± 0.00
	Recovery Step	0.12 ± 0.00	0.11 ± 0.00	0.11 ± 0.00	0.12 ± 0.00	0.11 ± 0.00
Flight Time	Preparatory step	0.09 ± 0.00 [#2345]	0.10 ± 0.00	0.10 ± 0.00	0.10 ± 0.00	0.10 ± 0.00
	Hurdle Step	0.35 ± 0.01	0.34 ± 0.01	0.34 ± 0.01	0.34 ± 0.01	0.34 ± 0.01
	Landing Step	0.09 ± 0.00	0.09 ± 0.00	0.09 ± 0.00	0.09 ± 0.00	0.09 ± 0.00
	Recovery Step	0.14 ± 0.00	0.14 ± 0.00	0.14 ± 0.00	0.14 ± 0.00	0.13 ± 0.00

Table 3. *Cont.*

		Hurdle 1	Hurdle 2	Hurdle 3	Hurdle 4	Hurdle 5
Step Length	Preparatory step	1.63 ± 0.01 [#2345]	1.76 ± 0.02	1.78 ± 0.01	1.79 ± 0.01	1.79 ± 0.01
	Hurdle Step	3.18 ± 0.03	3.21 ± 0.02	3.33 ± 0.02 [#1245]	3.20 ± 0.02	3.22 ± 0.02
	Take-off distance	1.79 ± 0.02 [#2345]	1.93 ± 0.02	2.01 ± 0.02 [#24]	1.93 ± 0.02	1.97 ± 0.02
	Landing distance	1.40 ± 0.03 [¥3#245]	1.28 ± 0.03	1.32 ± 0.03 [#5]	1.27 ± 0.03	1.25 ± 0.03
	Landing Step	1.48 ± 0.02	1.45 ± 0.02	1.50 ± 0.01	1.49 ± 0.01	1.50 ± 0.02
	Recovery Step	1.93 ± 0.01 [¥4]	1.98 ± 0.01	1.95 ± 0.01	1.98 ± 0.01	1.91 ± 0.01 [*2¥4]
Step Width	Preparatory step	0.10 ± 0.01	0.10 ± 0.01	0.09 ± 0.01	0.10 ± 0.01	0.09 ± 0.01
	Hurdle Step	0.13 ± 0.01 [¥5#234]	0.07 ± 0.01	0.06 ± 0.01	0.06 ± 0.01	0.07 ± 0.01
	Landing Step	0.12 ± 0.01	0.10 ± 0.01	0.11 ± 0.01	0.11 ± 0.01	0.13 ± 0.01
	Recovery Step	0.20 ± 0.01	0.19 ± 0.01	0.20 ± 0.01	0.20 ± 0.01	0.23 ± 0.01 [*2]

Inter-hurdle difference: * $p < 0.05$; ¥ $p < 0.01$; # $p < 0.001$. Superscripts numbers indicate the hurdle-unit (1–5) where differences exist.

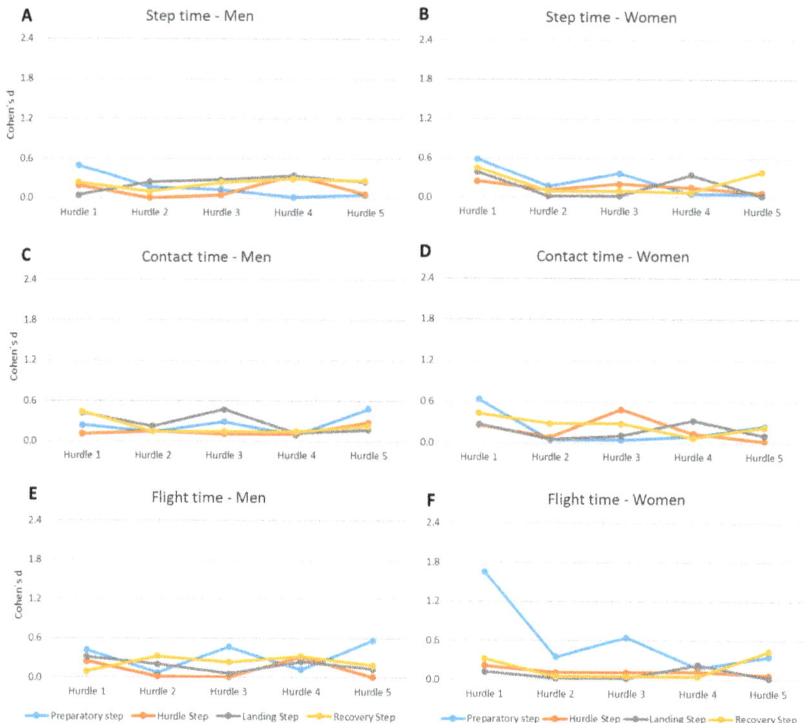

Figure 2. Magnitude of differences (Cohen's d) for the temporal parameters ((**A,B**) step time; (**C,D**) contact time; and (**E,F**) flight time) in relation to mean race values for men and women. An effect size greater than 0.6 or 1.2 was considered as moderate or large, respectively, by Hopkins et al. [25].

The differences in the spatial parameters with respect to the mean race values showing moderate, large, or very large effect sizes on step length in every step, except for the landing step, both for men and women. Moreover, the greatest values were achieved in the first and third hurdle (Figure 3A,B). The step width also showed moderate (men's recovery step), large (men's hurdle step), or very large (women's hurdle step) differences at the first hurdle (Figure 3C,D).

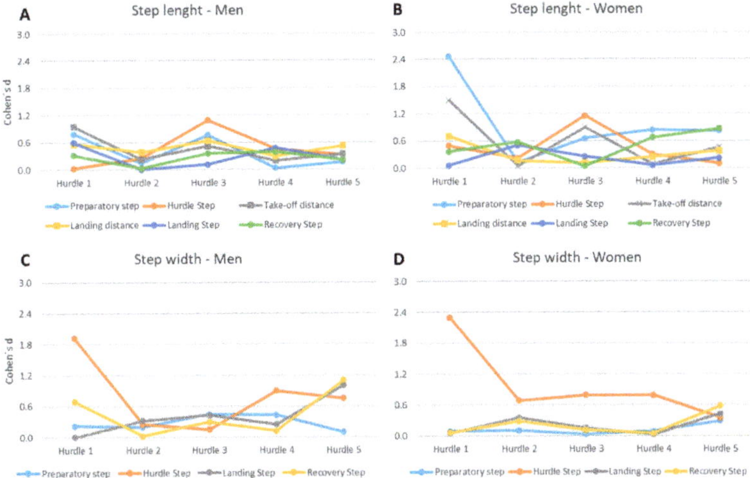

Figure 3. Magnitude of differences (Cohen's d) for the spatial parameters ((**A**,**B**) step length; (**C**,**D**) step width) in relation to mean race values for men and women. An effect size greater than 0.6 or 1.2 was considered as moderate or large, respectively, by Hopkins et al. [25].

3.3. Correlations of Kinematic Parameters with Hurdle-Unit Times

Step and flight times during the hurdle step showed very large relationships with the split times both for men (Figure 4A–E) and women (Figure 4B,F). Additionally, the step time during the recovery step showed a very large correlation with the split times for women (Figure 4B).

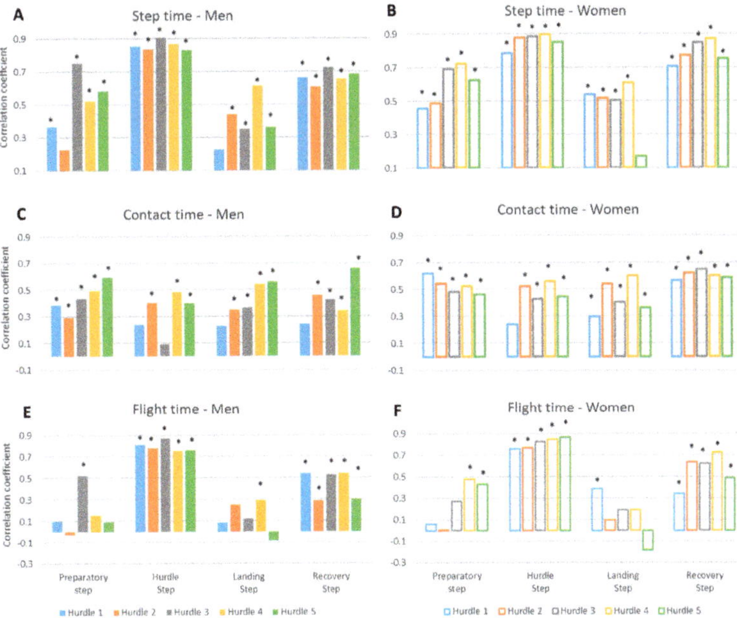

Figure 4. Pearson's correlation coefficient (r) between temporal parameters ((**A**,**B**) step time; (**C**,**D**) contact time; and (**E**,**F**) flight time) and the hurdle-unit splits times during the 44th Spanish Indoor and 12th IAAF World Indoor Championships. Correlations: * $p < 0.05$.

Every landing distance showed moderate to large (first and third hurdle) relationships with their respective split times in the men's race, while take-off distance showed moderate to large (fourth hurdle) correlations from the third to the fifth hurdle (Figure 5A). In the case of women athletes, the highest correlations were observed in the landing and recovery step of the last hurdle (moderate). The hurdle step width of men (Figure 5C) and women (Figure 5D) hurdlers showed moderate relationships with their respective split times in the first hurdle.

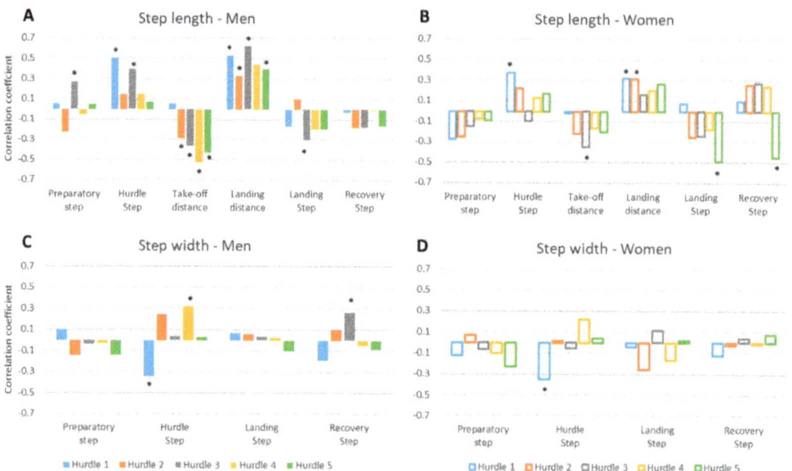

Figure 5. Pearson's correlation coefficient (r) between spatial parameters ((**A,B**) step length; and (**C,D**) step width) and the hurdle-unit split times during the 44th Spanish Indoor and 12th IAAF World Indoor Championships. Correlations: * $p < 0.05$.

4. Discussion

The present research examined the evolution of the kinematic parameters during the 60 m hurdle races of the 44th Spanish Indoor and 12th IAAF World Indoor Championships. Previous studies have focused on an isolated hurdle-unit analysis during the race, which could presumably have left a big knowledge-gap for coaches and athletes. The results of the present study indicate that each of the hurdle-units of the race presents a high correlation with the race result. Nevertheless, there were differences on some of the temporal and spatial parameters of the hurdle-unit subphases (especially the first and last ones).

4.1. Split Times

For the split times, the present research organized the hurdle-unit subphases from the preparatory to the recovery steps [9]. This hurdle-unit division allowed the inclusion of all performance criteria of the hurdle event in a more rational way following a deterministic model [17]. Split times calculated in the current investigation showed greater correlations with race results than values reported in previous studies [7,26], although, in all the cases, correlations were very large. Interestingly, correlation values increased from the first to the fourth hurdle in line with Tsionakos [6] in the 100/110 m event, whereas the values decreased in the fifth interval of the men's races. This may be due to the shorter distance between the last hurdle and the finish line, which could cause technical modifications (i.e., advance the trunk) in the last hurdle-unit phase to prepare for the run-in phase [27] or by the fatigue created by the greater relative height of the hurdles in men races [8,9,28].

Mean split times were slightly greater than the values of the finalists of international competitions [4–7,29–32], probably because of a greater sample size in the present research as it included both international and national standard level athletes. For both genders, there was a tendency in the first hurdle-unit to be the slowest of the race in line with previous studies [6]. However, in the case of men, the fifth (last) hurdle-unit was slower than the previous ones. This gender-dependent evolution in the split times coincides with that observed by Panoutsakopoulos et al. [7] and Kuitunen and Poon [26] and could be explained by the highest hurdle height of men. According to McDonald and Dapena [9], Salo et al. [11], and González-Frutos et al. [8], this height difference would represent a greater effort for men and would cause a loss of speed from the fourth hurdle on, whereas women can fasten up to the last hurdle.

4.2. Kinematic Differences between Hurdles

The first hurdle presented most of the kinematic differences throughout the race, even though the pattern was different from men to women. Male athletes presented a greater flight time in the hurdle step and a greater contact time on the landing step of the first hurdle, whereas female athletes presented a shorter step, contact, and flight times on the preparatory step. These differences could be explained by the relative differences in height and distance between hurdles in men and women hurdlers [9,28]. Indeed, the greater height of the hurdle in the male event could hinder the adjustments to the first hurdle with a possible greater take-off angle to avoid falling and its subsequent landing from a greater height [8,9,11]. For both genders, the take-off distance was the shortest and the step width was the greatest in the first hurdle step, which could be related to the need to avoid contact with the lead leg [29,30] by means of a lateral displacement. This could represent an important finding of the present research, as previous studies [9–12,15] have only focused on the study from the second to the fifth hurdle without providing information about the step width.

For the remaining hurdles, the longest take-off distance was observed in the third hurdle, which seems to be one of the most important performance indicators [8,9,11,15]. In the last (fifth) hurdle, male athletes increased their landing and recovery step length while female athletes presented a shorter recovery step length. These spatial differences in the last hurdle are in line with the temporal modifications mentioned above and suggest that gender differences, due to different distance and number of steps to the finish line, should be acknowledged by coaches when addressing this hurdle unit. Since previous studies have analyzed the second [10], third [11], fourth [9,12,15], or fifth hurdle [9,12] in isolation, our data clearly show that kinematics differences exist between hurdle units in a single race.

4.3. Correlations of Kinematic Parameters with Hurdle-Unit Times

The very large correlations of the step and flight times during the hurdle step with the hurdle-unit performance were in line with the very large correlations of these parameters with the race result [8]. These data reinforce the idea that shorter hurdle clearance times favor the maintenance of the horizontal velocity of the athletes [11]. Previous studies have not reported such large correlations [6] or have only observed high correlations in the men's race [26] or in some specific hurdle-units [7]. The differences with the current study might be due to a wider sample of athletes who were examined during the present research. For the female hurdlers, and as a difference to their male counterparts, very large correlations were observed between the recovery step and the split times. It is probable that the lower parabola observed for women [9] in the hurdle step, required the lengthening of the three inter-hurdle steps and highlighted the role of the recovery step. These gender differences somehow support the idea proposed by Mann [28] of increasing the distance between hurdles in the men's races and the hurdle height in the women's races. In the case of women, this would increase the importance of the technical ability [33], contributing to an upgrading of this event [34].

Regarding spatial variables, the correlations found between the take-off and landing distances with the split times of male hurdlers support the influence of the hurdle height on kinematics [8,9,11,34]. The fastest male hurdlers showed a clear ability to produce a longer take-off but shorter landing distances throughout the race. Furthermore, landing distances obtained greater correlations with the split times on the first half of the race, while take-off distances showed greater correlations with the split times in the second half of the race. Therefore, it could be understood that the lack of technical control in the landing distance during the first part of the race leads to fatigue in the second part, which might have a technical impact translating into a shorter take-off distance. Women hurdlers, on the other hand, presented the largest correlations between the split times in the landing and recovery step of the last hurdle. That is, a longer length of the landing and recovery after the last hurdle was related to a better split indicating the importance of the sprint ability [35]. Interestingly, for both genders, a greater step width over the first hurdle was related to a better split time. This had not been previously measured in competition, even though other studies focused on the approach run phase [36–38] and provides clear information on how the best hurdlers perform at the beginning of the race and negotiate the short take-off distance to the first hurdle.

4.4. Practical Applications and Research Limitations

According to the present results, coaches and practitioners should consider the key role of the first hurdle-unit and its technical demands with a greater step width and changes on the flight-contact times depending on gender. Also, the technical evolution throughout the race should be considered by maximizing the take-off distances at the beginning and minimizing the landing distances as the race progresses. Split times both in a training and competition setup should be recorded comprising the entire hurdle-unit structure (from preparatory to recovery steps).

Future research on the 60 m hurdle events could also consider the effect of contacting the hurdle during the flight phase on the kinematic parameters [39] or the differences in the race evolution according to the athletes' competitive level. Also, studies employing data recordings with a greater temporal resolution would be advisable to increase the uncertainty related to contact and flight times.

5. Conclusions

The results of this study provide solid evidence on the kinematic differences that competitive hurdlers present through the race. Split times improve from the first hurdle onwards and present gender-specific differences on the last (fifth) hurdle. In the first hurdle, men tend to increase their flight and contact times and women tend to decrease their contact and flight times. Also, both genders present shorter take-off distance, shorter landing distance, and greater step width than in the remainder of the race.

Faster split times were related to shorter step and flight times of the five hurdle steps for both genders. However, greater take-off distances for men in the first half of the race and shorter landing distances in the second half were related to faster performers. For women, longer recovery steps on the last hurdle were related to faster performers. The hurdle-unit split times from the preparatory to the recovery step present a more rational race division (according to a deterministic model) and can be compared with the split times of temporal studies.

From all of the above, coaches and athletes should implement their training programs considering the differences between men and women, to have an impact on the key variables according to the specific demands of each hurdle-unit phase.

Author Contributions: Conceptualization, P.G.-F., S.V., J.M., and E.N.; Data curation, P.G.-F.; Formal analysis, P.G.-F. and S.V.; Investigation, P.G.-F., S.V., and E.N.; Methodology, P.G.-F., S.V., J.M., and E.N.; Project administration, P.G.-F. and S.V.; Resources, E.N.; Software, E.N.; Supervision, E.N.; Validation, P.G.-F., J.M., and E.N.; Visualization, P.G.-F.; Writing—original draft, P.G.-F. and S.V.; Writing—review & editing, P.G.-F., S.V., J.M., and E.N. All authors have read and agreed to the published version of the manuscript.

Funding: This research received no external funding.

Acknowledgments: The author would like to acknowledge José Campos as the coordinator of the biomechanical studies developed during the 44th Spanish Indoor Championship and 12th IAAF World Indoor Championship, as well as the Royal Spanish Athletics Federation (RFEA) and the International Association of Athletics Federations (IAAF) for promoting, facilitating access and location for data collection, and managing the stay during the championships.

Conflicts of Interest: The authors declare no conflict of interest.

References

1. Schiffer, J. Hurdles (Part I). *New Stud. Athl.* **2006**, *21*, 71–104.
2. Schiffer, J. Hurdles (Part II). *New Stud. Athl.* **2006**, *21*, 97–122.
3. Brüggemann, G.-P.; Glad, B. Time Analysis of the 110 Metres and 100 Metres Hurdles. Scientific Research Project at the Games of the XXIVth Olympiad—Seoul 1988. *New Stud. Athlet.* **1990**, *5*, 91–131.
4. Muller, H.; Hommel, H. Biomechanical research project at the VIth world championships in athletics, Athens 1997. *New Stud. Athl.* **1997**, *12*, 43–73.
5. Graubner, R.; Nixdorf, E. Biomechanical analysis of the sprint and hurdles events at the 2009 IAAF world championships in athletics. *New Stud. Athlet.* **2011**, *26*, 19–53.
6. Tsiokanos, A.; Tsaopoulos, D.; Giavroglou, A.; Tsarouchas, E. Race pattern of women's 100-m hurdles: Time analysis of olympic hurdle performance. *Int. J. Kinesiol. Sports Sci.* **2017**, *5*, 56–64. [CrossRef]
7. Panoutsakopoulos, V.; Theodorou, A.S.; Kotzamanidou, M.C.; Fragkoulis, E.; Smirniotou, A.; Kollias, I.A. Gender and event specificity differences in kinematical parameters of a 60 m hurdles race. *Int. J. Kinesiol. Sports Sci.* **2020**, *20*, 668–682. [CrossRef]
8. González-Frutos, P.; Veiga, S.; Mallo, J.; Navarro, E. Spatiotemporal comparisons between elite and high-level 60 m hurdlers. *Front. Psychol.* **2019**, *10*, 2525. [CrossRef]
9. McDonald, C.; Dapena, J. Linear kinematics of the men's 110-m and women's 100-m hurdles races. *Med. Sci. Sports Exerc.* **1991**, *23*, 1382–1391.
10. McLean, B. The biomechanics of hurdling: Force plate analysis to assess hurdling technique. *New Stud. Athl.* **1994**, *9*, 55–58.
11. Salo, A.; Grimshaw, P.N.; Marar, L. 3-D biomechanical analysis of sprint hurdles at different competitive levels. *Med. Sci. Sports Exerc.* **1997**, *29*, 231–237. [CrossRef]
12. Coh, M. Biomechanical analysis of Colin Jackson's hurdle clearance technique. *New Stud. Athl.* **2003**, *18*, 37–45.
13. Park, Y.J.; Ryu, J.K.; Ryu, J.S.; Kim, T.S.; Hwang, W.S.; Park, S.K.; Yoon, S. Kinematic analysis of hurdle clearance technique for 110-m men's hurdlers at IAAF world championships, Daegue 2011. *Korean J. Sport Biomech.* **2011**, *21*, 529–540. [CrossRef]
14. Ryu, J.K.; Chang, J.K. Kinematic analysis of the hurdle clearance technique used by world top class women's hurdler. *Korean J. Sport Biomech.* **2011**, *21*, 131–140. [CrossRef]
15. Coh, M.; Iskra, J. Biomechanical studies of 110 m hurdle clearance technique. *Sport Sci.* **2012**, *5*, 10–14.
16. Čoh, M.; Bončina, N.; Štuhec, S.; Mackala, K. Comparative biomechanical analysis of the hurdle clearance technique of Colin Jackson and Dayron Robles: Key Studies. *Appl. Sci.* **2020**, *10*, 3302. [CrossRef]
17. Chow, J.W.; Knudson, D.V. Use of deterministic models in sports and exercise biomechanics research. *Sport Biomech.* **2011**, *10*, 219–233. [CrossRef]
18. Ho, C.S.; Chang, C.Y.; Lin, K.C. The wearable devices application for evaluation of 110 meter high hurdle race. *J. Hum. Sport Exerc.* **2020**, *15*, 34–42. [CrossRef]
19. Ito, A.; Ishikawa, M.; Isolehto, J.; Komi, P.V. Changes in the step width, step length, and step frequency of the world's top sprinters during 100 metres. *New Stud. Athl.* **2006**, *21*, 35–39.
20. Cala, A.; Veiga, S.; García, A.; Navarro, E. Previous cycling does not affect running efficiency during a triathlon world cup competition. *J. Sport Med. Phys. Fitness* **2009**, *49*, 152–158.
21. Abdel-Aziz, Y.I.; Karara, H.M. Direct linear transformation from comparator coordinates into space coordinates in close range photogrammetry. In *Proceedings of the Symposium on Close Range Photogrammetry*; The American Society of Photogrammetry: Falls Church, VA, USA, 1971; pp. 1–18.
22. Allard, P.; Blanchi, J.P.; Aíssaqui, R. Bases of three-dimensional reconstruction. In *Three Dimensional Analysis of Human Movement*; Allard, P., Stokes, I.A.F., Bianchi, J.P., Eds.; Human Kinetics: Champaign, IL, USA, 1995; pp. 19–40.
23. Veiga, S.; Cala, A.; Mallo, J.; Navarro, E. A new procedure for race analysis in swimming based on individual distance measurements. *J. Sports Sci.* **2013**, *31*, 159–165. [CrossRef]
24. Cohen, J.A. Power primer. *Psychol. Bull.* **1992**, *112*, 155–159. [CrossRef] [PubMed]
25. Hopkins, W.; Marshall, S.; Batterham, A.; Hanin, Y. Progressive statistics for studies in sports medicine and exercise science. *Med. Sci. Sports Exerc.* **2009**, *41*, 3–13. [CrossRef]

26. Kuitunen, S.; Poon, S. Race pattern of 60-m hurdles in world-class sprint hurdles: A biomechanical analysis of World Indoor Championships 2010. In Proceedings of the International Conference on Biomechanics in Sports, Marquette, MI, USA, 19–23 July 2010.

27. Kugler, F.; Janshen, L. Body position determines propulsive forces in accelerated running. *J. Biomech.* **2009**, *43*, 343–348. [CrossRef] [PubMed]

28. Mann, R. Rules-related limiting factors in hurdling. *Track Coach* **1996**, *136*, 4335–4337.

29. Pollitt, L.; Walker, J.; Bissas, A.; Merlino, S. Biomechanical report for the IAAF world championships 2017: 110 m hurdles men's in 2017. In *IAAF World Championships Biomechanics Research Project*; International Association of Athletics Federations: London, UK, 2018.

30. Pollitt, L.; Walker, J.; Bissas, A.; Merlino, S. Biomechanical report for the IAAF world championships 2017: 100 m hurdles women's in 2017. In *IAAF World Championships Biomechanics Research Project*; International Association of Athletics Federations: London, UK, 2018.

31. Walker, J.; Pollitt, L.; Paradisis, G.P.; Bezodis, I.; Bissas, A.; Merlino, S. *Biomechanical Report for the IAAF World Indoor Championships 2018: 60 Metres Hurdles Men*; International Association of Athletics Federations: Birmingham, UK, 2019.

32. Walker, J.; Pollitt, L.; Paradisis, G.P.; Bezodis, I.; Bissas, A.; Merlino, S. *Biomechanical Report for the IAAF World Indoor Championships 2018: 60 Metres Hurdles Women*; International Association of Athletics Federations: Birmingham, UK, 2019.

33. Bedini, R. Technical ability in the women's 100m hurdles. *New Stud. Athl.* **2016**, *31*, 117–132.

34. Stein, N. Reflections on a change in the height of the hurdles in the women's sprint hurdles event. *New Stud. Athl.* **2000**, *15*, 15–20.

35. Mackala, K. Optimisation of perfomance through kinematic analysis of the different phases of the 100 meters. *New Stud. Athl.* **2007**, *22*, 7–16.

36. López del Amo, J.L.; Rodríguez, M.C.; Hill, D.W.; González, J.E. Analysis of the start to the first hurdle in 110 m hurdles at the IAAF world athletics championships Beijing 2015. *J. Hum. Sport Exerc.* **2018**, *13*, 504–517. [CrossRef]

37. Hücklekemkes, J. Model technique analysis sheets for the hurdles. Part IV: The women's 100 m hurdles. *New Stud. Athl.* **1990**, *5*, 33–58.

38. Bezodis, I.N.; Brazil, A.; von Lieres und Wilkau, H.C.; Wood, M.A.; Paradisis, G.P.; Hanley, B.; Tucker, C.B.; Pollitt, L.; Merlino, S.; Vazel, P.-J.; et al. World-Class male sprinters and high hurdlers have similar start and initial acceleration techniques. *Front. Sports Act. Living* **2019**, *1*, 23. [CrossRef]

39. Iwasaki, R.; Shinkai, H.; Ito, N. How hitting the hurdle affects performance in the 110 m hurdles. *ISBS Proc. Arch.* **2020**, *38*, 268–271.

Article

Biological Maturity Status, Anthropometric Percentiles, and Core Flexion to Extension Strength Ratio as Possible Traumatic and Overuse Injury Risk Factors in Youth Alpine Ski Racers: A Four-Year Prospective Study

Lisa Steidl-Müller [1,*], Carolin Hildebrandt [1], Martin Niedermeier [1], Erich Müller [2], Michael Romann [3], Marie Javet [3], Björn Bruhin [3] and Christian Raschner [1]

1   Department of Sport Science, University of Innsbruck, 6020 Innsbruck, Austria; carolin.hildebrandt@uibk.ac.at (C.H.); martin.niedermeier@uibk.ac.at (M.N.); christian.raschner@uibk.ac.at (C.R.)
2   Department of Sport and Exercise Science, University of Salzburg, 5020 Salzburg, Austria; erich.mueller@sbg.ac.at
3   Section for Elite Sport, Swiss Federal Institute of Sport Magglingen, 2532 Magglingen, Switzerland; Michael.romann@baspo.admin.ch (M.R.); marie.javet@baspo.admin.ch (M.J.); bjoern.bruhin@baspo.admin.ch (B.B.)
*   Correspondence: lisa.steidl-mueller@uibk.ac.at; Tel.: +43-(0)512-507-45904

Received: 30 September 2020; Accepted: 27 October 2020; Published: 29 October 2020

Abstract: The aim of the present study was to investigate prospectively the role of biological maturity status, anthropometric percentiles, and core flexion to extension strength ratios in the context of traumatic and overuse injury risk identification in youth ski racing. In this study, 72 elite youth ski racers (45 males, 27 females) were prospectively observed from the age of 10 to 14 years. Anthropometric parameters, biological maturity status, and core flexion to extension strength ratios were assessed twice per year. Type and severity of traumatic and overuse injuries were prospectively recorded during the 4 years. Generalized estimating equations were used to model the binary outcome (0: no injury; 1: ≥1 injury). Factors tested on association with injury risk were sex, relative age quarter, age, maturity group, puberty status, core flexion to extension strength ratio, height percentile group, and weight percentile group. In total, 104 traumatic injuries and 39 overuse injuries were recorded. Age (odds ratio (OR) = 3.36) and weight percentile group (OR = 0.38) were significant risk factors for traumatic injuries (tendency: pubertal status). No significant risk factor for overuse injuries was identified (tendency: maturity group, puberty status, height percentile group). Future studies should focus on identifying risk factors for overuse injuries; growth rates might be of importance.

Keywords: youth ski racing; injury risk; biological maturity; anthropometric characteristics; core flexion to extension strength ratio; talent development

---

1. Introduction

Sport specialization in the immature athlete has profound effects on short- and long-term injury risks and joint health [1]. Even though alpine ski racing is a late specialization sport and the peak performance in elite alpine ski racing is achieved between the ages of 26 (females) and 28 years (males), youth ski racers at the ages of 10 to 14 years already have high weekly volumes of skiing-specific (734.0 ± 232.7 min) and additionally athletic-specific training (201.1 ± 122.0 min) [2]. High training volumes at an early age combined with an athlete's long competitive life needs to be considered when implementing strategies to develop talents. In this context, a large number of specialized ski boarding schools for talented young athletes provide a fundamental basis for a future professional career. Research has shown that training load parameters did not represent injury risk factors in young ski racers [2], which is a result that emphasizes the importance of such well-planned training structures for young athletes.

Nevertheless, alpine ski racing is a sport with a high risk of injury in elite athletes at World Cup and European Cup levels [3] but also at youth (10–14 years) [4] and adolescent levels (15–19 years) [5]. At the youth level, traumatic injury rates of 0.63/athlete per season, as well as overuse injury rates of 0.21/athlete per season were reported [4]. The knee was the most affected body part by both traumatic (36.5% of all traumatic injuries) and overuse injuries (82.3% of all overuse injuries). Up to now, studies prospectively investigating injury risk factors in youth ski racing were conducted only for one to two seasons [4,6–8]. Due to the short observation period, the small sample sizes because of the general small cohort of elite athletes in this age group, and the consequently small numbers of traumatic and overuse injuries, risk factors were usually identified for both traumatic and overuse injuries combined. However, due to the growing back overuse problems in alpine ski racing, it seems important to identify risk factors separately for traumatic and overuse injuries, respectively.

Growth-related factors such as biological immaturity contribute to overuse injuries [9]. The risk of sustaining overuse injuries is strongly intensified during the adolescent growth spurt [10]. Late maturing soccer players were at a higher risk for both overuse injuries [11] and severe traumatic injuries [12]. Additionally, late maturing youth ski racers were more vulnerable for more severe injuries (traumatic and overuse injuries combined) [4]. In this context, it seems important to prospectively assess youth athletes over a longer period during their pubertal years to better understand the role of biological maturity as a possible risk factor for both traumatic and overuse injuries. Additionally, further variables based on previous research might be considered in injury risk analyses. Firstly, anthropometric percentiles might be worth consideration for identifying possible risk factors among youth ski racers aged 10 to 14 years. Secondly, core strength was found to be a significant risk factor for injuries in youth ski racers [4], and for ruptures of the anterior cruciate ligament (ACL) in adolescent athletes [13]; in adolescent ski racers, core strength imbalances were significant risk factors for ACL injuries in male athletes [13]; however, core flexion to extension strength ratios have not yet been considered in injury risk identification in youth ski racers. Therefore, the aim of the present study was to investigate prospectively over a period of 4 seasons the role of biological maturity status, body weight, and body height percentiles and core flexion to extension strength ratios in the context of traumatic and overuse injury risk identification in elite youth ski racers from the ages of 10 years at the beginning of the observation period.

2. Materials and Methods

2.1. Study Design

The Institutional Review Board of the Department of Sport Science of the University of Innsbruck and the Board for Ethical Questions of the University of Innsbruck approved the study (#02/2014). A four-season prospective longitudinal study design was used to record traumatic and overuse injuries as well as anthropometric parameters, biological maturity status, and core flexion to extension strength ratios in a cohort of elite youth ski racers aged 10 years at the beginning of the study (until 14 years of age). All tests were performed in the laboratory of the Department of Sport Science, University of Innsbruck, by experienced researchers, at the same time of day (3 p.m.) and under standardized laboratory conditions (standardized shoes of the laboratory, same order of tests etc.) and using the same measurement systems to ensure repeatability and to limit confounding influences. The diverse parameters were tested twice per year: prior to the start of the competitive season (end of September) and at the end of the winter season prior to the preparation season (end of April). Traumatic and overuse injuries were recorded prospectively during the four years.

2.2. Participants

In total, 72 elite youth ski racers (45 males, 27 females), who were pupils of a very famous ski boarding school and who all competed at national levels in Austria, were included in the study. The athletes represent a cohort of very successful elite youth ski racers. No power calculations were performed, as all available pupils were included in the study. Inclusion criteria were as follows: (a) school attendance at the start of the study, including a medical screening (e.g., for musculoskeletal, circulatory, or neurological disorders) when entering the boarding school, (b) being free of acute injuries at the start of the study, and (c) being free of relevant chronic diseases. The pupils were prospectively observed and investigated from the age of approximately 10 years until the age of approximately 14 years, as they frequented the skiing-specific secondary modern school for 10- to 14-year-olds. At the start of the study, the mean age of the pupils was 10.6 ± 0.3 years. The mean weekly training volume of skiing-specific training was 727.3 ± 143.4 min, and the mean weekly training volume of athletic-specific training was 211.6 ± 76.2 min. The athletes participated in an average of 16.7 ± 9.2 races/season. The anthropometric characteristics of male and female athletes at the beginning of the study are presented in Table 1. All athletes, their parents, and coaches were informed about the study aims, risks, and benefits, and the parents of the athletes provided their written informed consent.

Table 1. Anthropometric characteristics of male and female athletes at the start of the study.

	Sex	
Anthropometric Characteristics	**Male**	**Female**
	M ± SD	**M ± SD**
Age [yrs]	10.6 ± 0.4	10.5 ± 0.3
Height [cm]	144.6 ± 5.5	142.5 ± 6.0
Weight [kg]	35.3 ± 5.1	34.0 ± 4.9
BMI [1] [kg/m^2]	16.8 ± 1.9	16.7 ± 1.7
Sitting height [cm]	77.5 ± 4.3	77.7 ± 4.6
APHV [2] [yrs]	13.5 ± 0.4	12.0 ± 0.4

[1] BMI: body mass index; [2] APHV: age at peak height velocity.

2.3. Data Collection

2.3.1. Anthropometric Parameters and Biological Maturity Status

The anthropometric characteristics of body height (0.5 cm; portable Stadiometer SECA 217; SECA, Hamburg, Germany), body weight (1N, Kistler force plate; Kistler Instrumente AG, Winterthur, Switzerland; calculated to the nearest 0.1 kg), and sitting height (0.5 cm; portable Stadiometer SECA 217; SECA, Hamburg, Germany; sitting height table) were assessed. Additionally, the leg length as difference between body height and sitting height (0.5 cm) and the body mass index (BMI; 0.1 kg/m^2) were calculated. The percentiles of body height and body weight were classified according to Braegger et al. [14]. Percentiles of body height and body weight were categorized in three groups (0: percentile 25 and below; 1: between percentile 25 and <75; 2: percentile 75 and above), which was done in accordance with Kromeyer-Hauschild et al. [15].

The biological maturity status was assessed using the non-invasive method of calculating the age at peak height velocity (APHV) [16]. These sex-specific prediction equations include the aforementioned anthropometric characteristics (except for BMI) as well as the calculated actual chronological age at the time of measurement. Then, based on these values, the maturity offset as the time before or after individual peak height velocity (PHV) was calculated. Then, the difference between the actual chronological age and the maturity offset was defined as APHV. Then, participants were categorized into three maturity groups according to the sex-specific APHV of the present sample [17,18]: early maturing, defined as APHV below sex-specific mean − standard deviation; normal maturing, defined as sex-specific mean APHV ± standard deviation; and late maturing: APHV higher than sex-specific mean + standard deviation. A negative maturity offset indicated that the athlete was pre PHV, whereas a positive maturity offset indicated post PHV [16]. In accordance with Faigenbaum et al. [19] and Lesinski et al. [20], puberty status was classified as pre-pubertal: maturity offset −1 and smaller; pubertal: maturity offset between −1 and 1; and post-pubertal: maturity offset 1 and larger. A previous study [18] confirmed the validity of this APHV method for 10- to 13-year-old youth ski racers. The difference between the calculated APHV and the actual age at the day of the occurrence of an injury was defined as maturity offset at time of injury occurrence. The birth months of the athletes were split into relative age quarters; relative age quarter 1 (Q1) included the birth months January, February, and March; Q2 included April, May, and June; Q3 included July, August, and September; and Q4 included October, November, and December.

2.3.2. Core Flexion to Extension Strength Ratio

The core flexion and extension strength test is part of the test battery for junior ski athletes as described by Raschner et al. [21], which was established in consultation with sports scientists, ski racing experts, and coaches. The maximal isometric core flexion and extension strength were measured using the slightly modified Back-Check (Dr. Wolff Sports & Prevention GmbH, Arnsberg, Germany). The athlete was in an upright standing position, with knees slightly flexed and pelvis stabilized; pads, which were connected to a force transducer, were set at sternum level (anteriorly and posteriorly). Against these cushioned pads, the athlete contracted maximally. The athletes conducted three trials each (flexion and extension), and the highest force of the three attempts each were recorded [21]. Then, the ratio of flexion to extension strength was calculated by dividing the flexion strength by the extension strength. The test–retest reliability analysis among high-level athletes identified intra class correlation coefficients (ICCs) of 0.77 (flexion) and 0.90 (extension) [22]. Subsequently, the ratio of core flexion to extension strength was categorized to 0 for a ratio ≤1 indicating higher strength in extensors or equal strength; and to 1 for a ratio >1 indicating higher strength in flexors.

2.3.3. Injury Registration

A sport-specific internet-based database was developed and used in the present study; a detailed description of the database was reported by Müller et al. [4]. Using this database, coaches of the ski boarding school recorded all relevant training data immediately after each skiing-specific and athletic training session. In case of absence of an athlete due to traumatic injury or overuse injury, a member of the study team contacted the coaches, physiotherapists, and/or physicians of the ski boarding school to get detailed information. Based on the information, all traumatic and overuse injuries that caused absence from training for at least one day were then registered by the member of the study team using the internet-based database. A detailed medical report was provided in case of an injury that required medical attention. The date of injury, the severity of injury, the circumstance of the injury, the affected body part, the type of injury were recorded, among other factors.

A traumatic injury was defined as an injury with a sudden onset based on time-loss definition [23]. Based on the injury surveillance consensus paper of the International Olympic Committee [24], the type of traumatic injury and the affected body part were defined. The injury severity was defined according to Fuller et al. [25]: An injury was classified as minimal with a time-loss of 1–3 days, extending to mild (4–7 days), moderate (8–28 days), severe (>28 days), or career ending. Based on the definition by Clarsen et al. [26], an overuse injury was defined as any physical complaint without a single identifiable event being responsible.

2.4. Statistical Analyses

Statistical analyses were done using SPSS v. 26 (IBM, New York, NY, USA). Risk factors for traumatic and overuse injuries were analyzed separately. The eight test time points resulted in repeated measurements (up to eight) in the athletes. Ordinary binary regression analysis is expected to produce biased estimates of the regression analysis. Therefore, generalized estimating equations (GEE) were used to model the binary outcome in which 0 represents no injury, and 1 represents at least one injury. GEE was preferred compared to ordinary binary regression analysis since GEE accounts for the inter-correlation of repeated observations in the same subjects over time and was used in the injury context previously [27]. A working correlation matrix with exchangeable structures was used. The first step of the primary statistical analysis consisted of a series of simple GEE models using all potential associated factors separately. In detail, the factors tested on association with injury risk were sex, relative age quarter, age, maturity group, puberty status, core flexion to extension strength ratio group, height percentile group, and weight percentile group. The odds ratio (OR) including 95% confidence intervals (95% CI) was calculated for each factor. In a second step of the primary statistical analyses, a multiple GEE model with all significant associated factors in the simple analysis was conducted. Quasi-information criterion (QIC) was given for the multiple GEE model. A lower QIC value indicates a better model fit [28].

As an additional analysis, a series of multiple binary regression analyses with stepwise backward elimination was performed separately for each test time point. The regression analysis is regarded as secondary and as a supplement to the GEE, since it cannot consider repeated observations across athletes. *p*-values < 0.05 were considered as significant and *p*-values ranging from 0.05 to 0.10 were considered as tendencies.

3. Results

3.1. Injury Chracteristics

In total, *n* = 482 observations from 72 ski racers were available with 104 traumatic injuries (45 separate athletes) and 39 overuse injuries (25 separate athletes). Average injury incidence across all time points was 0.22 injuries/athlete for traumatic and 0.08 injuries/athlete for overuse injuries (Figure 1). Most traumatic injuries were moderate (41.3%), followed by mild (23.1%), minimal (20.2%), and severe (15.4%). Most overuse injuries were mild (33.3%), followed by moderate (30.6%), minimal (22.2%), and severe (13.9%). None of the injuries was career ending. The predominant injury location of traumatic injuries was in the lower extremities (66.3% of all injuries), followed by upper extremities (19.2%), trunk (8.7%), and head (5.8%). Among the traumatic injuries located at the lower extremities, 44.7% were moderate injuries, and 18.4% were minimal, mild and severe injuries each. Overuse injuries were located in the lower extremities (82.4%) or in the trunk (17.6%). Among the overuse injuries located at the lower extremities, most of them were classified as moderate injuries (34.4%), followed by mild (31.3%), minimal (21.8%) and severe (12.5%). Twenty-two athletes had a traumatic (eight athletes had an overuse) injury on more than one observation. In 15 observations, more than one traumatic injury was found, resulting in 87 observations with at least one traumatic injury. In two observations, more than one overuse injury was found, resulting in 36 observations with at least one overuse injury.

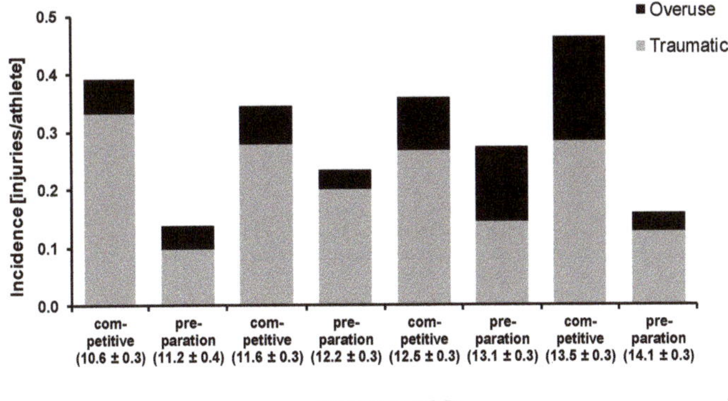

Figure 1. Incidence of traumatic (gray) and overuse (black) injuries per athlete by test time point.

3.2. Factors Associated with Traumatic Injuries

Out of all factors tested on association with traumatic injuries using simple GEE analysis, age and weight percentile group were significant (Table 2). Compared to the age of 14 years, being 10 years old was associated with a significantly higher risk for traumatic injury (OR = 3.36). Compared to being in the weight percentile of 75 and above, being in the weight percentile 25 and below was associated with a significantly lower risk for traumatic injury (OR = 0.38). Puberty status showed a tendency for a higher injury risk for pre-pubertal athletes compared to post-pubertal athletes. The other potential factors including sex, relative age variables, core flexion to extension strength ratio, and height were non-significant.

Table 2. Potential factors associated with traumatic injuries according to the simple general estimating equation analysis (*n* = 482 observations).

Variable	% Observations with Injuries	N Observations with Injuries	N Observations	*p*	OR [1]	95% CI lb [2]	95% CI ub [3]
Sex							
Female	16.2%	31	191	0.655	0.86	0.43	1.69
Male	19.2%	56	291		1 (ref)		
Relative age quarter							
1st	15.7%	24	153	0.692	0.79	0.25	2.52
2nd	20.1%	30	149	0.873	1.10	0.34	3.51
3rd	19.2%	20	104	0.965	1.03	0.29	3.63
4th	17.1%	13	76		1 (ref)		
Age [years]							
10	**28.1%**	**16**	**57**	**0.021**	**3.63**	**1.22**	**10.81**
11	17.4%	19	109	0.174	2.06	0.73	5.86
12	20.0%	26	130	0.080	2.46	0.90	6.73
13	15.6%	21	135	0.213	1.88	0.70	5.09
14	9.8%	5	51		1 (ref)		
Maturity group [4]							
Early	19.7%	15	76	0.670	1.26	0.43	3.73
Average	18.2%	60	330	0.739	1.16	0.48	2.78
Late	15.8%	12	76		1 (ref)		
Puberty status [5]							
Pre-pubertal	21.8%	39	179	0.073	1.86	0.94	3.67
Pubertal	16.7%	37	222	0.348	1.36	0.71	2.61
Post-pubertal	13.6%	11	81		1 (ref)		
Core flexion to extension strength ratio group [6]							
Ratio 1 and below	16.1%	38	236	0.287	0.76	0.45	1.26
Ratio > 1	19.5%	41	210		1 (ref)		
Height percentile group [7]							
25 and below	17.2%	20	116	0.403	0.68	0.27	1.70
>25 to <75	17.3%	51	295	0.401	0.72	0.34	1.54
75 and above	23.4%	15	64		1 (ref)		
Weight percentile group [8]							
25 and below	**11.0%**	**13**	**118**	**0.034**	**0.38**	**0.15**	**0.93**
>25 to <75	18.9%	54	286	0.333	0.69	0.32	1.47
75 and above	25.7%	18	70		1 (ref)		

[1] OR: odds ratio, [2] 95%CI lb: 95% confidence interval lower bound, [3] 95%CI ub: 95% confidence interval upper bound, [4] groups according to age at peak height velocity, [5] according to distance to age at peak height velocity, [6] missing observations = 36, [7] missing observations = 7, [8] missing observations = 8, bold values indicate statistically significant values.

According to the multiple GEE analysis, both age and weight percentile group remained significantly associated with traumatic injury risk (Table 3), QIC = 446.9. Adding the variable puberty status did not improve the model according to the QIC = 449.7. According to the series of multiple linear regression analysis with stepwise backward elimination, the weight percentile group was the only remaining significant factor at time points 3 and 4. No significant factors remained for the other time points.

Table 3. Results of the multiple general estimating equation analysis of factors associated with traumatic injuries based on n = 474 observations.

Variable	% Observations with Injuries	p	OR [1]	95%CI lb [2]	95%CI ub [3]
Age [years]					
10	28.1%	**0.018**	**3.72**	1.26	11.04
11	17.4%	0.148	2.15	0.76	6.03
12	20.0%	0.066	2.55	0.94	6.93
13	15.6%	0.220	1.86	0.69	5.02
14	9.8%		1 (ref)		
Weight percentile group					
25 and below	11.0%	**0.027**	**0.37**	0.15	0.89
>25 to <75	18.9%	0.375	0.71	0.33	1.52
75 and above	25.7%		1 (ref)		

[1] OR: odds ratio, [2] 95%CI lb: 95% confidence interval lower bound, [3] 95%CI ub: 95% confidence interval upper bound, bold values indicate statistically significant values.

3.3. Potential Factors Associated with Overuse Injuries

None of the factors tested on association with overuse injuries was significant (Table 4). Statistical trends were found for maturity group, puberty status, and height percentile group, $p < 0.10$, where a non-significantly higher injury risk was found for athletes being early maturing, post-pubertal, and larger.

Table 4. Potential factors associated with overuse injuries according to the simple general estimating equation analysis (n = 482 observations).

Variable	% Observations with Injuries	N Observations with Injuries	N Observations	p	OR [1]	95%CI lb [2]	95%CI ub [3]
Sex							
Female	7.3%	14	191	0.952	0.97	0.42	2.28
Male	7.6%	22	291		1 (ref)		
Relative age quarter							
1st	7.2%	11	153	0.875	0.91	0.30	2.82
2nd	7.4%	11	149	0.956	0.97	0.28	3.30
3rd	7.7%	8	104	0.998	1.00	0.30	3.27
4th	7.9%	6	76		1 (ref)		
Age [years]							
10	8.8%	5	57	0.600	1.50	0.33	6.82
11	3.7%	4	109	0.391	0.55	0.14	2.14
12	5.4%	7	130	0.865	0.88	0.20	3.79
13	12.6%	17	135	0.199	2.12	0.67	6.66
14	5.9%	3	51		1 (ref)		
Maturity group [4]							
Early	15.8%	12	76	0.091	2.61	0.86	7.93
Average	5.8%	19	330	0.757	0.84	0.28	2.55
Late	6.6%	5	76		1 (ref)		
Puberty status [5]							
Pre-pubertal	4.5%	8	179	0.069	0.37	0.12	1.08
Pubertal	8.6%	19	222	0.449	0.71	0.29	1.72
Post-pubertal	11.1%	9	81		1 (ref)		
Core flexion to extension strength ratio group [6]							
Ratio 1 and below	8.1%	19	236	0.274	1.45	0.75	2.81
Ratio > 1	7.1%	15	210		1 (ref)		

Table 4. *Cont.*

Variable	% Observations with Injuries	N Observations with Injuries	N Observations	*p*	OR [1]	95%CI lb [2]	95%CI ub [3]
Height percentile group [7]							
25 and below	6.9%	8	116	0.110	0.43	0.15	1.21
>25 to <75	6.4%	19	295	0.088	0.44	0.17	1.13
75 and above	14.1%	9	64		1 (ref)		
Weight percentile group [8]							
25 and below	4.2%	5	118	0.140	0.39	0.11	1.36
>25 to <75	8.0%	23	286	0.616	0.77	0.27	2.16
75 and above	11.4%	8	70		1 (ref)		

[1] OR: odds ratio, [2] 95%CI lb: 95% confidence interval lower bound, [3] 95%CI ub: 95% confidence interval upper bound, [4] groups according to age at peak height velocity, [5] according to distance to age at peak height velocity, [6] missing observations = 36, [7] missing observations = 7, [8] missing observations = 8.

4. Discussion

4.1. Incidence of Traumatic and Overuse Injuries

In total, 104 traumatic injuries occurred during the four-year period (63.9% reported by males); 45 separate athletes (62.5% of all athletes) were affected. An average injury incidence of 0.22 injuries/athlete per season (September until April or April until September) was calculated, which is lower compared to previous studies in youth ski racing (0.46 traumatic injuries/athlete per season) [7]. However, it has to be considered that in the present study, the incidence was calculated for half-year-periods (competitive season and preparation season). Nevertheless, a relatively low incidence of traumatic injuries was found compared with adolescent ski racers (injury risk: 0.67) [5] or elite ski racers at the World Cup level (0.36 injuries/athlete/season) [3]. However, when comparing the competitive seasons (September–April) of the four-year periods (10–14-year-old athletes) with the preparation seasons (April–September), the incidence of traumatic injuries was obviously higher during the competitive season for all age groups. This finding is in line with a study in adolescent ski racing, in which Hildebrandt and Raschner [5] reported higher rates of traumatic injuries during the winter seasons compared with the summer seasons for both males and females.

In total, only 39 overuse injuries were reported during the four-year period (64.1% reported by males); 25 separate athletes (34.7% of all athletes) were affected. A very low overuse injury rate of 0.08 injuries/athlete per season (September until April or April until September) was calculated, which is comparable to the rate found in a previous study in youth ski racing (0.13 overuse injuries/athlete/season) [7]. In adolescent ski racing (>50% of the athletes during 2-season study period) [5] and elite ski racing (>1/3 of the top-40 ranked Slalom athletes) [29], higher rates of overuse injuries were identified. Hildebrandt and Raschner [5] found a higher risk for overuse injuries during the summer season, while a higher risk for traumatic injuries was present during the winter season in adolescent ski racers. In the present study, a lower incidence for traumatic injuries during the preparation season was identified, but the incidence of overuse problems was approximately the same. When looking at descriptive results, a slightly higher incidence of overuse problems was identified in the competitive seasons and preparation season of athletes aged between 12.5 and 13.5 years. However, it only can be speculated that in the older athletes, the training loads are higher, which might lead to more overuse problems the older the athletes are, or overuse problems are developed with time and are therefore, more reported in older athletes, as the study of Hildebrandt and Raschner [5] shows.

4.2. Severity of Traumatic and Overuse Injuries and Affected Body Parts

Most traumatic injuries were moderate (41.3%), which means a time-loss of training of 8–28 days, whereas only 15.4% were severe (time-loss of >28 days). These findings are in line with previous studies in youth ski racing (44.9% moderate; 13.1% severe) and adolescent ski racing (15–19 years; 46% moderate; 41% severe) [5], whereas at the World Cup level, most injuries were classified as severe injuries (>36%) [30]. However, at the World Cup level, injury registration was performed retrospectively, which might lead to the assumption that not all minimal and moderate injuries were reported, as it was done in the present prospective study. With respect to injury location, the present results confirm previous findings in youth ski racing [6], adolescent ski racing [5], and elite ski racing [3]: most traumatic injuries (66.3%) affected the lower extremities; most of them were located at the knee. Based on these findings, it can be concluded that also in youth ski racing, the knee is the most vulnerable body part by traumatic injuries.

Most overuse injuries were mild (33.3%), followed by moderate (30.6%), minimal (22.2%), and only 13.9% were categorized as severe. Only 15% of the overuse problems were recurring, which is a finding that is comparable with a previous study in youth ski racing, in which 23.1% of the reported overuse injuries were recurrent [4]. This finding indicates that the youth athletes had sufficient time to recover, and the pressure to return to training and competition or train despite symptoms was not that high. In line with a previous study in youth ski racing [4], the lower extremities were the most affected body parts also by overuse injuries; in the present study, 82.4% affected the lower extremities, which is comparable to the 82.3% affecting the knee in the reported study [4]. Growing athletes suffer from unspecified but self-limiting knee problems; the growth rate of the apophysis is slower than that of the epiphyseal plate, leading to problems of the interface between bone and ligaments; most overuse injuries were classified as typical apophysitis, such as Osgood–Schlatter disease [31]. Based on the present and previous findings, it can be concluded that in young ski racers, overuse problems are more relevant in the lower extremities, whereas the older the athletes are, the more overuse problems in the trunk become more relevant [5].

4.3. The Role of Chronological and Relative Age in Injury Risk Identification

Chronological age seems to be a relevant factor in traumatic injury risk identification. A statistically significant higher risk for traumatic injuries was found in 10-year-old athletes (OR = 3.36). The highest rate of injured observations was found during the competitive season after the first investigation of the athletes, when they were in mean 10.6 ± 0.3 years old. This finding might be explained by the different preconditions of the young ski racers before frequenting the ski boarding school. After having passed the entrance exam for the school, which normally takes place in the fall, they start with their dual career (school and ski racing) in September (first examination of the athletes in present study). Some athletes had already trained with experienced coaches and were already well-prepared, whereas others did not have any planned training sessions except for participating in ski races; thus, an inhomogeneous group of athletes might start with the first school year. However, all of them have then the same training loads of skiing-specific and athletic-specific training. It might be assumed that these loads are too much for some athletes during the first months in the new school; together with the new life in the boarding school and psychological stress to perform well, it might lead to the high traumatic injury rates in this age group. This result is in line with a study in youth soccer, in which an association between traumatic injuries and chronological (younger) age was found [32]. In this context, in the future, it might be necessary to adapt the training loads based on the diverse preconditions of the pupils when entering the school.

Chronological age might be more relevant in identifying athletes at higher risk for sustaining traumatic injuries than relative age. In line with a previous study in youth ski racing [4], in which no difference in frequencies of traumatic injuries between the four relative age quarters was found, the relative age quarter was not identified to be a significant risk factor in the present study. However, the fourth relative age quarter was considered as a reference in the statistical analyses, which means that the other three quarters were compared with Q4. As it was also assumed by Müller et al. [4], relatively younger athletes, who might be expected to be at a higher injury risk based on previous findings in different sports (a.o., [32]), may have counteracted their relative age disadvantage if they were more mature and possibly at a higher level of physical fitness. Thus, they might present the same physical conditions as their relatively older counterparts and therefore, they might not be at a higher injury risk.

4.4. The Role of Biological Maturity in Injury Risk Identification

Although not significant in the present dataset, pre-pubertal athletes might be more at risk for traumatic injuries compared to post-pubertal athletes, as at least a tendency was observed. This finding seems plausible, since Kemper et al. [33] showed that anthropometric characteristics such as rapid growth rates and BMI increase seem to identify youth athletes at high injury risk; this seems particularly true between the year before PHV and the year of PHV [9,11]. Additionally, Caine et al. [34] found the adolescent growth spurt to be a time of an increased risk for sports injuries; an association between PHV and peak fracture rate was found. Jayanthi et al. [35] additionally revealed that growth plate tissue may be more vulnerable to injury during periods of rapid growth, which means that athletes are more susceptible to injuries during rapid phases of growth, which might be true for pre-pubertal as well as pubertal athletes but not for post-pubertal athletes. Growth spurts might lead to changes in the ratios of muscle strength to body or limb mass and moments of inertia; soft tissues may sustain increased stress and strain [36]. Individual monitoring of anthropometrics and corresponding growth spurts is necessary to determine an increased risk of injury occurrence [36]. Regular assessments of anthropometric characteristics, e.g., all three months, should be performed in order to be able to consider growth rates in future studies, which is especially important for the age group of athletes prior to or at the beginning of puberty [37,38].

In contrast to traumatic injuries, a tendency was present that post-pubertal athletes might be more prone to overuse injuries. Previous findings in youth ski racing identified *lighter* athletes at a higher injury risk [6]. Thus, the present results revealing a significant higher injury risk in athletes being in the higher weight percentile group compared to the lower percentile group (OR = 0.38) was surprising and cannot really be explained. It should be noted in this context that unlike previous studies, the present GEE analysis does not exclusively compare heavier versus lighter athletes (due to repeated measurements across athletes). This approach may consider intra-individual fluctuations over the four seasons, e.g., comparing not only between separate athlete groups, but also between different time points, where athletes fall into different weight percentiles.

Previous studies in soccer [11,12] and youth ski racing [6] showed that late maturing athletes were at a higher risk for sustaining traumatic and overuse injuries, whereas in the present study, no significant difference in injury risk for both types of injuries was identified between early, normal, and late maturing athletes. However, in the study of Müller et al. [6], an association was identified only between APHV and injury severity; namely, athletes with lower APHV values, which means that they are earlier maturing, were more likely to miss fewer days of training due to injuries. In the present study, injury severity was not considered in identifying risk factors, which means that minimal, mild, moderate, and severe injuries were included combined. However, the results are in line with those of Müller et al. [6] showing no difference in injury risk in general (injured or not) between different maturity groups. With respect to overuse injuries, the unexpected finding that a tendency was present for *early* maturing athletes being at a higher risk was found. Even though this result cannot really be explained, as previous studies in soccer showed a higher risk of sustaining overuse injuries for late maturing athletes [11], it leads to the assumption that traumatic and overuse injuries should be considered separately when talking about injury risk among athletes of differing maturity statuses.

4.5. Gender-Specific Differences in Injury Risk

Previous findings on the role of sex in injury risk included that female adolescent ski racers (15–19 years) were at a twofold higher risk of sustaining a rupture of the ACL than their male counterparts (relative risk of 2.3) [13]; similar findings were found in older studies for World Cup athletes [39,40]. However, other studies in elite ski racers revealed that male athletes were at a higher risk for injuries in general and for time-loss injuries than females [29,41]. Spörri et al. [3] concluded that the influence of sex might depend on the type of injury, as females might be at a higher risk for ACL injuries and injuries at the knee in general. In other types of sport, such as basketball and soccer, a fourfold to sixfold higher risk in sustaining an ACL injury was found for females compared to males, independent of the age of the athletes [42,43]. However, in the present study, sex did not play a role in injury risk identification of both traumatic and overuse injuries. This might be explained by the fact that all injuries were considered combined (separated by traumatic and overuse injuries), and specific injuries, such as ACL injuries, were not analyzed separately. In the present sample, only three ACL ruptures occurred; two of them affected female athletes. A slightly higher percentage of males were affected by traumatic injuries (19.2%) compared to females (16.2%), whereas approximately 7% of both males and females were affected by overuse injuries. These findings are in line with a previous study in youth ski racing [4], in which no sex-specific differences were found in the occurrence of traumatic injuries; it was hypothesized that this might be associated with similar training-load exposure, as skiing-specific and athletic-specific training sessions were conducted together for males and females [4], which was also the case in the present study. In addition, Raschner et al. [13] argued that during alpine ski racing events, the influence of external factors, such as weather and snow conditions, overrule gender-specific risk factors such as anatomical differences or hormonal influences. It might be hypothesized that anatomical and hormonal differences might be more relevant after puberty.

Thus, it can be concluded that in the age group of 10- to 14-year-old ski racers, sex does not represent a critical factor for sustaining injuries. This might be explained by the fact that in this age group, males and females do not present great differences in physical performance and therefore, they might be similarly prepared for the demands in ski racing. Additionally, in the ski boarding school, males and females participate in the same training sessions together, and therefore, they have the same training loads and contents; they participate in the same races with the same difficulties. Only males and females starting at the FIS age participate in diverse races with different conditions/difficulties.

4.6. The Role of Core Flexion to Extension Strength Ratio in Injury Risk Identification

One study in adolescent ski racing revealed core strength imbalances as an ACL injury risk factor in male athletes [13]. Male athletes who sustained an ACL injury showed either much higher mean values (1.10), indicating that their core flexor muscles were too strong, or lower values (0.65), indicating weak trunk extensor muscles [13]. Insufficient core strength, inappropriate recruitment of the core muscles, and imbalances of these muscles might contribute to low back pain (LBP) [44], which represents a severe overuse problem in alpine ski racing in elite athletes [3] and adolescent athletes [5]. Even though LBP is not such a severe overuse problem in youth ski racers, as most of the overuse injuries were located in the lower extremities (82.4%) and only 17.6% were located in the trunk in the present study, a well-balanced core strength seems important in an injury prevention context. However, the present results did not reveal a significant association between core strength ratios and injury risk according to the GEE analysis. In this context, the used threshold of the core flexion to extension strength ratio of >1 and ≤1 has to be critically considered, as reference values of "good" or "weak" or "high risk" trunk strength ratios of alpine ski racers do not exist in the literature. In the present study, the ratio was defined only as stronger flexors or stronger extensors.

5. Limitations

The main limitation of the present study was that despite prospectively monitoring athletes over a period of four years, still, a low number of injuries was reported, especially of overuse injuries. This fact is of course desirable for the athletes and underlines the well-prepared and planned talent development work done in the ski boarding school, but it makes it difficult to have sufficiently powered datasets. Furthermore, it is unclear whether the used coding of diverse variables was appropriate, as for example, the threshold value of the core strength ratios of >1 versus 1 and smaller. However, the main strength of the present study was the long period for a prospective study: namely, four years.

Additionally, the GEE analysis is a relatively new way of identifying risk factors in youth ski racing by combining several seasons and considering the data structure of repeated measurements. A new aspect of the present study was considering traumatic and overuse injuries separately, because potentially different mechanisms and consequently different risk factors are assumed. As some significant risk factors for traumatic injuries were identified (chronological age, weight percentile group; tendency: pubertal status), no significant risk factor was found for overuse injuries. Therefore, future studies including other parameters, such as growth rates, are necessary in order to be able to identify risk factors for overuse injuries as well.

6. Conclusions

The present study was the first prospective study investigating possible risk factors for traumatic and overuse injuries in youth alpine ski racers over a longer period (4 years). Anthropometric percentiles, biological maturity status, core flexion to extension strength ratio, as well as sex, relative age, and chronological age were included as possible injury risk factors. With respect to traumatic injuries, only two injury risk factors were identified: age and weight percentile. Puberty status showed a tendency for a higher injury risk for pre-pubertal athletes. The analyses did not reveal a significant risk factor for overuse injuries; maturity group, puberty status and height percentile showed only tendencies in the analysis. Based on these results, it can be concluded that injury prevention programs in youth ski racing should consider chronological age and biological maturity. Additionally, traumatic and overuse injury risk should be considered separately. In future studies, growth rates should be considered in order to consider the impact of growth.

Author Contributions: Conceptualization, L.S.-M., C.H., M.N., E.M., M.R., M.J., B.B. and C.R.; methodology, L.S.-M. and C.H.; formal analysis, M.N. and L.S.-M.; investigation, L.S.-M., C.H. and C.R.; resources, C.R.; data curation, L.S.-M. and C.H.; writing—original draft preparation, L.S.-M., E.M. and M.N.; writing—review and editing, L.S.-M., C.H., M.N., E.M., M.R., M.J., B.B. and C.R.; visualization, M.N. and L.S.-M.; project administration, L.S.-M. All authors have read and agreed to the published version of the manuscript.

Funding: This research received no external funding. The authors want to thank the University of Innsbruck for the funding of the publication.

Acknowledgments: The authors want to thank all pupils and the coaches for participation in this study.

Conflicts of Interest: The authors declare no conflict of interest.

References

1. Lansdown, D.A.; Rugg, C.M.; Feeley, B.T.; Pandya, N.K. Single Sport Specialization in the Skeletally Immature Athlete: Current Concepts. *J. Am. Acad. Orthop. Surg.* **2020**, *28*, e752–e758. [CrossRef]
2. Hildebrandt, C.; Oberhoffer, R.; Raschner, C.; Müller, E.; Fink, C.; Steidl-Müller, L. Training load characteristics and injury and illness risk identification in elite youth ski racing: A prospective study. *J. Sport. Health. Sci.* **2020**. [CrossRef]
3. Spörri, J.; Kröll, J.; Gilgien, M.; Müller, E. How to prevent injuries in alpine ski racing: What do we know and where do we go from here? *Sports Med.* **2017**, *47*, 599–614. [CrossRef]
4. Müller, L.; Hildebrandt, C.; Müller, E.; Oberhofer, R.; Raschner, C. Injuries and illnesses in a cohort of elite youth alpine ski racers and the influence of biological maturity status and relative age: A two-season prospective study. *Open Access J. Sports Med.* **2017**, *8*, 113–122. [CrossRef]
5. Hildebrandt, C.; Raschner, C. Traumatic and overuse injuries among elite adolescent alpine skiers: A two-year retrospective analysis. *Int. Sport Med. J.* **2013**, *14*, 245–255.
6. Müller, L.; Hildebrandt, C.; Müller, E.; Fink, C.; Raschner, C. Long-term athletic development in youth alpine ski racing: The effect of physical fitness, ski racing technique, anthropometrics and biological maturity status on injuries. *Front. Physiol.* **2017**, *8*, 656. [CrossRef]
7. Steidl-Müller, L.; Hildebrandt, C.; Müller, E.; Raschner, C. Relationship of Changes in Physical Fitness and Anthropometric Characteristics over One Season, Biological Maturity Status and Injury Risk in Elite Youth Ski Racers: A Prospective Study. *Int. J. Environ. Res. Public Health* **2020**, *17*, 364. [CrossRef]

8. Steidl-Müller, L.; Hildebrandt, C.; Müller, E.; Fink, C.; Raschner, C. Limb symmetry index in competitive alpine ski racers: Reference values and injury risk identification according to performance level. *J. Sport Health Sci.* **2018**, *7*, 405–415. [CrossRef]

9. DiFiori, J.P.; Benjamin, H.J.; Brenner, J.S.; Gregory, A.; Jayanthi, N.; Landry, G.L.; Luke, A. Overuse injuries and burnout in youth sports: A position statement from the American medical society for sports medicine. *Br. J. Sports Med.* **2014**, *48*, 287–288. [CrossRef]

10. Fort-Vanmeerhaeghe, A.; Romero-Rodriguez, D.; Montalvo, A.M.; Kiefer, A.W.; Lloyd, R.S.; Myer, G.D. Integrative neuromuscular training and injury prevention in youth athletes. Part I: Identifying risk factors. *Str. Cond. J.* **2016**, *38*, 36–48. [CrossRef]

11. Van der Sluis, A.; Elferink-Gemser, M.T.; Brink, M.S.; Visscher, C. Importance of peak height velocity timing in terms of injuries in talented soccer players. *Int. J. Sports Med.* **2015**, *36*, 327–332. [CrossRef]

12. Le Gall, F.; Carling, C.; Reilly, T. Biological maturity and injury in elite youth football. *Scand. J. Med. Sci. Sports* **2007**, *17*, 564–572. [CrossRef]

13. Raschner, C.; Platzer, H.P.; Patterson, C.; Werner, I.; Huber, R.; Hildebrandt, C. The relationship between ACL injuries and physical fitness in young competitive ski racers: A 10-year longitudinal study. *Br. J. Sports Med.* **2012**, *46*, 1065–1071. [CrossRef]

14. Braegger, C.; Jenni, O.G.; Konrad, D.; Molinari, L. Neue Wachstumskurven für die Schweiz. *Paediatrica* **2011**, *22*, 9–11.

15. Kromeyer-Hauschild, K.; Wabitsch, M.; Kunze, D.; Geller, F.; Geisz, H.C.; Hesse, V.; von Hippel, A.; Jaeger, U.; Johnsen, D.; Korte, W.; et al. Percentiles of body mass index in children and adolescents evaluated from different regional German studies. *Monatsschrift Kinderheilkd.* **2001**, *149*, 807–818. [CrossRef]

16. Mirwald, R.L.; Baxter-Jones, A.D.G.; Bailey, D.A.; Beunen, G.P. An assessment of maturity from anthropometric measurements. *Med. Sci. Sports Exerc.* **2002**, *34*, 689–694. [CrossRef]

17. Sherar, L.B.; Baxter-Jones, A.D.G.; Faulkner, R.A.; Russell, K.W. Do physical maturity and birth date predict talent in male youth ice hockey players? *J. Sports Sci.* **2007**, *25*, 879–886. [CrossRef]

18. Müller, L.; Müller, E.; Hildebrandt, C.; Kapelari, K.; Raschner, C. Die Erhebung des biologischen Entwicklungsstandes für die Talentselektion—Welche Methode eignet sich? *Sportverletz. Sportsc.* **2015**, *29*, 56–63. [CrossRef]

19. Faigenbaum, A.D.; Lloyd, D.G.; Oliver, J.L. *Essentials of Youth Fitness*; Human Kinetics: Champaign, IL, USA, 2020.

20. Lesinski, M.; Schmelcher, A.; Herz, M.; Puta, C.; Gabriel, H.; Arampatzis, A.; Laube, G.; Büsch, D.; Granacher, U. Maturation-, age-, and sex-specific anthropometric and physical fitness percentiles of German elite young athletes. *PLoS ONE* **2020**, *15*, e0237423. [CrossRef]

21. Raschner, C.; Müller, L.; Patterson, C.; Platzer, H.P.; Ebenbichler, C.; Luchner, R.; Lembert, S.; Hildebrandt, C. Current performance testing trends in junior and elite Austrian alpine ski, snowboard and ski cross racers. *Sport Orthop. Sport Traumatol.* **2013**, *29*, 193–202. [CrossRef]

22. Platzer, H.P.; Raschner, C.; Patterson, C. Performance-determining physiological factors in the luge start. *J. Sports Sci.* **2009**, *27*, 221–226. [CrossRef] [PubMed]

23. Brooks, J.H.; Fuller, C.W. The influence of methodological issues on the results and conclusions from epidemiological studies of sports injuries: Illustrative examples. *Sports Med.* **2006**, *36*, 459–472. [CrossRef]

24. Junge, A.; Engebretsen, L.; Alonso, J.M.; Renström, P.; Mountjoy, M.; Aubry, M.; Dvorak, J. Injury surveillance in multi-sport events: The International Olympic Committee approach. *Br. J. Sports Med.* **2008**, *42*, 413–421. [CrossRef]

25. Fuller, C.W.; Ekstrand, J.; Junge, A.; Andersen, T.E.; Bahr, R.; Dvorak, J.; Hägglund, M.; Mc Crory, P.; Meeuwisse, W.H. Consensus statement on injury definitions and data collection procedures in studies of football (soccer) injuries. *Clin. J. Sport Med.* **2006**, *16*, 97–106. [CrossRef]

26. Clarsen, B.; Myklebust, G.; Bahr, R. Development and validation of a new method for the registration of overuse injuries in sports injury epidemiology. *Br. J. Sports Med.* **2013**, *47*, 495–502. [CrossRef]

27. Clarsen, B.; Bahr, R.; Heymans, M.W.; Engedahl, M.; Midtsundstad, G.; Rosenlund, L.; Thorsen, G.; Myklebust, G. The prevalence and impact of overuse injuries in five Norwegian sports: Application of a new surveillance method. *Scand. J. Med. Sci. Sports* **2015**, *25*, 323–330. [CrossRef]

28. Pan, W. Akaike's information criterion in generalized estimating equations. *Biometrics* **2001**, *57*, 120–125. [CrossRef]

29. Jahnel, R.; Spörri, J.; Kröll, J.; Müller, E. Prevalence of overuse problems in World Cup alpine skiers—An explorative approach. In *6th International Congress on Science and Skiing—Book of Abstracts*; Müller, E., Kröll, J., Lindinger, S., Pfusterschmied, J., Stöggl, T., Eds.; Department of Sport Science and Kinesiology, University of Salzburg: Salzburg, Austria, 2013; p. 143.

30. Bere, T.; Florenes, T.W.; Nordsletten, L.; Bahr, R. Sex differences in the risk of injury in World Cup alpine skiers: A 6-year cohort study. *Br. J. Sports Med.* **2013**, *48*, 36–40. [CrossRef]

31. Caine, D.; DiFiori, J.; Maffulli, N. Physeal injuries in children's and youth sport: Reason for concerns? *Br. J. Sports Med.* **2006**, *40*, 749–760. [CrossRef]

32. Rommers, N.; Rössler, R.; Goossens, L.; Vaeyens, R.; Lenoir, M.; Witvrouw, E.; D'Hondt, E. Risk of acute and overuse injuries in youth elite soccer players: Body size and growth matters. *J. Sci. Med. Sport* **2020**, *23*, 246–251. [CrossRef]

33. Kemper, G.L.J.; van der Sluis, A.; Brink, M.S.; Visscher, C.; Frencken, W.G.P.; Elferink-Gemser, M.T. Anthropometric Injury Risk Factors in Elite-standard youth soccer. *Int. J. Sports Med.* **2015**, *36*, 1112–1117. [CrossRef]

34. Caine, D.; Purcell, L.; Maffulli, N. The child and adolescent athlete: A review of three potentially serious injuries. *Sports Sci. Med. Rehabil.* **2014**, *6*, 22. [CrossRef]

35. Jayanthi, N.A.; LaBella, C.R.; Fischer, D.; Pasulka, J.; Dugas, L.R. Sports-specialized intensive training and the risk of injury in young athletes. *Am. J. Sports Med.* **2015**, *43*, 794–801. [CrossRef]

36. Hawkins, D.; Metheny, J. Overuse injuries in youth sports: Biomechanical considerations. *Med. Sci. Sports Exerc.* **2001**, *33*, 1701–1707. [CrossRef]

37. Baxter-Jones, A.D.G.; Thompson, A.M.; Malina, R.M. Growth and Maturation in Elite Young Female Athletes. *Sports Med. Arthrosc. Rev.* **2002**, *10*, 42–49. [CrossRef]

38. Balyi, I.; Way, R.; Collin, R. *Long-Term Athlete Development*, 1st ed.; Human Kinetics Publishers: Champaign, IL, USA, 2013; p. 199.

39. Stevenson, H.; Webster, J.; Johnson, R.; Beynnon, B. Gender differences in knee injury epidemiology among competitive alpine ski racers. *Iowa Orthop. J.* **1998**, *18*, 64–66. [PubMed]

40. Eckeland, A.; Dimmen, S.; Lystad, H.; Aune, A.K. Completion rates and injury in alpine races during the 1994 Olympic Winter Games. *Scand. J. Med. Sci. Sports* **1996**, *6*, 287–290. [CrossRef]

41. Flørenes, T.W.; Bere, T.; Nordsletten, L.; Heir, S.; Bahr, R. Injuries among male and female World Cup alpine skiers. *Br. J. Sports Med.* **2009**, *43*, 973–978. [CrossRef]

42. Agel, J.; Arendt, E.; Bershadsky, B. Anterior cruciate ligament injury in national collegiate athletic association basketball and soccer: A 13-year review. *Am. J. Sports Med.* **2005**, *33*, 524–530. [CrossRef]

43. Renstrom, P.; Ljungqvist, A.; A Arendt, E.; Beynnon, B.D.; Fukubayashi, T.; E Garrett, W.; Georgoulis, T.; E Hewett, T.; Johnson, R.P.; Krosshaug, T.; et al. Non-contact ACL injuries in female athletes: An International Olympic Committee current concepts statement. *Br. J. Sports Med.* **2008**, *42*, 394–412. [CrossRef]

44. Bergstrøm, K.A.; Brandseth, K.; Fretheim, S.; Tvilde, K.; Ekeland, A. Back injuries and pain in adolescents attending a ski high school. *Knee Surg. Sports Traumatol. Arthrosc.* **2004**, *12*, 80–85. [CrossRef] [PubMed]

Publisher's Note: MDPI stays neutral with regard to jurisdictional claims in published maps and institutional affiliations.

Article

Effects of Differential Jump Training on Balance Performance in Female Volleyball Players

Philip X. Fuchs [1,2,*], Andrea Fusco [1,2,*], Cristina Cortis [2] and Herbert Wagner [1]

[1]   Department of Sport and Exercise Science, University of Salzburg, Schlossallee 49, Hallein-Rif, 5400 Salzburg, Austria; spowi@sbg.ac.at
[2]   Department of Human Sciences, Society and Health, University of Cassino e Lazio Meridionale, Viale dell'Università, 03043 Cassino FR, Lazio, Italy; c.cortis@unicas.it
*   Correspondence: philip.fuchs@sbg.ac.at (P.X.F.); andrea.fusco@unicas.it (A.F.)

Received: 15 July 2020; Accepted: 24 August 2020; Published: 26 August 2020

Featured Application: The presented training concept can be applied to combine two seemingly contradictory desired effects that are associated with increased jump performance and injury prevention in an effective and efficient way without adding excess physiological stress.

Abstract: The purpose of this study was to determine whether coordinative jump training that induces neuromuscular stimuli can affect balance performance, associated with injury risk, in elite-level female volleyball players. During the competitive season, the balance performance of 12 elite female players (highest Austrian division) was obtained via a wobble board (WB; 200 Hz) placed on an AMTI force plate (1000 Hz). Three identically repeated measurements defined two intervals (control and intervention phases), both comparable in duration and regular training. The intervention included 6 weeks of differential training (8 sessions of 15–20 min) that delivered variations in dynamics around the ankle joints. Multilevel mixed models were used to assess the effect on postural control. WB performance decreased from 27.0 ± 13.2% to 19.6 ± 11.3% during the control phase and increased to 54.5 ± 16.2% during the intervention (β = 49.1 ± 3.5; $p < 0.001$). Decreased sway area [cm^2] (β = −7.5 ± 1.6; $p < 0.001$), anterior–posterior (β = −4.1 ± 0.4; $p < 0.001$) and mediolateral sway [mm] (β = −2.7 ± 0.6; $p = 0.12$), and mean velocity [$mm \cdot s^{-1}$] (β = −9.0 ± 3.6; $p < 0.05$) were observed during the intervention compared with the control phase. Inter-limb asymmetry was reduced (β = −41.8 ± 14.4; $p < 0.05$). The applied training concept enhanced balance performance and postural control in elite female volleyball players. Due to the low additional physiological loads of the program and increased injury risk during the competitive season, we recommend this intervention for supporting injury prevention during this period.

Keywords: postural control; center of pressure; in-season intervention; injury risk; ankle sprain prevention

---

1. Introduction

Sport-specific performance training is required in all sports to achieve a high international level. In volleyball, an important performance criterion correlating with competition level and a main objective in volleyball conditioning programs is to increase jump height [1]. This is commonly achieved through various types of strength and power training [2,3]. On the other hand, volleyball is known for its severe risk of ankle injuries during jumping [4] that cannot be prevented by strength training alone [5]. Therefore, traditional performance training should be accompanied by additional injury prevention programs [4]. The most frequent injury is ankle sprains [4], which are associated with chronic ankle instability and an increased chance of recurrence [5]. Neuromuscular impairment due to traumatized mechanoreceptors of the ligaments and muscles in the ankle is expected to cause the recurrence of sprains [5]. Unsurprisingly, neuromuscular and proprioceptive training has shown positive effects and is recommended to prevent ankle injuries [5,6]. Such training consists of balance exercises on stable and unstable platforms or the combination of, e.g., balance, plyometric, and sport-specific exercises [6]. The objective is to improve proprioception and neuromuscular responses. Thus, mechanoreceptors and stabilizing muscles around the ankle sense and process the risk of balance loss early and can react in time to prevent falls and injuries.

Recent technical-coordinative jump training [7] may have targeted both performance and prevention aspects during volleyball spike jumping. The concept was based on differential training [8,9], which applies movement variability and allows for adaptations in neuromuscular activation patterns [10]. Differential training has been shown to provide good transferability of training effects on sport-specific jump performance when compared with traditional concepts [11]. The program focused on spike jump performance determinants [12], centering around approach velocity, feet position, and velocity conversion strategy through the dominant leg (i.e., side of the striking arm). The conversion strategy is associated with neuromuscular activation patterns in the lower limbs [13]. Therefore, the emphasis and implementation of numerous variations in these movement characteristics is expected to constantly create proprioceptive and neuromuscular stimuli for ankle stabilization. The effect of such differential jump training on prevention can be operationalized via balance assessment, since reduced balance performance can be associated with an increased risk of injury [14]. Both injury prevention and balance performance are affected by the proprioceptive and neuromuscular systems that stabilize and maintain balance [5,6,15,16]. Therefore, a transfer of effects on balance performance during such differential jump training can be presumed.

For balance assessment, quantitative posturography via force platforms is considered the gold standard and superior to functional balance tests with respect to sensitivity and objectivity [17]. The most frequently analyzed characteristic is the excursion of the center of pressure (CoP) [18]. However, there are many variables that can be calculated from the CoP. A systematic review of 32 published articles involving CoP analyses recommended including both distance (e.g., sway area) and time–distance (e.g., mean velocity) based variables [19]. Its authors advised against the usage of minimal, maximal, and peak-to-peak values as they represent severe data reduction, great variance, and low reliability [19]. It seems unclear whether to alternatively prioritize analyses on fractal dimensions or on resultant horizontal data [19]. The drawbacks of posturography via CoP are the numerous variables and uncertainty about the best choice among variables to reflect balance performance [19]. Moreover, the high costs, complex handling, and space required for the equipment often complicate its usage in clinical settings [17].

A handy and validated alternative with fair to excellent reliability is found in computerized wobble boards (WBs) [20]. WBs are unstable platforms frequently used in clinical and therapeutic settings. They can be equipped with, e.g., accelerometers to collect the WB tilt angle that can then be processed by software and displayed on a screen in real time. Besides their practical handling, the strength of WBs is the single output variable of balance performance (i.e., time spent at ~0° tilt). Moreover, classical assessment of the CoP when standing on a force platform can also be applied to standing on a WB [21]. Poor correlation between WB performance and the outcome of one of the most frequently used validated balance tests, the Y Balance Test [20], can be explained by the complexity of the underlying mechanisms of postural control [14] and different test-specific skills of postural control [20]. Since WBs are a common tool in sprain prevention programs [5] and are suitable for balance assessment in individuals with chronic ankle instability [16], their usage is reasonable for the context of the current study (i.e., injury prevention in high-risk individuals).

The objective of this study was to determine the effect of a differential-training-based jump intervention on postural control in high-level female volleyball players. It was hypothesized that a 6-week differential jump training regimen that induces proprioceptive and neuromuscular stimuli for ankle stabilization would improve WB balance performance and CoP characteristics.

2. Materials and Methods

2.1. Participants

A female volleyball team (n_{team} = 12) from the highest league in Austria participated in this study (age: 22.8 ± 3.7 years; training experience: 11.8 ± 3.8 years; body height: 1.78 ± 0.09 m; mass: 69.9 ± 9.4 kg; body mass index: 22.0 ± 1.9 kg·m^{-2}; spike jump height: 0.44 ± 0.09 m). Depending on player positions and associated jumping skills, players were categorized as spikers (n_1 = 6), blockers (n_2 = 2), or setters and libera (n_3 = 4). The individual's preferred striking arm during the volleyball spike defined the dominant side. In accordance with the Declaration of Helsinki, the study was approved by the local institutional research ethics committee (approval code: 29/2014). All players were informed about the risks, benefits, and procedure of the study. They signed a consent form and reported being free of injuries at the beginning of the investigation. The sample size was determined by the size of the team and the accessibility of such high-level athletes.

2.2. Study Design and Training Intervention

Three identical testing sessions were conducted during the competitive season (i.e., 1. Control, 2. Post-Control = Pre-Intervention, 3. Post-Intervention). Post-Control and Post-Intervention testing took place 5.5 ± 0.5 and 4.5 ± 0.5 days after the most recent match, respectively. The first interval (6.5 weeks) defined the control phase. The second interval (6 weeks) included an intervention with no systematic differences in the regular training scheme, training loads (8.8 ± 0.3 vs. 7.7 ± 0.2 h per week), and competition loads (both phases: 1 international, 7 national matches). This approach to assess training effects by comparing control and intervention phases has been used before [7,22]. A control group was not feasible for ethical reasons and due to the limited numbers of accessible elite female volleyball players [22].

The intervention consisted of eight sessions of differential training [8,9], 15–20 min per session, and focused on the volleyball-specific spike jump movement. Differential training implements movement variations to allow for individual adaptations towards an individual optimum. The concept was described by its founder [8]; the methodology and positive effects across sports, ages, and levels have been reported [23–25]. The objective of the applied training was to induce coordinative adaptations in previously identified performance determinants [12]. Variations in approach speed (thus, dynamics), feet positioning (thus, joint angles), and the distribution of weight and time-related power development through both legs (thus, neuronal activation patterns) create neuromuscular stimuli for stabilizing muscles around the ankles. In contrast to traditional balance training, these variations exploit system fluctuations [26] at the kinematic and kinetic levels. For instance, the athletes were instructed to vary the length of the penultimate step, the ankle joint angle at planting, and the pressure perceived at the left and right feet during planting. The exercises followed two concepts: (1) the athletes alternatingly performed two opposing, detrimental extremes, reducing the discrepancy between both variations after each pair of trials; and (2) starting from one detrimental extreme, the players approached an individual optimum. A detailed description of the program and a full list of variations are available [7].

2.3. Data Collection and Processing

Three identical repeated measurements were made, following a validated protocol and instructions [20].

The participants performed test trials for familiarization with a validated mediolateral and anterior–posterior tilting WB equipped with tri-axis accelerometers (GSJ Service, Rome, Italy) [20]. Subsequently, three trials per leg (30 s per trial) of a single-leg stance on the WB, hands at the hips, were measured at 200 Hz. The sequence was blocked for legs and randomized for participants. In real time, validated software (GSJ Service, Rome, Italy) [20] displayed a motion marker based on the WB tilt angle and circles around a central area (~0° WB tilt) on a screen in front of the participants. Participants aimed to keep the motion marker within the central area for as long as possible over the full span of each trial. To prevent fatigue, there was approximately one minute of break time between trials.

Simultaneously, a force platform (120 × 60 cm; AMTI, Watertown, MA, USA) recorded ground reaction forces at 1000 Hz. The CoP was calculated in Visual3D (C-Motion, Inc., Rockville, MD, USA). Based on residual analyses [27] of the current data and a recommendation for CoP filtering [19], a fourth-order zero-lag Butterworth low-pass filter at 10 Hz was applied (Supplementary Materials).

2.4. Criterion Variables and Definitions

WB performance was defined as the time for which the WB was kept at ~0° tilt, divided by the trial duration [%]. In agreement with a CoP review [19], ground reaction forces were used to compute the sway area (i.e., 95% confidence ellipse; see Figure 1), mediolateral and anterior–posterior sway (i.e., fractal standard deviation), and mean velocity of sway (i.e., total CoP path distance, divided by trial duration). As a characteristic associated with injury risk, the inter-limb asymmetry [%] was also calculated [28]. Development during the phases was defined as the difference between two measurements. Development in WB performance and asymmetry is displayed as percentage points (e.g., a rise from 20% to 40% equals a development of +20% points). For all variables, outliers [mean ± 2 times the standard deviation (SD)] within participants were removed.

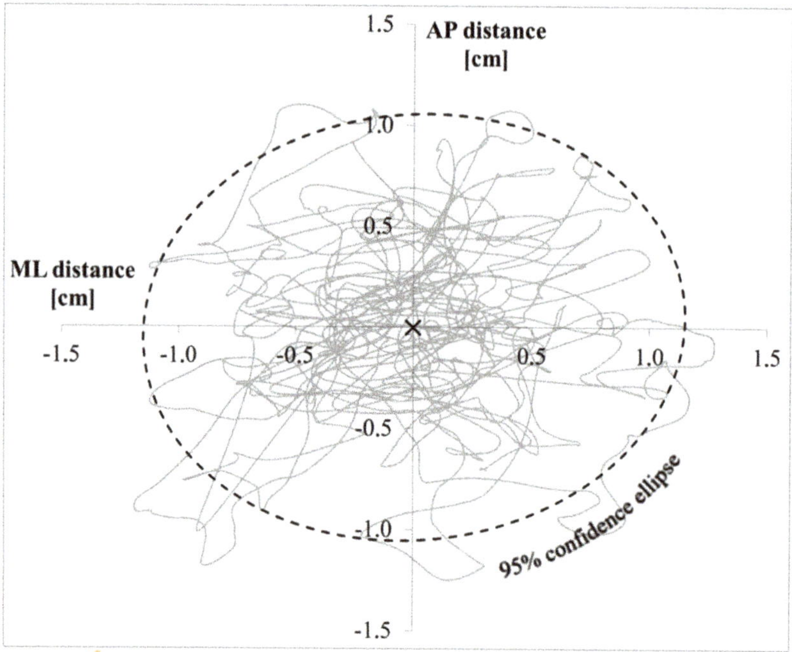

Figure 1. Exemplary path of the center of pressure of one participant during one trial and the sway area defined as a 95% confidence ellipse. Note: AP, anterior–posterior; ML, mediolateral.

2.5. Statistical Analyses

PASW Statistics (version 18.0; SPSS Inc, Chicago, IL, USA) was used for statistical analyses and Office Excel 2019 (Microsoft Corporation, Redmond, WA, USA) was used for visualization. For WB performance, as the primary interest of this investigation, linear multilevel mixed model analyses accounting for the repeated effect of trials were applied. The superiority of such an approach over traditional analysis of variance has been explained by reviews, technical reports, and empirical assessments with real data [29–31], and the reasons include the way of handling missing data and interindividual variance. Since both were observed in the current data, this approach was suitable for the specific set of data and purpose. Such a model predicts a dependent variable (i.e., WB performance) in consideration of fixed and random effects of intercepts and independent variables (i.e., measurement session, phase, leg used during the trial, player position). A random intercept was considered for participants and fixed and random effects for independent variables. A fixed intercept was defined as a pre-intervention measurement to allow for comparison with control and post-intervention measurements. Effects were included one after another, and the fit of the derived models was assessed for effects and specifications of covariances via chi-square (X^2) statistics based on -2 log-likelihood and degrees of freedom (df) [30,31]. The covariance structures for independent variables were defined after assessment of the correlation matrix and confirmed by the fit of the model when compared to alternative models. The covariance structure of the random intercept was specified as the variance component for all models. Depending on the data type, collinearity was checked via Pearson or Spearman correlation analyses, and none was found. Among models of comparable fit, the one with the smallest number of parameters was accepted as the final model. All analyses were performed using maximum-likelihood estimation [31]. The same statistical approach was applied for comparison of the developments between the control and intervention phases, with development during the control phase as a fixed intercept. Following the same procedure as for WB performance, the most successful single-predictor models with variable-specific specification of the covariance structure were calculated for each other dependent variable (i.e., inter-limb asymmetry, sway area, mediolateral sway, anterior–posterior sway, and CoP velocity). The performances at all measurement sessions are presented as mean ± SD and 95% confidence intervals (CIs). The results are presented as *F*-value, degree of freedom (df), estimate (β), and standard error (SE). p-values below 0.05 were considered significant.

3. Results

One ankle sprain occurred during the control phase, with no effect on data collection and presentation. One participant missed the first measurement; removal from analyses was not required thanks to the statistical approach.

The performances at the three measurement sessions, covariance specifications of the final models, and the effects of measurements are presented in Table 1.

Table 1. Means ± standard deviations (SDs), 95% confidence intervals (CIs), *F*-statistics for the main effect of measurements, estimates (β), and standard error (SE) for effects between single measurements.

	Control Phase		Intervention Phase					
	Control		Post-Control = Pre-Intervention		Post-Intervention				
Variable	Mean ± SD	β	Mean ± SD	β	Mean ± SD	*F*	*df*	*p*
	95% CI	SE	95% CI	SE	95% CI			
WB performance [%] †	27.02 ± 13.18 18.17 – 35.87	(−5.20) (2.71)	19.60 ± 11.31 12.00 – 27.19	42.87 3.05	59.49 ± 16.16 48.64 – 70.35	88.03	2, 14.17	<0.001
Sway area [cm²] §	8.03 ± 3.28 5.82 – 10.23	6.35 1.01	10.91 ± 3.42 8.61 – 13.21	−6.63 1.31	8.02 ± 2.89 6.01 – 9.96	23.41	2, 2.90	<0.05
AP sway [mm] †	7.04 ± 1.61 5.96 – 8.12	(1.03) (0.66)	9.18 ± 1.93 7.88 – 10.47	(−0.79) (0.81)	7.09 ± 1.47 6.10 – 8.08	1.53	2, 3.33	0.34
ML sway [mm] ‡	6.02 ± 1.23 5.19 – 6.85	0.98 0.29	6.34 ± 0.98 5.69 – 7.00	−1.24 0.30	5.82 ± 1.12 5.06 – 6.58	8.41	2, 9.58	<0.01
CoP velocity [mm·s⁻¹] †	68.54 ± 6.86 63.93 – 73.15	(3.51) (2.18)	67.88 ± 8.12 62.42 – 73.34	−5.85 2.11	64.08 ± 7.88 58.78 – 69.37	4.77	2, 14.47	<0.05
Asymmetry [%] ¥	23.64 ± 17.33 11.24 – 36.03	(6.38) (6.76)	32.69 ± 23.30 16.02 – 49.36	−25.89 6.23	10.15 ± 8.26 4.24 – 16.07	9.62	2, 52.33	<0.001

Note: Control, Post-Control, Pre-Intervention, and Post-Intervention denote measurement timings; WB, wobble board; AP, anterior–posterior; ML, mediolateral; CoP, center of pressure; *df*, degrees of freedom for the numerator and denominator, separated by a comma; β is converted so that negative values indicate decreased values from the previous to the subsequent measurement; β and SE relate to two adjacent measurements and are in parentheses if the effect was not significant ($p > 0.05$); Covariance structure specification: †, antedependence first order; §, unstructured; ‡, factor analytic first order; ¥, heterogeneous autoregressive first order.

The best model with measurement as the only predictor for WB performance included a fixed and a random intercept, a fixed slope for measurement, and an error. It detected a main effect for measurements ($F(2,58.24) = 173.83$, $p < 0.001$), decrease from control to pre-intervention ($\beta = -8.86$, SE $= 2.36$, $p < 0.001$), and increase from pre- to post-intervention ($\beta = 40.69$, SE $= 2.29$, $p < 0.001$). The random intercept for participants yielded 65.95% of this model's overall variance ($p < 0.05$).

The final model with the best fit consisted of 18 parameters ($X^2 = 29.29$, $\Delta df = 13$, $p < 0.01$, compared with the initial model). It was expressed by the following equation, where Y_{is} (WB performance), X_{is} (measurement), Z_{is} (leg), and ε_{is} (error) vary as functions of individual observations (i; level 1 variable) and participants (s; level 2 variable). b_0 and u_{0s} represent a fixed and a random intercept, respectively; b_1, b_2, and b_3 define the gradients.

$$Y_{is} = (b_0 + u_{0s}) + b_1 X_{is} + b_2 Z_{is} + b_3 X_{is} Z_{is} + \varepsilon_{is}$$

Besides the effects of measurements (Table 1), no effect was found for leg ($F(1,26.10) = 0.33$, $p = 0.57$), but an effect was found for the interaction of measurement by leg ($F(2,20.19) = 6.98$, $p < 0.01$). The random intercept for participants yielded 0.41% of this model's overall variance ($p < 0.05$).

Additional fixed and random effects did not produce significantly improved models (best alternative model including player position as a random effect, compared with the final model: $X^2 = 1.14$, $\Delta df = 1$, $p = 0.28$).

Model-based effects between the control and intervention phases are displayed in Figure 2.

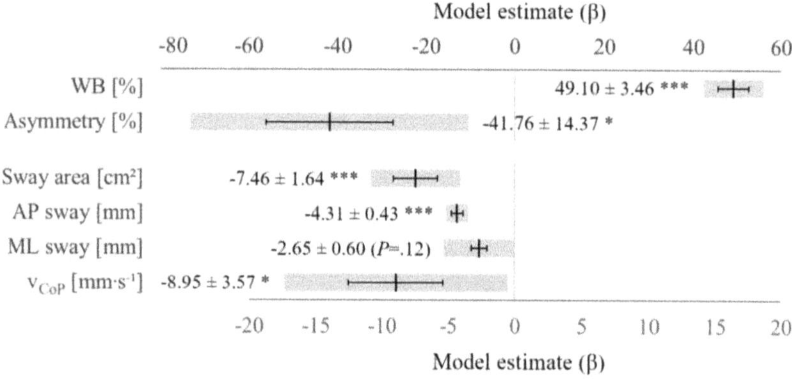

Figure 2. Discrepancies between development during the control and intervention phases, displayed as the model estimate (β) ± standard error and 95% confidence interval (i.e., grey bars). Note: WB, wobble board performance; Asymmetry, inter-limb asymmetry; AP, anterior–posterior; ML, mediolateral; v_{CoP}, mean velocity of the center of pressure; * $p < 0.05$; *** $p < 0.001$.

The final model for WB performance over the phases consisted of three parameters and was expressed by the following equation, where Y_i (WB performance), X_i (phase), and ε_i (error) vary as functions of individual observations (i). b_0 represents a fixed intercept and b_1 represents the gradient.

$$Y_i = b_0 + b_1 X_i + \varepsilon_i$$

The inclusion of random intercept, phase, leg, player position, and interactions as fixed and/or random effects did not deliver improved models (best models for each additional effect in comparison with the final model: $0 \leq X^2 \leq 8.06$, $1 \leq \Delta df \leq 20$, $0.09 \leq p \leq 1$). In these alternative models, no effects were found for leg ($F(1,46) = 2.87$, $p = 0.10$), leg by phase ($F(2,46) = 1.53$, $p = 0.23$), position ($F(1,46) = 2.24$, $p = 0.14$), or position by phase ($F(2,46) = 1.48$, $p = 0.24$).

For all other criterion variables, the final models included significant random intercepts for participants.

4. Discussion

The objective of this study was to investigate the effects of sport-specific differential jump training on balance performance and associated injury risk in elite female volleyball players. A mixed model analysis was deemed suitable because it is superior to traditional statistical approaches in the current scenario [29–31]. The first reason for this was increased power, as mixed models are capable of handling missing data instead of removing a participant from analyses. This applied in the current data since one participant missed the first measurement session. Second, mixed models differentiate between interindividual variance and random error variance. Ignoring occurrent interindividual variability leads to over- or underestimation of statistical significance in repeated measures analysis of variance; in fact, this generates inaccurate results and can affect the overall interpretation [31]. The significant effects of the random intercept for participants indicated interindividual variance [29,31] in the current study and supported the usage of mixed models. In the simple model for WB performance, the random intercept explained two-thirds of the overall variance. This is a great but not exceptional amount (e.g., values of >50% are systematically reported in sleep deprivation) and underlines the relevance of accounting for interindividual variance [31]. The reduced (yet significant) variance in the more complex model is probably due to the added effects explaining variance that was previously absorbed by the random intercept.

The results showed decreased balance performance during the control phase. This was not surprising as decreased performance during the competitive season is known in volleyball and is explained by increased physiological loads [7,22]. Sway area and ML sway increased significantly, which is considered detrimental for postural control. Also, WB performance, AP sway, mean CoP velocity, and inter-limb asymmetry tended ($0.08 \leq p \leq 0.36$) to develop unfavorably. During the intervention, all variables improved significantly (except for AP sway, only indicating ($p = 0.35$) a beneficial development).

WB performance is a valid single outcome reflecting postural control [20], and it was found to be decreased in subjects with chronic ankle instability, which is associated with increased risk of ankle sprains [16]. Previous data showed that healthy limbs scored 19.6–21.8 s (i.e., 65.33–72.66%) in WB performance [16]. In the current study, WB performance was comparably low during the control ($27.02 \pm 13.18\%$) and pre-intervention ($19.60 \pm 11.31\%$) measurements. This indicates and corroborates the assertion that volleyball players are at high risk for ankle sprains [4], since reduced balance performance is associated with increased ankle instability [16] and injury incidence [6]. After the 6-week intervention, WB performance ($59.49 \pm 16.16\%$) almost reached the level of the previously reported healthy limbs [16]. Interestingly, there is some evidence that it takes 8–10 weeks of neuromuscular training to reduce injury risk to a baseline level [5]. Therefore, it can be presumed that a prolonged intervention of the current type may have closed the gap between the currently achieved WB performance and the baseline level of healthy limbs.

CoP variables were considered insightful in addition to WB performance and were selected in agreement with a review on CoP analyses [19]. This review recommends the consideration of multiple variables because reports are controversial, no single gold standard is apparent, and various variables may reflect different aspects of postural control [19]. In the case of conflicting findings, individual assessment of the CoP variables allowed for specific interpretation of single characteristics. However, the developments of all CoP variables were in line with the change in WB performance, showing detrimental trends during the control phase and beneficial changes during the intervention phase. Based on this, we cannot conclude that specific CoP variables were more suitable for WB performance than others. Moreover, the data do not indicate that specific CoP variables reflect different aspects of postural control on WB. Despite mean velocity being the most reliable CoP variable [19], its values should be interpreted with caution. Mean velocity easily gives the misleading impression of reflecting adaptability (i.e., quick adjustment of CoP to keep the center of mass within the base of stability). However, the mean velocity is calculated as the total path of the CoP normalized by time [19]. Therefore, it reflects total sway and does not provide additional information about adaptability. No recommendation was found in the literature to properly quantify adaptability in balance performance.

Decreased inter-limb asymmetry is considered desirable. Therefore, a detrimental trend during the control phase and a beneficial change during the intervention phase were observed. The recommended equation to calculate asymmetry includes the smaller of the two values from both legs as the denominator [28]. Thus, asymmetry values rapidly approach 100% if the smaller value is close to 0, even when only a small absolute discrepancy between both values exists. Such values usually do not occur in the analyses of net peak vertical ground reaction forces where this equation was applied previously [28]. However, this problem was identified in one case during the current study and should be considered in future investigations. This data point was excluded from further analyses. Removal of this data point did not change the statistical outcome.

The results of the comparison of developments during both phases (Figure 2) are in line with the analyses across the three measurement sessions (Table 1). Throughout all variables, beneficial trends were observed during the intervention when compared with the control phase. ML sway was the only variable where this trend did not reach statistical significance ($p = 0.12$). However, the full range of the 95% CI was found to be on the negative (i.e., beneficial) side of the scale. Therefore, it can be stated with 95% confidence for this study that the intervention induced a beneficial effect on ML sway compared with the control phase. Considering the large sizes of the model estimates in comparison with the baseline values during the control and pre-intervention measurements, the intervention produced a great, beneficial effect on postural control and inter-limb asymmetry.

The lack of a main effect of leg on performance implies that any leg-specific factors contributing to balance performance (e.g., previous injuries, chronic instability) were distributed equally to both legs. This result indicates that, in general, both legs performed comparably. However, the training program specifically stressed the usage of the dominant leg for approach velocity conversion, and an interaction effect of measurement by leg was observed. Probably, the focus of the jump training on the dominant leg induced different stimuli and resulted in different adaptations in the two legs.

For player position, no main or interaction effects were found. This suggests that, first, position was not a crucial factor in balance performance for volleyball players and, second, training stimuli do not need to be position-specific to achieve improvements.

Neuromuscular prevention training may only be effective in individuals with previous injuries [5]. Volleyball players are known to be a high-risk group with frequent ankle injuries [4]. The current data suggest that the participants' balance performance prior to intervention was poor, even in comparison with individuals with chronic ankle instability [16]. Therefore, the same effects of training may not be expected in individuals with no history of injuries. However, transferability of the current findings to other high-risk sports and previously injured individuals seems reasonable.

The primary focus of the training program was to improve sports-specific jump performance via coordinative adaptations [7]. Jump height increased by 11.9%, and promising effects on biomechanical determinants were reported in detail [7]. Improvements in balance performance and ankle stabilization were expected side-effects thanks to neuromuscular stimuli that accompanied the implemented movement variations in the program. The positive effects of the differential jump training program on both jump and balance performance support its potential for combining two effects that are usually targeted separately. A repeated effect for balance performance may be expected, as a review on ankle sprains [5] suggested increased risk and training effects after reoccurring injuries. After every ankle sprain, the injury risk can be reduced to baseline risk via neuromuscular training within 8–10 weeks [5]. However, the repeated effect of the same program was not tested in the current study.

The longitudinal design of this study could be discussed as a limitation. This is a common difficulty; previously published investigations [7,22] applied the same study design and, thus, these data serve as a suitable reference. The control and intervention phases were comparable in terms of factors that were deemed potentially influential (except for the intervention program itself). All factors were reported and accounted for during the statistical analyses and interpretations. We are not aware of other factors that contributed to the training effects in a phase-specific fashion. A cross-sectional approach was not feasible for ethical reasons and due to the accessibility of such high-level female athletes.

To achieve the observed improvement in spike jump height, biomechanical understanding of specific performance determinants is required, which could be a challenge for coaches [7]. This is not the case regarding the effect on balance performance. The current training program did not target certain balance determinants specifically. The applied movement variations were expected to affect the dynamics around the ankle joints and thus induce neuromuscular stimuli for ankle stabilization. Therefore, it can be concluded that these nonspecific stimuli are the essential factor to generate the observed adaptations in postural control. This makes the program practical for a range of coaches, given that neuromuscular stimuli are delivered. It may be presumed that the great coordinative challenge in ballistic balance during the jumping variations has a positive effect on the static balance performance.

A longer intervention period (8–10 weeks) may result in even larger effects [5]. Preventive effects were not directly measured in this study but can be expected based on the observed effect on postural control and the associated injury risk [6], especially for interventions conducted immediately after sprains [5] and in high-risk individuals. Due to its low physiological loads and complementary effects in jump height and postural control, the program is feasible during the competitive season for the whole team. It is also recommended for individuals who cannot fully participate in regular training after an ankle sprain.

5. Conclusions

The current program was a differential jump training program that successfully provided complementary improvements in sport-specific jump performance and postural control, which is an indicator of injury risk and especially relevant for high-risk target groups (e.g., volleyball players). The program adds no severe physiological loads and requires no compromises in regular training. This suggests its potential for implementation during the competitive season to enhance performance and injury prevention without the risk of overtraining.

Supplementary Materials: The following are available online at http://www.mdpi.com/2076-3417/10/17/5921/s1, Data S1: data sheet, containing the data used in this study.

Author Contributions: Conceptualization, P.X.F., A.F., C.C. and H.W.; methodology, P.X.F., A.F., C.C. and H.W.; validation, P.X.F., A.F., C.C. and H.W.; formal analysis, P.X.F. and A.F.; investigation, P.X.F. and A.F.; resources, C.C. and H.W.; writing—original draft preparation, P.X.F.; writing—review and editing, P.X.F., A.F., C.C. and H.W.; visualization, P.X.F. and A.F.; supervision, C.C. and H.W.; project administration, P.X.F.; All authors have read and agreed to the published version of the manuscript.

Funding: This research received no external funding.

Acknowledgments: We want to thank the players and the coaches of the PSV Salzburg for participation in this training intervention study. Special thanks also to Kathryn A. Hill for technical support allowing us to advance the analytical measures used in this study.

Conflicts of Interest: The authors declare no conflict of interest.

References

1. Ziv, G.; Lidor, R. Vertical jump in female and male volleyball players: A review of observational and experimental studies. *Scand. J. Med. Sci. Sports* **2010**, *20*, 556–567. [CrossRef] [PubMed]
2. Powers, M.E. Vertical Jump Training for Volleyball. *Strength Cond. J.* **1996**, *18*, 18–23. [CrossRef]
3. Sheppard, J.M.; Cronin, J.B.; Gabbett, T.J.; McGuigan, M.R.; Etxebarria, N.; Newton, R.U. Relative importance of strength, power, and anthropometric measures to jump performance of elite volleyball players. *J. Strength Cond. Res.* **2008**, *22*, 758–765. [CrossRef] [PubMed]

4. Bahr, R.; Bahr, I. Incidence of acute volleyball injuries: A prospective cohort study of injury mechanisms and risk factors. *Scand. J. Med. Sci. Sports* **1997**, *7*, 166–171. [CrossRef] [PubMed]

5. Verhagen, E.; Bay, K. Optimising ankle sprain prevention: A critical review and practical appraisal of the literature. *Br. J. Sport Med.* **2010**, *44*, 1082–1088. [CrossRef]

6. Hübscher, M.; Zech, A.; Pfeifer, K.; Hänsel, F.; Vogt, L.; Banzer, W. Neuromuscular Training for Sports Injury Prevention: A Systematic Review. *Med. Sci. Sports Exer.* **2010**, *42*, 413–421. [CrossRef]

7. Fuchs, P.X.; Fusco, A.; Bell, J.W.; von Duvillard, S.P.; Cortis, C.; Wagner, H. Effect of Differential Training on female volleyball spike jump technique and performance. *Int. J. Sports Physiol.* **2020**, *15*, 1019–1025. [CrossRef]

8. Schöllhorn, W.I. Applications of systems dynamic principles to technique and strength training. *Acta Acad. Olymp. Est.* **2000**, *8*, 67–85.

9. Schöllhorn, W.I.; Hegen, P.; Davids, K. The Nonlinear Nature of Learning—A Differential Learning Approach. *Open Sports Sci. J.* **2012**, *5*, 100–112. [CrossRef]

10. Haudum, A.; Birklbauer, J.; Kröll, J.; Müller, E. Constraint-led changes in internal variability in running. *J. Sports Sci. Med.* **2012**, *11*, 8–15.

11. Pfeiffer, M.; Jaitner, T. Sprungkraft im Nachwuchstraining Handball: Training und Diagnose. *Z. Angew. Train.* **2003**, *10*, 86–95.

12. Fuchs, P.X.; Fusco, A.; Bell, J.W.; von Duvillard, S.P.; Cortis, C.; Wagner, H. Movement characteristics of volleyball spike jump performance in females. *J. Sci. Med. Sports* **2019**, *22*, 833–837. [CrossRef] [PubMed]

13. Fuchs, P.X.; Menzel, H.-J.K.; Guidotti, F.; Bell, J.; von Duvillard, S.P.; Wagner, H. Spike jump biomechanics in male versus female elite volleyball players. *J. Sports Sci.* **2019**, *37*, 2411–2419. [CrossRef] [PubMed]

14. Brown, C.N.; Mynark, R. Balance deficits in recreational athletes with chronic ankle instability. *J. Athl. Train.* **2007**, *42*, 367–373. [PubMed]

15. Horak, F.B. Postural orientation and equilibrium: What do we need to know about neural control of balance to prevent falls? *Age Ageing* **2006**, *35*, ii7–ii11. [CrossRef]

16. Fusco, A.; Giancotti, G.F.; Fuchs, P.X.; Wagner, H.; Varalda, C.; Cortis, C. Wobble board balance assessment in subjects with chronic ankle instability. *Gait Posture* **2019**, *68*, 352–356. [CrossRef]

17. Mancini, M.; Horak, F.B. The relevance of clinical balance assessment tools to differentiate balance deficits. *Eur. J. Phys. Rehab. Med.* **2010**, *46*, 239–248.

18. Crétual, A. Which biomechanical models are currently used in standing posture analysis? *Neurophysiol. Clin.* **2015**, *45*, 285–295. [CrossRef]

19. Ruhe, A.; Fejer, R.; Walker, B. The test–retest reliability of centre of pressure measures in bipedal static task conditions—A systematic review of the literature. *Gait Posture* **2010**, *32*, 436–445. [CrossRef]

20. Fusco, A.; Giancotti, G.F.; Fuchs, P.X.; Wagner, H.; Varalda, C.; Capranica, L.; Cortis, C. Dynamic balance evaluation: Reliability and validity of a computerized wobble board. *J. Strength Cond. Res.* **2020**, *34*, 1709–1715. [CrossRef]

21. Janura, M.; Bizovska, L.; Svoboda, Z.; Cerny, M.; Zemkova, E. Assessment of postural stability in stable and unstable conditions. *Acta Bioeng. Biomech.* **2017**, *19*, 89–94. [CrossRef]

22. Newton, R.U.; Rogers, R.A.; Volek, J.S.; Häkkinen, K.; Kraemer, W.J. Four weeks of optimal load ballistic resistance training at the end of season attenuates declining jump performance of women volleyball players. *J. Strength Cond. Res.* **2006**, *20*, 955–961.

23. Coutinho, D.; Santos, S.; Conçalves, B.; Travassos, B.; Wong, D.P.; Schöllhorn, W.; Sampaio, J. The effects of an enrichment training program for youth football attackers. *PLoS ONE* **2018**, *13*, e0199008. [CrossRef]

24. Savelsbergh, G.J.; Kamper, W.J.; Rabius, J.; De Koning, J.J.; Schöllhorn, W. A new method to learn to start in speed skating: A differencial learning approach. *Int. J. Sports Psychol.* **2010**, *41*, 415–427.

25. Wagner, H.; Müller, E. The effects of differential and variable training on the quality parameters of a handball throw. *Sports Biomech.* **2008**, *7*, 54–71. [CrossRef]

26. Schöllhorn, W.I.; Beckmann, H.; Davids, K. Exploiting system fluctuations. Differential training in physical prevention and rehabilitation programs for health and exercise. *Medicina* **2010**, *46*, 365–373. [CrossRef]

27. Sinclair, J.; Taylor, P.J.; Hobbs, S.J. Digital filtering of three-dimensional lower extremity kinematics: An assessment. *J. Hum. Kinet.* **2013**, *39*, 25–36. [CrossRef]

28. Bishop, C.; Read, P.; Lake, J.; Chavda, S.; Turner, A. Interlimb asymmetries: Understanding how to calculate differences from bilateral and unilateral tests. *Strength Cond. J.* **2018**, *40*, 1–6. [CrossRef]

29. McCulloch, C.; Searle, S. *Generalized, Linear, and Mixed Models*; John Wiley and Sons: New York, NY, USA, 2000. [CrossRef]
30. Linear Mixed-Effects Modeling in SPSS. Available online: https://www.spss.ch/upload/1126184451_Linear%20Mixed%20Effects%20Modeling%20in%20SPSS.pdf (accessed on 30 June 2020).
31. Van Dongen, H.P.A.; Olofsen, E.; Dinges, D.F.; Maislin, G. Mixed-Model Regression Analysis and Dealing with Interindividual Differences. *Method Enzymol.* **2004**, *384*, 139–171. [CrossRef]

 *applied sciences*

Article

Association between the On-Plane Angular Motions of the Axle-Chain System and Clubhead Speed in Skilled Male Golfers

Morgan V. Madrid, Marco A. Avalos, Nicholas A. Levine, Noelle J. Tuttle, Kevin A. Becker and Young-Hoo Kwon *

Biomechanics Laboratory, Texas Woman's University, Denton, TX 76204, USA; mmadrid1@twu.edu (M.V.M.); mavalos1@twu.edu (M.A.A.); nlevine@twu.edu (N.A.L.); ntuttle@twu.edu (N.J.T.); kbecker1@twu.edu (K.A.B.)
* Correspondence: ykwon@twu.edu; Tel.: +1-940-898-2598

Received: 1 August 2020; Accepted: 18 August 2020; Published: 19 August 2020

Abstract: The on-plane rotations of the inclined axle-chain system on the functional swing plane (FSP) can represent the angular motions of the golfer–club system closely. The purpose of this study was to identify key performance factors in golf through a comprehensive investigation of the association between the angular motion characteristics of the axle-chain system and clubhead speed in skilled golfers. Sixty-six male golfers (handicap ≤ 3) performed full-effort shots in three club conditions: driver, 5-iron, and pitching wedge. Swing trials were captured with an optical motion capture system, and the hip/shoulder lines, upper lever, club, and wrist angular positions/velocities were calculated. Time, angular position, range of rotation, and peak angular velocity parameters were extracted and their correlation coefficients (Pearson and Spearman) to actual and normalized clubhead speeds were computed ($p < 0.05$). Higher clubhead speed was associated with shorter downswing phases, larger rotation ranges (hip/shoulder lines, and upper lever), larger hip–shoulder separation at impact, delayed transitions (hip line and upper lever), faster rotations (backswing, downswing, and impact), and larger angular velocity losses (hip line and upper lever) with additional club- and body-specific correlations. Clubhead speed was not well associated with wrist cock angles/ranges, X-factors/stretches, and timings of the downswing peak.

Keywords: double pendulum; kinematic sequences; delayed release; transition phase

---

1. Introduction

The golf swing is an angular motion-dominant sport movement in which a high clubhead speed is generated primarily by the angular motions of the golfer–club system. To develop a high clubhead speed, the golfer must (1) generate large external moments by using the ground reaction forces (GRFs) and moments (GRMs), and (2) transfer angular momentum sufficiently from the body to the club [1]. To maximize the clubhead speed, it is imperative to not only use the ground effectively/efficiently, but also move the body/club in an orchestrated manner during the swing. The legs are primarily involved in the golfer–ground interaction, while the trunk/arms are responsible for generation of a fast-angular motion of the club. Han and colleagues [2] modeled the upper part of the golfer–club system (i.e., pelvis and above) as an axle-chain system consisting of an inclined multi-segment axle (trunk) and an open chain. The open chain was further modeled as a functional double-pendulum (FDP) system [3] in which the thorax, shoulder girdles, and arms form the upper lever (UL), while the club constitutes the lower lever (LL). In the FDP model, the hub of the open chain is located at the mid-section of the trunk through which the functional swing plane (FSP) [4] typically passes.

Effective central-to-peripheral (i.e., proximal-to-distal) angular momentum transfer that leads to a high clubhead speed can be achieved by well-orchestrated angular motions of the axle-chain system in which the kinematic sequences (i.e., backswing, transition, and downswing sequences) play key roles [2]. Interest in the kinematic sequence in golf has grown ever since it was initially investigated in striking and throwing motions [5] and subsequently in other sport motions [6–8]. The end outcome of good proximal-to-distal kinematic sequences is maximization of the speed of the distal end of the kinematic chain (i.e., clubhead) through effective summation of the segmental angular velocities of the elements of the axle-chain system [2]. A proximal-to-distal kinematic sequence typically comes with more efficient use of the muscles (i.e., stretch-shortening cycle (SSC)) along with minimization of the moment of inertia of the chain for each joint motion.

Summation of speed in golf has been established in the perspective of angular velocity peaks of the segments [9–13] or linear velocity peaks of the joints/segments [14–16]. Full or partial proximal-to-distal sequences of angular velocity peaks have been reported in protocols examining the difference between clubs [10], skill levels [9,13], and hip–shoulder torsional separation styles [2]. Early generation of pelvis rotation during the backswing and downswing phases [17] along with large ranges of hip [18] and shoulder [19] motions contribute to higher clubhead speed. The wrist joint, the most distal joint in the axle-chain system, has also been investigated for the role it plays in increasing clubhead speed. More wrist cock [20] and delayed onset of wrist motion (also known as delayed release) [21,22] have been associated with increased clubhead speed. The summation of speed principle also establishes that, as the motion sequence is nearing completion, the system follows a proximal-to-distal deceleration pattern with the proximal segments slowing down as the acceleration increases in the next distal segment [9,16].

In addition to the backswing and downswing sequences based on the summation of speed principle, the backswing-to-downswing transition sequence also has received attention in the perspective of SSC [9,23]. Since the distal end of the axle-chain system is not constrained, a continuous backswing-to-downswing transition motion tends to lead to a proximal-to-distal transition sequence within the axle-chain system. A complete proximal-to-distal transition sequence can naturally induce SSC-style activation of the muscles that are responsible for generating angular motions of the axle-chain system. In a study involving a group of elite male golfers, Han and colleagues [2] identified four hip–shoulder separation styles and reported that the hip–shoulder separation pattern during the backswing influenced the backswing-to-downswing transition sequence, resulting in partial or full proximal-to-distal sequences.

Another group of swing concepts that are related to orchestrated behavior of the axle-chain system and have received substantial attention are the X-factor (i.e., hip–shoulder separation angle) and the X-factor stretch [24,25]. Investigators have compared X-factor parameters among different golfer groups or swing conditions and reported significant inter-group or inter-condition differences: e.g., skill groups [13,24,26,27], ball speed groups [28], genders [29], effort levels [30], and training stages [31]. While correlation/regression studies involving large heterogeneous samples have reported significant associations between X-factor parameters and club/ball speed [20,28], it is likely that these associations were practically driven by the heterogeneity in the samples. Correlation studies involving homogeneous golfer samples (e.g., skilled male golfers) in fact revealed no significant association between X-factor/stretch parameters and clubhead speed [2,3]. In elite male golfers, however, it was reported that the overall hip–shoulder separation pattern during the backswing did affect the backswing-to-downswing transition pattern/sequence [2].

While consistency in shot distance and direction is the most essential aspect of golf performance [4], the importance of shot distance itself increases with the level of competitiveness [3,9]. Shot distance is directly correlated to the clubhead speed at impact [21], and a golfer's ability to generate a high clubhead speed with sufficiently accurate direction control depends on well-orchestrated angular motions of the axle-chain system [2,3,22] as well as the level of interaction with the ground [1]. Clubhead speed in fact can serve as an effective indicator of the quality of the swing mechanics employed in a golfer's swing. In order to truly identify the key angular motion characteristics that contribute to clubhead speed positively, however, it is crucial to eliminate the influence of the heterogeneity in the sample by employing a homogeneous group of skilled golfers instead of a heterogeneous sample of golfers of varying skill levels. The angular motions of the axle-chain system can be accurately represented by the planar angular motions of the body lines (hip line (HL) and shoulder line (SL)) and the levers (UL and LL) of the axle-chain system on the FSP [2]. The purpose of this study, therefore, was to identify key performance factors in golf swings through a comprehensive investigation of the association between the on-plane angular motion characteristics of the axle-chain system and clubhead speed in a group of elite male golfers. It was hypothesized that (1) temporal parameters, on-plane rotation ranges, and peak angular velocities of the axle-chain system would be significantly correlated to clubhead speed, and (2) proximal-to-distal kinematic sequences would be observed during the backswing, transition, and downswing in a group of skilled male golfers.

2. Materials and Methods

2.1. Participants

A total of 66 right-handed male skilled golfers (touring professionals, collegiate players, and teaching professionals; handicap $\leq$ 3) participated in this study. The average ($\pm$ SD) mass, height, age, and driving clubhead speed were 84.4 $\pm$ 9.0 kg, 182.1 $\pm$ 6.1 cm, 29.4 $\pm$ 7.4 years, and 49.4 $\pm$ 2.1 m/s (110.4 $\pm$ 4.8 mph), respectively. The age range of the golfers was 19–50 years. At the time of data collection, participants reported being free of any serious injuries that could affect their maximum-effort swing performance. Golfers who consistently presented outliers due to their unusual timing characteristics (e.g., extremely slow backswing) were excluded from the study. The study protocol was approved by Texas Woman's University's Institutional Review Board, and informed consent was obtained from each golfer prior to participation.

2.2. Trial Conditions

Participants performed maximal-effort swings until seven successful trials were collected in each of the following three club conditions: driver, 5-iron, and pitching wedge. Shots were performed indoors with foam balls against a wall (located 10.6 m away from the center of the ball mat and marked with a vertical target line) in a motion analysis laboratory. Success of a shot was determined by the shot characteristics, such as direction and launch angle judged from the ball impact location on the target wall, and golfer's perceived solidness of contact.

Each participant performed a self-selected warm-up prior to data collection and completed the trials with his own clubs/shoes. Trials were captured in the order of pitching wedge, 5-iron, and driver. Practice shots were allowed at the beginning of the data capture session and between the club conditions. The first five trials (out of the seven captured trials) that were free of marker labeling issues were used in further processing and analysis.

2.3. Experimental Setup and Procedure

Marker trajectories were captured by a 10-camera real-time optical motion capture system (Oqus, Qualisys, Gothenburg, Sweden) operated at 500 Hz. Cameras were calibrated with a manufacturer-provided calibration wand before each data capture session. The ball mat (Hank Haney Profinity Practice Hitting Mat) was placed within the calibrated control volume. In the driver condition, rubber tees of various heights were used so that golfers could choose their preferred height.

Participants were required to be shirtless while wearing spandex shorts for minimization of the marker motion artifacts. Static trials (T-pose, ball mat, and club) were captured with seventy-six retro-reflective markers (spherical or hemispherical; 10-mm in diameter) placed on the participant's body, shoes, club, ball, and the ball mat [2]. Once the static trials were captured, 15 static-only markers (4 clubface markers, the ball marker, and 10 body markers) were removed before proceeding with the swing trials. The T-pose static trial was captured once for each participant, while the ball mat static trial and the club static trial were captured once for each club condition.

Captured trials were labeled by using Qualisys Tracking Manager (QTM, Qualisys, Gothenburg, Sweden). Labeled trials were saved in the c3d format (http://www.c3d.org) for subsequent processing.

2.4. Data Reduction and Processing

Captured marker coordinates were imported into Kwon3D Motion Analysis Suite (Version 5, Visol Inc., Seoul, Korea) for processing and analysis. Imported position data were upsampled to 1000 Hz. A 4th-order zero-phase lag Butterworth low-pass filter was used in filtering the marker coordinates. The residual plot method outlined by Winter [32] (pp. 42–50) was used in assessing the frequency contents of the marker trajectory data, and markers were eventually clustered into three groups, each with different cutoff frequencies (8, 15, and 30 Hz) [2].

Secondary points (including joint centers) were defined and computed from the filtered marker coordinates and measured anthropometric parameters [2]. The Tylkowski–Andriacchi hybrid method was used in locating the hip joint centers, while the mid-point method was used in locating the shoulder joint centers [4,33]. The mid-trunk point was defined as the centroid of the xiphoid process and T12 (i.e., the spinous process of the 12th thoracic vertebra). The mid-hand point was defined as the centroid of the hand centers. The HL and the SL were defined by the vectors drawn from the left joint center to the right for the hips and shoulders, respectively. The UL was defined as the line drawn from the mid-trunk point to the mid-hand point, whereas the LL (club) was defined as the line drawn from the mid-hand point to the clubhead point.

2.5. Variable Computation

Twelve swing event frames were identified in each trial based on the club and lead arm (left arm) positions: BA (breakaway), MB (mid-backswing), LBA (late backswing, arm-based), LB (late backswing), EPR (end of pelvis rotation), TB (top of backswing), EDA (early downswing, arm-based), ED (early downswing), MD (mid downswing), BI (ball impact), MF (mid-follow-through), and MC (maximum cocking) [2]. BA is the instant the clubhead moves more than 3 cm away from the initial address position. EPR is the instant the direction of rotation of the pelvis changes from backward to downward. The ZYX (longitudinal–anteroposterior–mediolateral) rotation sequence was used in computing the orientation angles of the pelvis (rotation, lateral tilt, and anteroposterior tilt), and EPR was detected based on the first orientation angle. TB is the instant the club's direction of rotation changes from backward to downward in the frontal view. BI is the instant the clubface comes into contact with the ball. At the "mid" events, the club shaft is parallel to the ground in the frontal view. "Early" or "late" events are when the club shaft becomes vertical before or after the corresponding "mid" event. At the "arm-based" events, the lead arm is parallel to the ground. MC is the instant the wrist (the link between the upper lever and the club) is cocked the most on the FSP. BI was used as the zero-time event and all event times were expressed relative to BI. Five phases were defined from the events: backswing (BA–TB), downswing (TB–BI), transition (EPR–TB), early downswing (TB–EDA), and late downswing (EDA–BI).

The position and normal vector of the FSP were calculated from the clubhead trajectory (MD–MF) following the method outlined by Kwon et al. [4]. A group of on-plane angles of the axle-chain system (i.e., HL, SL, UL, club, and wrist cock) were defined on the FSP (Figure 1) as outlined by Han et al. [2], and ranges of rotation (ROR) and angular velocities were computed from the angular positions. The angle between the HL and the SL was used as the X-factor (Figure 1). The increases in the X-factor beyond EPR observed during the transition (EPR–TB) and the downswing (TB to maximum) were labeled as the backswing and downswing X-factor stretches, respectively. The total stretch was computed as the sum of the backswing and downswing stretches.

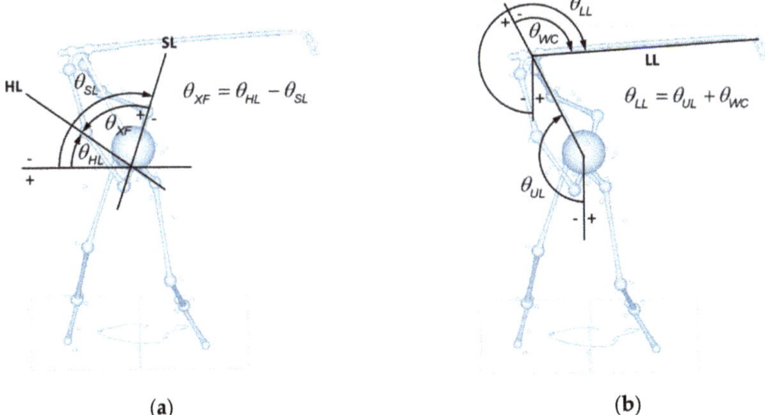

(a)   (b)

Figure 1. The on-plane angles of the axle-chain system defined on the functional swing plane. The hip line (HL) and shoulder line (SL) angles are relative to the horizontal line (**a**), while the upper lever (UL) and lower lever (LL; club) angles are relative to the vertical line (downward) (**b**). The wrist cock (WC) angle is the relative angular position of the LL to the UL. Counterclockwise angles yield positive values in these angles. The X-factor is the angle formed by the HL and the SL. A more backward-rotated (more negative) SL position relative to the HL yields a positive X-factor.

The instants the on-plane angular velocities of the body parts/club reached their respective peak values during backswing/downswing were identified as the peak points (Figure 2). The instants their on-plane rotations changed direction from backward to downward were identified as the transition points. All time data were expressed relative to BI. The actual clubhead speed (ACHS) was normalized to the body height (BH) to obtain the normalized clubhead speed (NCHS).

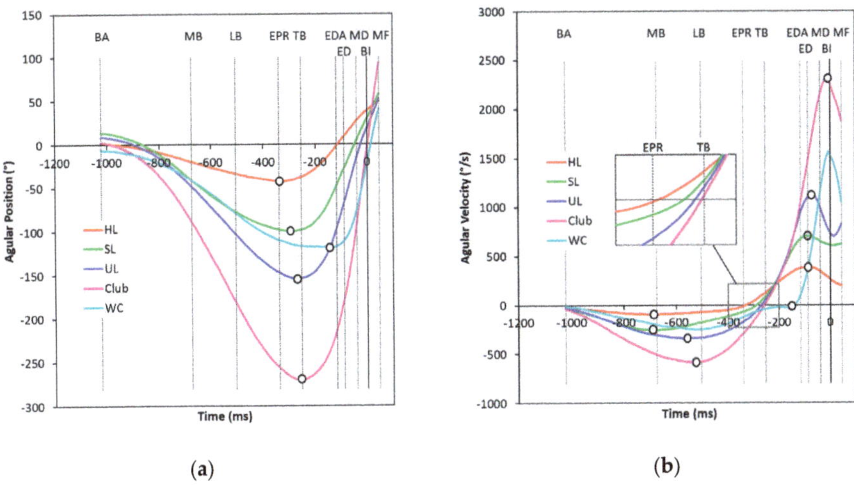

(a)   (b)

Figure 2. Ensemble-average on-plane angular position (**a**) and angular velocity (**b**) patterns of the axle-chain system (driver; $n = 66$). Transition points and angular velocity peaks (backswing and downswing) are marked. Negative angular positions mean backward-rotated or cocked. Abbreviations: HL—hip line, SL—shoulder line, UL—upper lever, and WC—wrist cock angle.

The following temporal/kinematic parameters were extracted from the trials for statistical analysis:

- Event times (relative to BI): BA, MB, LBA, LB, EPR, TB, EDA, ED, MD, and MC;
- Phase times: backswing (BA–TB), downswing (TB–BI), transition (EPR–TB), early downswing (TB–EDA), and late downswing (EDA–BI);
- Angular positions (address, transition, and BI) and RORs (backward and downward): UL, HL, and SL;
- Wrist cock angles (address, EDA, BI, and maximum) and ranges (cocking and uncocking);
- X-factors (address, EPR, TB, BI, and maximum) and range;
- X-factor stretches (backswing, downswing, and total);
- Times of backswing angular velocity peaks: HL, SL, UL, club, and wrist;
- Transition times: HL, SL, UL, club, and wrist;
- Times of downswing angular velocity peaks: HL, SL, and UL;
- Peak backswing angular velocities: HL, SL, UL, club, and wrist;
- Peak downswing angular velocities: HL, SL, UL, club, and wrist;
- Angular velocities at BI: HL, SL, and UL;
- Peak-to-BI angular velocity decreases: HL, SL, and UL.

2.6. Statistical Analysis

The average values of five repeated trials were used in the statistical analysis. The correlations between the select temporal/kinematic parameters and the clubhead speed (actual and normalized) were computed ($\alpha = 0.05$). Normality of the temporal/kinematic parameters was tested by using the Shapiro–Wilk test. When deviation from normality was caused by a small number (1 or 2) of outliers, the outliers were removed, and Pearson's correlation coefficient was calculated. When substantial deviation from normality was observed, Spearman's correlation coefficient was calculated instead. $r = 0.1, 0.3$, and 0.5 were used as the thresholds for small, moderate, and large correlations [34].

In addition, three two-way repeated-measure ANOVAs ($\alpha = 0.017$) were conducted for the kinematic-sequence time data (peak and transition points), one in each club condition, with two within-subject factors: SEQUENCE (backswing, transition, and downswing) and BODY (HL, SL, UL, and club). Post-hoc analysis was conducted if the BODY factor effect and/or the two-way interaction was significant. IBM SPSS Statistics (Version 25) was used for all statistical analyses.

3. Results

The average ACHS scores of the participants were 49.4 ± 2.1 m/s (110.4 ± 4.8 mph), 42.1 ± 1.8 m/s (94.2 ± 4.0 mph), and 38.2 ± 1.9 m/s (85.5 ± 4.3 mph) for the driver, 5-iron, and pitching wedge, respectively. The average NCHS scores were 26.8 ± 1.3 BH/s, 22.7 ± 1.1 BH/s, and 20.6 ± 1.2 BH/s, respectively. Tables 1–4 present the average values of the temporal/kinematic parameters and their correlations to clubhead speeds (ACHS and NCHS) along with the counts of outliers removed.

3.1. Event and Phase Times

Among the swing events, the downswing events (TB, EDA, ED, and MD) showed significant ($p < 0.05$) correlations (moderate to large) to both ACHS and NCHS across all club conditions (Table 1). Among the backswing events, MB and LB were characterized by significant correlations (small to moderate) to ACHS in all club conditions with one exception (MB/driver). More delayed onset was associated with higher clubhead speed. Downswing phase times (total, early downswing, and late downswing) showed significant negative correlations (moderate to large) to NCHS across all club conditions (Table 1). The downswing and late downswing times also revealed significant correlations to ACHS in all club conditions. Shorter phase time was associated with higher clubhead speed.

Table 1. Correlation coefficients of the event and phase times (in ms) to clubhead speed ($n = 66$).

Parameter	Club								
	Driver			5-Iron			Pitching Wedge		
	$M \pm SD$	ACHS	NCHS	$M \pm SD$	ACHS	NCHS	$M \pm SD$	ACHS	NCHS
	Event time (relative to BI)								
BA	−1022 ± 105	0.226	0.172	−999 ± 97	0.223	0.221	−992 ± 92	0.401 *	0.390 *
MB	−672 ± 63	0.248 *	0.146	−670 ± 59	0.213	0.166	−672 ± 58	0.343 *	0.307 *
LBA	−515 ± 48	0.162	0.121	−512 ± 45	0.126	0.136	−511 ± 48	0.218	0.234
LB	−503 ± 49	0.287 *	0.183	−493 ± 44	0.248 *	0.184	−491 ± 44	0.316 *	0.282 *
EPR	−336 ± 25 [#]	0.177	0.239	−334 ± 26 [#]	0.143	0.230	−338 ± 26 [1]	0.258 *	0.405 *
TB	−253 ± 26	0.481 *	0.446 *	−256 ± 22 [1]	0.393 *	0.405 *	−266 ± 25 [1]	0.441 *	0.495 *
EDA	−118 ± 9	0.531 *	0.475 *	−113 ± 9	0.595 *	0.612 *	−117 ± 13 [#]	0.720 *	0.654 *
ED	−88 ± 6	0.810 *	0.735 *	−91 ± 7	0.736 *	0.773 *	−97 ± 8 [1]	0.761 *	0.811 *
MD	−39 ± 2	0.745 *	0.644 *	−37 ± 2 [#]	0.737 *	0.747 *	−38 ± 3	0.802 *	0.808 *
MC	−183 ± 71 [#]	0.174	0.199	−183 ± 75 [#]	0.161	0.232	−207 ± 85 [#]	0.176	0.218
	Phase time								
BS	769 ± 100	−0.112	−0.064	741 ± 88	−0.125	−0.108	724 ± 78	−0.284 *	−0.250 *
DS	253 ± 26	−0.481 *	−0.446 *	256 ± 22 [1]	−0.393 *	−0.405 *	266 ± 25 [1]	−0.441 *	−0.495 *
Transition	83 ± 31	0.274 *	0.168	77 ± 31	0.224	0.126	72 ± 30	0.162	0.070
Early DS	136 ± 25	−0.318 *	−0.302 *	144 ± 23	−0.211	−0.261 *	151 ± 26	−0.199	−0.268 *
Late DS	118 ± 9	−0.531 *	−0.475 *	113 ± 9	−0.595 *	−0.612 *	117 ± 13 [#]	−0.720 *	−0.654 *

Abbreviations: ACHS—actual clubhead speed, NCHS—normalized clubhead speed, BA—breakaway, MB—mid backswing, LBA—late backswing, arm-based, LB—late backswing, EPR—end of pelvis rotation, TB—top of backswing, EDA—early downswing, arm-based, ED—early downswing, MD—mid downswing, BI—ball impact, MC—maximum cocking, BS—backswing, and DS—downswing. [#] Nonparametric (Spearman's) correlation coefficient; [ ] Number of outliers removed; * Statistically significant ($p < 0.05$).

3.2. On-Plane Angular Positions and Ranges

The transition (maximum backward-rotated) positions and RORs (backward and downward) of the HL, SL, and UL showed significant correlations (small to large) to NCHS across all club conditions (Table 2). The angular position at BI of the HL also showed significant correlations (moderate) to NCHS across all club conditions. The angular position at BI and downward ROR of the HL and the transition position and RORs (backward and downward) of the SL/UL revealed significant correlations (small to moderate) to ACHS. More backward-rotated transition position, more downward-rotated position at BI, and larger RORs were associated with higher clubhead speed.

In the on-plane motion of the wrist, no consistent correlation profile was identified (Table 2). The wrist cock angle at EDA showed significant correlations (small to moderate) to clubhead speeds (ACHS and NCHS) in the driver condition only. More wrist cocking at EDA was associated with higher clubhead speed. Among the X-factors, the values at TB and BI were found to be significantly correlated (small to moderate) to ACHS across all club conditions (Table 2). The X-factor at BI was also correlated (moderate) to NCHS in two club conditions (iron and wedge). More HL–SL separations at these events were associated with higher clubhead speed. No consistent significant correlation was observed in the X-factor stretches.

3.3. Kinematic Sequence Time Parameters

In terms of the time of backswing angular velocity peak, only the SL revealed significant correlations (small to moderate) to both ACHS and NCHS across all club conditions (Table 3). The UL also showed significant correlations (small) to ACHS in all club conditions except the pitching wedge. Delayed onset of the angular velocity peaks during the backswing was associated with higher clubhead speed. The times of transition of the HL, UL, and the club showed significant correlations (small to large) to NCHS across all club conditions. The UL and the club also showed significant correlations (moderate) to ACHS across all club conditions. In these cases, delayed transitions were associated with higher clubhead speed. The timings of angular velocity peaks in the downswing were characterized by a lack of significant correlation to clubhead speed.

Table 2. Correlation coefficients of the angular positions and ranges (in °) to clubhead speed ($n = 66$).

Parameter	Club								
	Driver			5-Iron			Pitching Wedge		
	$M \pm SD$	ACHS	NCHS	$M \pm SD$	ACHS	NCHS	$M \pm SD$	ACHS	NCHS
Hip line angle									
At address	3 ± 5	0.158	0.269 *	2 ± 4	0.171	0.204	2 ± 4	0.131	0.169
Transition	−42 ± 8	−0.010	−0.266 *	−38 ± 7	0.046	−0.270 *	−34 ± 7	−0.189	−0.354 *
At BI	39 ± 8	0.325 *	0.315 *	31 ± 7	0.382 *	0.373 *	27 ± 7	0.422 *	0.400 *
Backward ROR	45 ± 8	0.096	0.409 *	40 ± 7	0.051	0.397 *	37 ± 7	0.274 *	0.470 *
Downward ROR	81 ± 10	0.278 *	0.472 *	69 ± 9	0.253 *	0.495 *	62 ± 9	0.469 *	0.584 *
Shoulder line angle									
At address	15 ± 5	0.233	0.300 *	12 ± 4	0.080	0.213	12 ± 4	0.070	0.206
Transition	−100 ± 8	−0.254 *	−0.339 *	−96 ± 9	−0.161	−0.338 *	−93 ± 9 [1]	−0.362 *	−0.374 *
At BI	30 ± 5	0.180	0.192	23 ± 4 #	0.173	0.126	21 ± 3 [1]	0.096	0.134
Backward ROR	115 ± 11	0.287 *	0.422 *	108 ± 10	0.173	0.379 *	104 ± 10	0.351 *	0.444 *
Downward ROR	130 ± 9	0.306 *	0.439 *	119 ± 9	0.241	0.412 *	114 ± 9	0.403 *	0.454 *
Upper lever angle									
At address	9 ± 4	0.280 *	0.260 *	6 ± 3	0.020	0.171	5 ± 1	0.055	0.206
Transition	−154 ± 9	−0.333 *	−0.394 *	−148 ± 8	−0.335 *	−0.400 *	−142 ± 9	−0.398 *	−0.410 *
At BI	19 ± 5	−0.064	0.015	12 ± 4	−0.193	−0.025	11 ± 4	−0.276 *	−0.139
Backward ROR	164 ± 10	0.386 *	0.437 *	154 ± 9	0.312 *	0.420 *	147 ± 10	0.387 *	0.443 *
Downward ROR	173 ± 9	0.299 *	0.408 *	160 ± 8	0.262 *	0.414 *	152 ± 8	0.316 *	0.395 *
Wrist cock angle									
At address	−6 ± 4	−0.133	−0.175	−9 ± 4	−0.091	−0.157	−10 ± 4	−0.147	−0.210
Max. cocking	−121 ± 10	−0.172	−0.101	−114 ± 8 [1]	−0.078	0.030	−110 ± 8 [1]	−0.066	−0.018
At BI	−20 ± 5	0.204	0.068	−20 ± 5	0.135	0.026	−20 ± 5	0.185	0.112
At EDA	−117 ± 9 #	−0.362 *	−0.265 *	−111 ± 8	−0.133	−0.188	−105 ± 7 [1]	−0.109	−0.131
Cocking range	115 ± 11	0.104	−0.019	106 ± 10	0.006	−0.030	100 ± 10	−0.029	−0.031
Uncocking range	101 ± 11	0.255 *	0.119	95 ± 10	0.105	0.049	90 ± 10	0.119	0.111
X-factor									
At address	−12 ± 6	−0.067	−0.036	−10 ± 5	0.072	0.004	−10 ± 5	0.042	−0.028
At EPR	56 ± 8	0.237	0.156	56 ± 8	0.228	0.169	56 ± 8	0.258 *	0.199
At TB	61 ± 9	0.265 *	0.175	60 ± 8	0.247 *	0.160	59 ± 8	0.270 *	0.183
Maximum	62 ± 8	0.218	0.159	61 ± 8	0.219	0.151	60 ± 8	0.263 *	0.182
At BI	9 ± 8	0.249 *	0.235	8 ± 7	0.295 *	0.309 *	6 ± 7	0.391 *	0.372 *
Range	74 ± 9 #	0.306 *	0.121	71 ± 8 #	0.164	0.142	69 ± 8	0.226	0.192
X-factor stretch									
Backswing	5 ± 3 #	0.070	0.020	4 ± 3 #	0.066	−0.015	3 ± 3 #	0.110	0.068
Downswing	1 ± 2 #	−0.258 *	−0.063	1 ± 1 #	−0.193	−0.060	1 ± 1 #	−0.009	0.025
Total	6 ± 4 #	−0.130	−0.066	5 ± 3 #	−0.017	−0.038	3 ± 3 #	0.038	0.026

Abbreviations: ACHS—actual clubhead speed, NCHS—normalized clubhead speed, ROR—range of rotation, EPR—end of pelvis rotation, TB—top of backswing, EDA—early downswing, and BI—ball impact. # Nonparametric (Spearman's) correlation coefficient; [ ] Number of outliers removed; * Statistically significant ($p < 0.05$).

Table 3. Correlation coefficients of the kinematic-sequence time parameters (in ms; relative to BI) to clubhead speed ($n = 66$).

Parameter	Driver			5-Iron			Pitching Wedge		
	$M \pm SD$	ACHS	NCHS	$M \pm SD$	ACHS	NCHS	$M \pm SD$	ACHS	NCHS
Backswing peak									
Hip line	−657 ± 98 [#]	0.065	0.084	−646 ± 94	0.093	0.132	−644 ± 93	0.184	0.232
Shoulder line	−675 ± 75	0.366 *	0.249 *	−665 ± 75	0.299 *	0.272 *	−668 ± 77	0.350 *	0.337 *
Upper lever	−572 ± 64 [#]	0.281 *	0.214	−565 ± 61 [#]	0.273 *	0.168	−565 ± 57 [1]	0.155	0.099
Club	−536 ± 71	0.243 *	0.100	−553 ± 72	0.194	0.089	−569 ± 74	0.152	0.112
Wrist (cocking)	−539 ± 93	0.217	0.063	−570 ± 105	0.169	0.068	−584 ± 100	0.140	0.071
Time difference	SL/HL → UL → club			SL/HL → UL/club			SL/HL → club/UL		
Transition									
Hip line	−327 ± 24 [#]	0.215	0.294 *	−321 ± 25 [#]	0.211	0.318 *	−323 ± 29 [#]	0.315 *	0.418 *
Shoulder line	−286 ± 20 [#]	0.117	0.129	−288 ± 25 [#]	0.121	0.086	−294 ± 29 [#]	0.280 *	0.326 *
Upper lever	−266 ± 19	0.442 *	0.419 *	−265 ± 15 [1]	0.386 *	0.382 *	−273 ± 19 [1]	0.479 *	0.514 *
Club	−258 ± 25	0.499 *	0.420 *	−262 ± 23	0.429 *	0.464 *	−270 ± 24 [1]	0.416 *	0.447 *
Wrist (uncocking)	−164 ± 63 [#]	0.182	0.206	−160 ± 65 [#]	0.217	0.277 *	−172 ± 71 [#]	0.288 *	0.330 *
Time difference	HL → SL → UL → club			HL → SL → UL → club			HL → SL → UL/club		
Downswing peak									
Hip line	−87 ± 18	−0.014	−0.104	−76 ± 23 [#]	−0.037	−0.146	−66 ± 27 [#]	−0.202	−0.306 *
Shoulder line	−84 ± 25 [#]	0.184	0.082	−83 ± 20 [#]	0.143	0.067	−75 ± 28 [#]	−0.006	−0.046
Upper lever	−73 ± 6	0.083	0.046	−70 ± 6	0.140	0.060	−68 ± 6	0.177	0.069
Time difference	HL/SL → UL → club			SL → UL → club HL → club			SL/UL/HL → club		

Abbreviations: ACHS—actual clubhead speed, NCHS—normalized clubhead speed, and BI—ball impact. [#] Nonparametric (Spearman's) correlation coefficient; [ ] Number of outliers removed; * Statistically significant ($p < 0.05$).

The two-way repeated ANOVAs of the kinematic-sequence time parameters revealed significant SEQUENCE*BODY interactions in all club conditions: driver (Huynh–Feldt $F = 44.384$, $p < 0.001$, partial $\eta = 0.406$), 5-iron (Huynh–Feldt $F = 39.744$, $p < 0.001$, partial $\eta = 0.383$), and pitching wedge (Huynh–Feldt $F = 37.251$, $p < 0.001$, partial $\eta = 0.368$). The most consistent proximal-to-distal sequence was observed in the transition phase (Table 3). Both the driver and 5-iron conditions showed full proximal-to-distal sequences, while the pitching wedge condition showed a partial sequence with the UL and club transitions not significantly separated.

In the backswing sequence, the timings of the HL and SL angular velocity peaks were not significantly separated with the tendency of the SL peak appearing earlier than the HL peak (Table 3; Figure 2b). The timing of the UL peak was significantly separated from the club peak only in the driver condition. The SDs of the backswing angular velocity peak times were substantially larger than those of the other sequences. In the downswing sequence, however, only the separation between the club and the rest of the axle-chain system (SL/HL/UL) was notable, which was consistent across all club conditions (Table 3; Figure 2b). The downswing angular velocity peaks of the body parts occurred in a narrow time window with inconsistent sequences, although some statistically significant separations were observed (e.g., SL vs. UL in the driver and 5-iron conditions).

3.4. Angular Velocity Parameters

The peak backswing angular velocities of the HL, SL, and UL were characterized by significant correlations (small to large, mostly moderate) to both ACHS and NCHS across all club conditions (Table 4). Faster backward rotations were associated with higher clubhead speed. All peak downswing angular velocities of the axle-chain system were significantly correlated to ACHS and NCHS across all club conditions. Faster downward rotations were associated with higher clubhead speed. At impact, the angular velocities of the SL and UL were significantly correlated (small to moderate) to both ACHS and NCHS with one exception (UL/driver vs. NCHS). Faster rotations at impact were associated with higher clubhead speed. The peak-to-BI angular velocity decrease in the HL and UL was significantly correlated (small to large) to NCHS across all club conditions. In addition, the HL also revealed significant correlations (moderate to large) to ACHS. More losses in the angular velocity were associated with higher clubhead speed.

Table 4. Correlation coefficients of the angular velocity parameters (in °/s) to clubhead speed ($n = 66$).

Parameter	Club								
	Driver			5-Iron			Pitching Wedge		
	M ± SD	**ACHS**	**NCHS**	*M ± SD*	**ACHS**	**NCHS**	*M ± SD*	**ACHS**	**NCHS**
Backswing peak									
Hip line	−96 ± 22 [1]	−0.314 *	−0.435 *	−86 ± 21	−0.275 *	−0.451 *	−79 ± 19	−0.464 *	−0.539 *
Shoulder line	−259 ± 56	−0.332 *	−0.311 *	−247 ± 53	−0.263 *	−0.334 *	−239 ± 51	−0.401 *	−0.424 *
Upper lever	−349 ± 47	−0.330 *	−0.308 *	−337 ± 44	−0.322 *	−0.341 *	−329 ± 43	−0.409 *	−0.413 *
Club	−610 ± 87	−0.186	−0.128	−576 ± 83	−0.240	−0.183	−568 ± 77	−0.296 *	−0.237
Wrist cocking	−289 ± 65 [#]	0.003	0.068	−268 ± 62 [#]	−0.070	−0.012	−257 ± 59	−0.097	−0.055
Downswing peak									
Hip line	394 ± 62	0.447 *	0.583 *	340 ± 61	0.398 *	0.579 *	305 ± 63	0.541 *	0.620 *
Shoulder line	711 ± 63	0.461 *	0.580 *	662 ± 60	0.398 *	0.603 *	622 ± 63	0.618 *	0.716 *
Upper lever	1142 ± 67	0.552 *	0.696 *	1077 ± 69	0.546 *	0.756 *	1020 ± 72 [1]	0.685 *	0.802 *
Club	2364 ± 115	0.906 *	0.787 *	2341 ± 108	0.846 *	0.799 *	2251 ± 122	0.874 *	0.867 *
Wrist uncocking	1615 ± 138	0.539 *	0.506 *	1627 ± 127	0.419 *	0.465 *	1563 ± 131	0.494 *	0.558 *
At BI									
Hip line	262 ± 63	−0.150	−0.115	259 ± 53	−0.026	0.030	247 ± 51	0.155	0.185
Shoulder line	610 ± 65	0.265 *	0.299 *	572 ± 62	0.258 *	0.290 *	553 ± 63	0.365 *	0.383 *
Upper lever	725 ± 84	0.324 *	0.220	705 ± 80	0.441 *	0.315 *	680 ± 79	0.494 *	0.395 *
Peak−to−BI decrease									
Hip line	−123 ± 66 [2]	−0.476 *	−0.552 *	−81 ± 62 [#]	−0.417 *	−0.507 *	−58 ± 55 [#]	−0.517 *	−0.575 *
Shoulder line	−101 ± 80 [#]	−0.148	−0.182	−89 ± 71 [#]	−0.083	−0.150	−69 ± 65 [#]	−0.212	−0.283 *
Upper lever	−417 ± 100	−0.102	−0.287 *	−372 ± 94	−0.026	−0.285 *	−336 ± 91 [#]	−0.183	−0.373 *

Abbreviations: ACHS—actual clubhead speed, and NCHS—normalized clubhead speed. [#] Nonparametric (Spearman's) correlation coefficient; [ ] Number of outliers removed; * Statistically significant ($p < 0.05$).

4. Discussion

The main purpose of this study was to investigate the association between the on-plane angular motion parameters of the axle-chain system and clubhead speed in three different club conditions. The on-plane angular positions, RORs, and angular velocities of the HL, SL, UL, club, and wrist were calculated along with a host of temporal parameters (event times and peak/transition times). The clubhead speed was normalized to the body height and the temporal/kinematic parameters were correlated to both the actual (ACHS) and normalized (NCHS) clubhead speeds. The coefficients of variation (COVs) of the ACHS were 4.3%, 4.2%, and 5.0% for driver, 5-iron, and pitching wedge, respectively, while those of the NCHS were 4.9%, 4.8%, and 5.9%. The NCHS and the pitching wedge condition tended to show larger COVs than the ACHS and the driver/5-iron conditions, respectively.

The backswing-to-downswing time ratio was approximately 3.0 to 1 in the driver condition with a decreasing trend in the shorter clubs (2.9 to 1 for the 5-iron and 2.7 to 1 for the pitching wedge). As the club got shorter, the backswing time decreased while the downswing time increased (Table 1). The association between NCHS and event/phase times was expected, as faster motion naturally leads to shorter phases and delayed onset of events closer to BI (Table 1). Higher NCHS was associated with shorter downswing phase times (total, early downswing, and late downswing). One noteworthy finding was that the NCHS was associated with the downswing event times but not with the backswing event times. The timing of EPR, the beginning of the transition phase, showed significant deviations from normal distribution in the longer club conditions (Table 1). This suggests that the way golfers start the transition phase is highly dependent on the individual and the swing style employed.

In a study involving 74 male elite golfers, Han and colleagues [2] identified five distinct swing styles based on the X-factor pattern and the SL acceleration characteristic. The LBS (large backswing stretch) style, characterized by the longest transition phase and a complete (and well-spaced) proximal-to-distal transition sequence, showed an almost 1:1:1 (109:117:119 ms) time ratio between the transition, early downswing, and late downswing phases. The LDS (large downswing stretch) style scored an average time ratio of 1:2.5:2 (63:153:119 ms) with the transition phase being substantially shorter than the early downswing phase. Simultaneous transitions of the SL, UL, and the club were observed in the LDS style. The average phase times in the current study (83 ms, 136 ms, and 118 ms for the transition, early downswing, and late downswing phases, respectively) were similar to those of the LSA (late shoulder acceleration) style (82 m, 137 m, and 119 ms, respectively) reported by Han and associates [2]. Further explanatory analysis revealed significant negative correlations between the phase times of the early downswing phase and the transition phase (r = -0.584, -0.571, and -0.526 for driver, 5-iron, and pitching wedge, respectively), which means a longer transition phase is associated with a shorter early downswing phase. The findings of the study by Han and colleagues [2] and the current study collectively imply that the swing mechanics employed in the transition phase can be a key determinant of the biomechanical robustness of a golf swing. Therefore, more in-depth investigations on various transition strategies employed by skilled golfers are warranted.

The pitching wedge condition was characterized by additional significant correlations between the NCHS and the backswing event/phase times. Higher clubhead speed was associated with a shorter backswing phase and delayed onset of the backswing events. This suggests that the swing mechanics used in the pitching wedge condition could be somewhat different from that used in the longer club conditions. More in-depth investigations of the mechanics involved in full and partial wedge swings are necessary.

Higher NCHS was associated with larger on-plane angular motions (i.e., more backward-rotated transition positions and larger backward/downward RORs) of the HL, SL, and UL (Table 2; Figure 2a). Additional body-specific (e.g., HL angular position at BI) and club-specific (e.g., driver at address) significant correlations were observed in the factors contributing to larger RORs. The driver-specific additional correlations at address appeared to be related to the ball position, as the ball is typically located closer to the lead foot in the driver condition when compared to other club conditions. The HL-specific additional correlations also suggest that the pelvis position at impact can be useful in assessing the golfer's impact posture.

In general, the clubhead speed was not well associated with the wrist-cock angles/ranges (Table 2). Club-specific correlations of wrist-cock angle to clubhead speed, however, were observed in the driver condition at EDA. Unlike iron/wedge swings, one of the main goals in a driver swing is to maximize the distance. The GRF moment generated by the combined GRF about the center of mass (COM) in the frontal plane and the pivoting moment generated by individual foot GRFs about the combined center of pressure (COP) in the horizontal plane typically reach their peak values at or slightly before EDA [1]. The relative position of the club to the COM at EDA directly affects the moment of inertia of the golfer–club system and the resulting angular acceleration caused by the golfer–ground interaction moments. For a faster angular acceleration of the body–club system, therefore, it is imperative to keep the club closer to the body by maintaining the lag (relatively large wrist-cock angle) near EDA.

It is noteworthy that the X-factor/stretch parameters overall lack association to clubhead speed while the angular position/range and velocity parameters of the HL and the SL individually do exhibit significant associations. Among the X-factors at various time points, only that at BI showed significant correlations (in 5-iron and pitching wedge). Higher NCHS was associated with larger HL–SL torsional separation at BI (i.e., more downward-rotated HL position than the SL). Promotion of the pelvis lateral tilt with the left hip positioned higher at BI and suppression of early opening of the shoulders can promote more favorable impact conditions. The NCHS was not associated with any X-factor stretch parameters (backswing, downswing, and total). It was apparent that normalization of the clubhead speed weakened the association between the X-factor/stretch parameters and clubhead speed (Table 2).

The X-factor range and stretch parameters were also characterized by significant deviations from normal distribution across all club conditions, suggesting swing style-dependent HL–SL torsional separations. Although not statistically significant, there was a tendency that higher NCHS was generally associated with larger backswing stretch than smaller downswing stretch (Table 2). Han and colleagues [2] reported that the LDS swing style, showing the largest downswing stretch among the styles, was characterized by a partial proximal-to-distal transition sequence with notable separation between the pelvis and the rest (HL→SL/UL/club). The LBS style, showing the largest backswing stretch, on the contrary, was characterized by a well-spaced, complete proximal-to-distal transition sequence (HL→SL→UL→club) with an elongated transition phase. While skilled male golfers overall showed proximal-to-distal transition sequences, the separation between the UL and club got obscured in the pitching wedge condition (Table 3). These findings suggest that more deviations from a full proximal-to-distal sequence may be observed in the shorter club conditions in some swing styles (e.g., LDS).

In a study involving 18 skilled male golfers, Kwon et al. [3] reported no significant correlation between the NCHS and the FSP-based X-factor/stretch parameters including the value at BI for the driver condition. The maximum X-factor and the downswing/total stretches reported in their study (59°, 1°, and 8°, respectively) were comparable to those from the current study (61°, 1°, and 6°, respectively; Table 2). In a study involving a large heterogeneous group of golfers ($n = 100$, three ball speed groups), Myers et al. [28] reported significant correlations of ball speed to X-factor at TB ($r = 0.55$) and maximum X-factor ($r = 0.54$). In another study involving an even larger heterogeneous group (308 golfers; 266 males and 42 females; 8.4 ± 8.4 handicap), Chu et al. [20] reported a significant regression that associated ball velocity with six kinematic parameters measured at TB: lateral bend, superior-inferior shift velocity, X-factor, leading arm angle, wrist hinge, and leading knee flexion ($R^2 = 0.437$). However, it is likely that these significant correlation/regression results were largely driven by the heterogeneity in the sample [35], evidenced by the reported significant inter-group differences in the ball speed and X-factor among various ball velocity [28] and skill [13] groups.

In terms of the kinematic-sequence time parameters, higher NCHS was associated with delayed onset of SL's backswing peak and delayed transitions of all bodies except SL (Table 3). The NCHS, however, was not associated with the timings of the downswing peaks in any consistent way. This lack of association in the downswing may be explained by the interplay of two conflicting necessities: (1) faster swing tends to delay the downswing peaks due to shorter downswing phase time; (2) the body has to reach peak angular velocities early enough to allow a sufficient time for the angular momentum transfer from the body to the club [2].

The backswing sequences were characterized by the separation between the angular velocity peaks of the axle and the chain (i.e., SL/HL vs. UL/club) (Table 3). The SL tended to reach the angular velocity peak earlier than the HL did in the backswing. Timings of the backswing peaks were characterized by large SDs when compared to the transition and downswing peak times due to the slower backswing motion and the large SDs in the backswing phase time (Tables 1 and 4). The transition sequences were more consistent than the other sequences, showing full or partial proximal-to-distal sequences (Table 3). Longer clubs (driver and 5-iron) exhibited full proximal-to-distal sequences while the pitching wedge condition lost the separation between the UL and club transition points. The timings of the downswing angular velocity peaks of the HL, SL, and UL were not well separated from each other and the sequences were not consistent across the club conditions (Table 3) [2]. The separation of the downswing peaks between the body (HL/SL/UL) and the club was obvious, but the peaks of the axle segments and the UL occurred within narrow time windows (14, 13, and 9 ms on average for the driver, 5-iron, and pitching wedge, respectively). In the driver condition, the angular velocity peaks of the axle segments (HL and SL) on average occurred at ED (Tables 1 and 3). As the club gets shorter, ED tends to occur earlier while the angular velocity peaks of the axle segments occur later (i.e., closer to BI).

The peak angular velocities (downswing in particular) showed the strongest association with the NCHS (Table 4). Higher NCHS was associated with faster backward rotations of the body (HL, SL, and UL), but no association was observed with the angular speed of the club and the wrist. Club and wrist angular velocities during the backswing can easily be manipulated by the wrist action. While LB occurred on average 12 ms, 19 ms, and 20 ms after LBA in the driver, 5-iron, and wedge conditions, respectively (Table 1), 23 (driver), 19 (5-iron), and 18 (pitching wedge) golfers (out of 66 each) actually showed the reversed sequence (LB→LBA). Early initiation of wrist cocking in the backswing using the wrist muscles causes the club and wrist to rotate faster/earlier and affects the way the body rotates. For example, Han and colleagues [2] reported that the STS (small total stretch) style exhibited a wrist-dominant swing pattern with early HL–SL separation and small X-factor stretches. It was speculated that the swing style-dependent variations weakened NCHS' association with the peak angular velocities of the club/wrist during the backswing.

Higher NCHS was also associated with faster downward rotations of all body parts of the axle-chain system, more slow-down of the HL and UL toward BI, and faster rotations of the SL and UL at BI (Table 4). Among the body parts, the UL revealed a unique correlation profile: i.e., the more it slows down after reaching the peak, and the faster the rotation still is at BI, the higher the NCHS is. The UL directly interacts with the club during the swing, and more slow-down in the UL rotation was found to be strongly associated with faster wrist uncocking (r = 0.758, 0.770, and 0.808 for the driver, iron, and pitching wedge, respectively). Effective slow-down of the UL rotation and speed-up of the wrist uncocking can be promoted with a delayed release [22].

Among the elements of the axle-chain system, the UL revealed significant correlations to NCHS across all club conditions in all four angular velocity parameter categories (with one exception of angular velocity at BI in the driver condition): backswing peak, downswing peak, velocity at BI, and peak-to-BI decrease. The SL showed significant correlations to clubhead speed in all categories except the peak-to-BI decrease. The HL showed significant correlations to clubhead speed in all categories except the angular velocity at BI.

From a supplementary correlation analysis, it was found in the HL and SL that the peak backswing angular velocity and the peak downswing angular velocity were correlated to each other significantly across all club conditions: r = 0.500, 0.597, and 0.591 for the driver, 5-iron, and pitching wedge, respectively, in the HL, and r = 0.339, 0.493, and 0.530, respectively, in the SL. This means faster backward rotation of the axle tends to result in faster downward rotation of the axle. The backswing–downswing association got weaker in the UL and club/wrist, and the driver condition revealed weaker associations than the iron and wedge conditions did, suggesting added complexity in the UL/club/wrist motions, particularly in the driver swing.

The overall findings of this study (correlations and kinematic sequences) and those of Han and colleagues [1,2] collectively highlight the benefits of a faster backswing motion for skilled golfers. The potential benefits of active (faster) backswing include the following: (1) a faster backward rotation of the axle leads to a more body-driven backswing along with large backward angular momentums of the axle segments; (2) a body-driven backswing promotes the LBS swing pattern by increasing the backswing X-factor stretch and elongating the transition phase; (3) the continuous HL–SL torsional separation during the transition phase that comes with the LBS swing style induces full and well-spaced proximal-to-distal transition sequence; (4) a continuous proximal-to-distal transition promotes SSC-style muscle actions during the transition and early downswing phases making the swing easier with more efficient use of muscles; (5) an active and continuous backswing-to-downswing transition requires the lower body to interact with the ground more actively, resulting in large golfer–ground interaction moments (i.e., the GRF moment in the frontal plane and the pivoting moment in the horizontal plane) [1]; (6) large golfer–ground interaction moments at TB enable fast acceleration of the downward rotation of the body from the beginning of the downswing, thus shortening the early downswing phase without a waste of time; (7) early acceleration of the downward rotation allows the axle and the UL to reach their angular velocity peaks early and secure a sufficient time for the body-to-club angular momentum transfer; (8) the LBS-style (body-driven) swing also promotes delayed release of the club (i.e., wrist uncocking).

5. Conclusions

The main purpose of this study was to investigate the association between the on-plane angular motion parameters of the axle-chain system and the clubhead speed in three different club conditions (driver, 5-iron, and pitching wedge) in a group of skilled male golfers. It was concluded that:

- Higher normalized clubhead speed is associated with shorter downswing phases (total, early downswing, and late downswing) and more delayed onset of downswing events (TB and thereafter) closer to impact. In addition, a shorter early downswing phase is associated with a longer transition phase.

- Higher normalized clubhead speed is associated with larger backward/downward on-plane rotations (hip/shoulder lines and upper lever) on the functional swing plane. Normalized clubhead speed is not significantly associated with the wrist-cock angles or wrist motion ranges. In the driver swing, more wrist cocking at EDA (keeping the club closer to the body) is associated with higher normalized clubhead speed.

- Higher normalized clubhead speed is associated with a larger X-factor at BI (with the shoulder line trailing the hip line). Normalized clubhead speed is not significantly associated with other X-factor/stretch parameters.

- Higher normalized clubhead speed is associated with more delayed onset of backswing angular velocity peaks (shoulder line) and more delayed backswing-to-downswing transitions (hip/shoulder lines, upper lever, and club). Normalized clubhead speed is not significantly associated with the timings of the backswing angular velocity peaks.

- Only the transition sequence exhibits consistent proximal-to-distal sequences with a minor deviation in the pitching wedge condition. The backswing and downswing sequences deviate from the proximal-to-distal sequence significantly and are inconsistent across club conditions. The backswing sequence revealed a dominant separation between the axle (hip/shoulder lines) and the chain (upper lever and club), while the downswing sequence exhibited a dominant separation between the body and the club. The downswing angular velocity peaks of the body (hip/shoulder lines and upper lever) typically occur within a narrow time window.

- Higher normalized clubhead speed is associated with faster backward rotations (hip/shoulder lines and upper lever). Higher normalized clubhead speed is also associated with faster downward rotations (hip/shoulder lines and upper lever), more slowdowns after reaching the peak (hip line and upper lever), and faster rotations at BI (shoulder line and upper lever). Faster backward rotations are associated with faster downward rotations (hip/shoulder lines).

- While larger/faster backward and downward rotations of the axle-chain system on the functional swing plane are the overarching pre-requisites of a high clubhead speed, the correlation and kinematic sequence profiles fundamentally highlight two key aspects of a biomechanically robust golf swing: body-driven active (fast) backswing and active/elongated backswing-to-downswing transition.

Author Contributions: Conceptualization, M.V.M., K.A.B. and Y.-H.K.; methodology, all; software, Y.-H.K.; validation, M.V.M. and Y.-H.K.; formal analysis, M.V.M., M.A.A., N.A.L. and N.J.T.; investigation, M.V.M., M.A.A., N.A.L. and N.J.T.; resources, M.V.M. and Y.-H.K.; data curation, M.V.M., M.A.A., N.A.L. and N.J.T.; writing—original draft preparation, M.V.M. and Y.-H.K.; writing—review and editing, all; supervision, K.A.B. and Y.-H.K.; project administration, M.V.M. and Y.-H.K. All authors have read and agreed to the published version of the manuscript.

Funding: This research received no external funding.

Acknowledgments: The authors would like to thank Qualisys AB for providing an optical motion capture system.

Conflicts of Interest: The authors declare no conflict of interest.

References

1. Han, K.H.; Como, C.; Kim, J.; Lee, S.; Kim, J.; Kim, D.K.; Kwon, Y.-H. Effects of the golfer-ground interaction on clubhead speed in skilled male golfers. *Sports Biomech.* **2019**, *18*, 115–134. [CrossRef]
2. Han, K.H.; Como, C.; Kim, J.; Hung, C.-J.; Hasan, M.; Kwon, Y.-H. Effects of pelvis-shoulders torsional separation style on kinematic sequence in golf driving. *Sports Biomech.* **2019**, *18*, 663–685. [CrossRef]
3. Kwon, Y.-H.; Han, K.H.; Como, C.; Lee, S.; Singhal, K. Validity of the X-factor computation methods and relationship between the X-factor parameters and clubhead velocity in skilled golfers. *Sports Biomech.* **2013**, *12*, 231–246. [CrossRef]
4. Kwon, Y.-H.; Como, C.S.; Han, K.; Lee, S.; Singhal, K. Assessment of planarity of the golf swing based on the functional swing plane of the clubhead and motion planes of the body points. *Sports Biomech.* **2012**, *11*, 127–148. [CrossRef]
5. Putnam, C.A. Sequential motions of body segments in striking and throwing skills: Descriptions and explanations. *J. Biomech.* **1993**, *26*, 125–135. [CrossRef]
6. Marshall, R.N.; Elliott, B.C. Long-axis rotation: The missing link in proximal-to-distal segmental sequencing. *J. Sports Sci.* **2000**, *18*, 247–254. [CrossRef]
7. Fuchs, P.X.; Lindinger, S.J.; Schwameder, H. Kinematic analysis of proximal-to-distal and simultaneous motion sequencing of straight punches. *Sports Biomech.* **2018**, *17*, 512–530. [CrossRef]
8. Wagner, H.; Pfusterschmied, J.; Duvillard, S.P.V.; Müller, E. Skill-dependent proximal-to-distal sequence in team-handball throwing. *J. Sports Sci.* **2012**, *30*, 21–29. [CrossRef]
9. Hellstrom, J. Competitive golf: A review of the relationships between playing results, technique and physique. *Sports Med.* **2009**, *39*, 723–741. [CrossRef]
10. Joyce, C.; Burnett, A.; Cochrane, J.; Ball, K. Three-dimensional trunk kinematics in golf: Between-club differences and relationships to clubhead speed. *Sports Biomech.* **2013**, *12*, 108–120. [CrossRef]
11. Lynn, S.K.; Frazier, B.S.; New, K.N.; Wu, W.F.W.; Cheetham, P.J.; Noffal, G.J. Rotational kinematics of the pelvis during the golf swing: Skill level differences and relationship to club and ball impact conditions. *Int. J. Golf Sci.* **2013**, *2*, 116–125. [CrossRef]
12. Vena, A.; Budney, D.; Forest, T.; Carey, J. Three-dimensional kinematic analysis of the golf swing using instantaneous screw axis theory, Part 2: Golf swing kinematic sequence. *Sports Eng.* **2011**, *13*, 125–133. [CrossRef]
13. Zheng, N.; Barrentine, S.W.; Fleisig, G.S.; Andrews, J.R. Kinematic analysis of swing in pro and amateur golfers. *Int. J. Sports Med.* **2008**, *29*, 487–493. [CrossRef] [PubMed]
14. Cheetham, P.J.; Rose, G.A.; Hinrichs, R.N.; Neal, R.J.; Mottram, R.E.; Hurrion, P.D.; Vint, P.F. Comparison of kinematic sequence parameters between amateur and professional golfers. In *Science and Golf V: Proceedings of the World Scientific Congress of Golf*; LPGA Foundation: Phoenix, AZ, USA, 2008.
15. Horan, S.A.; Kavanagh, J.J. The control of upper body segment speed and velocity during the golf swing. *Sports Biomech.* **2012**, *11*, 165–174. [CrossRef] [PubMed]
16. Tinmark, F.; Hellström, J.; Halvorsen, K.; Thorstensson, A. Elite golfers' kinematic sequence in full-swing and partial-swing shots. *Sports Biomech.* **2010**, *9*, 236–244. [CrossRef]
17. Burden, A.M.; Grimshaw, P.N.; Wallace, E.S. Hip and shoulder rotations during the golf swing of sub-10 handicap players. *J. Sports Sci.* **1998**, *16*, 165–176. [CrossRef]
18. Okuda, I.; Gribble, P.; Armstrong, C. Trunk rotation and weight transfer patterns between skilled and low skilled golfers. *J. Sports Sci. Med.* **2010**, *9*, 127–133.
19. Egret, C.; Dujardin, F.; Weber, J.; Chollet, D. 3-D kinematic analysis of the golf swings of expert and experienced golfers. *J. Hum. Motion Mov. Stud.* **2004**, *47*, 193–204.
20. Chu, Y.; Sell, T.C.; Lephart, S.M. The relationship between biomechanical variables and driving performance during the golf swing. *J. Sports Sci.* **2010**, *28*, 1251–1259. [CrossRef]
21. Nesbit, S.M. A three dimensional kinematic and kinetic study of the golf swing. *J. Sports Sci. Med.* **2005**, *4*, 499–519.
22. Sprigings, E.J.; Mackenzie, S.J. Examining the delayed release in the golf swing using computer simulation. *Sports Eng.* **2002**, *5*, 23–32. [CrossRef]
23. Hume, P.A.; Keogh, J.; Reid, D. The role of biomechanics in maximising distance and accuracy of golf shots. *Sports Med.* **2005**, *35*, 429–449. [CrossRef] [PubMed]

24. Cheetham, P.J.; Martin, P.E.; Mottram, R.E.; St. Laurent, B.F. The importance of stretching the 'X-Factor' in the downswing of golf: The 'X-Factor stretch'. In *Optimising Performance in Golf*; Thomas, P.R., Ed.; Australian Academic Press: Brisbane, QLD, Australia, 2001; pp. 192–199.
25. McLean, J. Widen the gap. *Golf Mag.* **1992**, *1992*, 49–53.
26. Cole, M.; Grimshaw, P. The X-factor and its relationship to golfing performance. *J. Quant. Anal. Sports* **2009**, *5*. [CrossRef]
27. McTeigue, M.; Lamb, S.R.; Mottram, R.; Pirozzolo, F. Spine and hip motion analysis during the golf swing. In *Science and Golf II: Proceedings of the World Scientific Congress of Golf*; Cochran, A.J., Farrally, M.R., Eds.; E & FN Spon: London, UK, 1994; pp. 50–58.
28. Myers, J.; Lephart, S.; Tsai, Y.S.; Sell, T.; Smoliga, J.; Jolly, J. The role of upper torso and pelvis rotation in driving performance during the golf swing. *J. Sports Sci.* **2008**, *26*, 181–188. [CrossRef]
29. Horan, S.A.; Evans, K.; Morris, M.R.; Kavanagh, J.J. Thorax and pelvis kinematics during the downswing of male and female skilled golfers. *J. Biomech.* **2010**, *43*, 1456–1462. [CrossRef]
30. Meister, D.W.; Ladd, A.L.; Butler, E.E.; Zhao, B.; Rogers, A.P.; Ray, C.J.; Rose, J. Rotational biomechanics of the elite golf swing: Benchmarks for amateurs. *J. Appl. Biomech.* **2011**, *27*, 242–251. [CrossRef]
31. Lephart, S.M.; Smoliga, J.M.; Myers, J.B.; Sell, T.C.; Tsai, Y.S. An eight-week golf-specific exercise program improves physical characteristics, swing mechanics, and golf performance in recreational golfers. *J. Strength Cond. Res.* **2007**, *21*, 860–869. [CrossRef]
32. Winter, D.A. *Biomechanics and Motor Control of Human Movement*, 3rd ed.; John Wiley & Sons: Hoboken, NJ, USA, 2005.
33. Bell, A.L.; Pedersen, D.R.; Brand, R.A. A comparison of the accuracy of several hip center location prediction methods. *J. Biomech.* **1990**, *23*, 617–621. [CrossRef]
34. Lee, D.K. Alternatives to P value: Confidence interval and effect size. *Korean J. Anesthesiol.* **2016**, *69*, 555–562. [CrossRef]
35. Ball, K.A.; Best, R.J. Different centre of pressure patterns within the golf stroke I: Cluster analysis. *J. Sports Sci.* **2007**, *25*, 757–770. [CrossRef] [PubMed]

Article

Validation and Application of Two New Core Stability Tests in Professional Football

Saioa Etxaleku [1], Mikel Izquierdo [2,*], Eder Bikandi [3], Jaime García Arroyo [4], Iñigo Sarriegi [5], Iosu Sesma [1] and Igor Setuain [1,2,*]

[1] TDN, Orthopedic Surgery and Advanced Rehabilitation Center, Clinical Research Department, Pol Mutilva C/V nave 3, 31192 Mutilva, Spain; setxaleku@tdnclinica.es (S.E.); sesmanoble@hotmail.com (I.S.)
[2] Navarrabiomed, Complejo Hospitalario de Navarra (CHN)-Universidad Pública de Navarra (UPNA), Navarra Institute for Health Research (IdiSNA), 31008 Pamplona, Spain
[3] Athletic Club, Mazarredo Zumarkalea, 23. 48009 Bilbao, Bizkaia, Spain; ederbikandi@gmail.com
[4] TEN Centre, C/San Julian nº 8 Majadahonda, 28220 Madrid, Spain; jaimegarroyo@gmail.com
[5] Southampton FC, Staplewood Football Development & Support Centre, Marchwood, Southampton SO40 4WR, UK; isarriegui@saintsfc.co.uk
* Correspondence: mikel.izquierdo@gmail.com (M.I.); igorsetuain@gmail.com (I.S.); Tel.: +34-948-417876 (I.S.)

Received: 1 July 2020; Accepted: 4 August 2020; Published: 8 August 2020

Abstract: The purpose of the first study was to validate two newly proposed core stability tests; Prone Plank test (PPT) and Closed Kinetic Chain test (CCT), for evaluating the strength of the body core. Subsequently, these tests were employed in a longitudinal prospective study implementing a core stability training program with a professional Spanish football team. For the validation study, 22 physically active men (*Tegner Scale 6–7*) performed three trials of the PPT and CCT tests in two different testing sessions separated by one week. In the longitudinal study, 13 male professional football players were equally evaluated (PPT and CCT) before and after the competitive session in which they completed a core training program. Intra-/intersession, and intertester, reliability was analyzed. PPT and CCT demonstrated excellent to good test–retest reliability and acceptable error measurement (ICCs for intratester and intrasession reliability ranged from 0.77 to 0.94 for the PPT, and 0.8–0.9 for the CCT) in all but one of the testing conditions (female tester for CCT test; ICC = 0.38). Significant improvements on core strength were found from pre to post evaluation in both the PPT ($p < 0.01$) and CCT ($p < 0.01$) after the implementation of a core training program in professional football players.

Keywords: core function; football; field test; reliability

1. Introduction

The "Core musculature" term refers to the combination of muscles that surround and comprise the lumbopelvic region, and act synergistically to stabilize the trunk and hip, contributing to the movement control and stability of more distal joints [1]. On the other hand, core stability, defined as the ability to control the position and motion of the trunk over the pelvis, governs optimal force production, transfer and application to more distal segments in many functional athletic activities [2].

The core muscles are considered to play a key role in sport performance as they seem to provide "proximal stability for distal mobility" [1,3] throughout the entire kinetic chain. Moreover, core stabilization precedes gross motor movement, as the central nervous system activates the trunk musculature before limb movement to provide the stability and stiffness of the trunk and pelvis in anticipation of the forces produced at this level [3].

"Core stability" is accomplished through a complex interaction of neuromuscular coordination, proprioception, strength and endurance of the trunk and hip musculature. This inherent complexity has been the focus of a wide range of studies employing different measurement techniques for core stability evaluation in recent years [3]. In this way, there is a lack in the literature of a gold standard for core stability evaluation in sports. Therefore, the assessment of trunk stability in the field setting could become even more complex because it requires the combination of different variable measurements, such as force, endurance, etc. [4,5].

Considering this multi-characteristic approach, a wide variety of testing methods have been described to evaluate core stability performance, both in laboratory and in clinical settings. In the laboratory, the most common testing method to measure core strength has been developed by the use of isokinetic dynamometers, whereas trunk stability tests have been performed through the employment of custom-designed devices [6,7]. Beside these expensive and time-consuming tests, "clinical tests" are the preferred choice for evaluating core stability for "in the field conditions".

No gold standard for core stability evaluation has been established, until recently. These tests have also involved vague or unclear scoring systems and require expensive instrumentations or the development of individually customized equipment. The definition of a gold standard test for core stability evaluation has become challenging, as there is not a widely accepted conclusive definition for core stability. In fact, there are numerous muscles that participate in the body core's dynamic stabilization. Several reliable and easy to use methods have been proposed to assess core musculature function. These include the prone bridging test, the Biering–Sørensen test, the double leg lowering test, prone plank stability], the side plank bridging test], the front abdominal power test (FAPT), the side abdominal power test (SAPT) [3,8–12] and the flexion-rotation test [13]. Some of the limitations of these tests are as follows: (a) they isolate one aspect of this complex interaction, such as core strength or endurance; (b) they are linked to a subjective scoring or evaluation system; and (c) they asses a more complete picture of this interaction, but not in a functionally relevant position, aligned with the sport-specific core region's biomechanical and or functional requirements [3]. Furthermore, these tests have also reported vague or unclear scoring systems and/or require expensive equipment, or they require the development of individually customized measurement devices which limit the extrapolation of the results obtained to other populations.

The interaction among the lumbopelvic and hip muscles makes it difficult for researchers to develop a single test that could encompass all muscles and structures [12]. Considering the abovementioned methodological limitations, there is a clear need to develop a quick, simple, valid and reliable test to accurately evaluate the core stability of an athlete, due to this capacity's proven strong relationship with sport performance as well as with low back and lower limb injury risk [2,14–17].

In the present research, we aimed to describe two functional core stability tests that aimed to evaluate the athlete's ability to produce isometric force, from two core stability-challenging positions, through the use of a hand-held dynamometer. Secondly, we evaluated a professional football team before and after a standardized core training program in order to elucidate if a motor control-based progression would have any effect on the player's ability to produce isometric force in the previously validated tests.

2. Methods

2.1. Experimental Approach to the Problem

This investigation consisted of two consecutively performed studies.

Study I: (Cross-sectional validation study)

A preliminary validation study of the two new core functional strength tests was performed. The core function testing positions were derived from traditional core-conditioning exercises that address both closed [2] (Closed Kinetic Chain test; CCT) and open kinetic chain efforts [12] (Prone Plank Test; PPT) (Figure 1A and 1B respectively). These two tests aimed to assess the force exertion capacity in those two positions with a hand-held dynamometry. The intra- and intertester reliability of the measurements was evaluated for both tests.

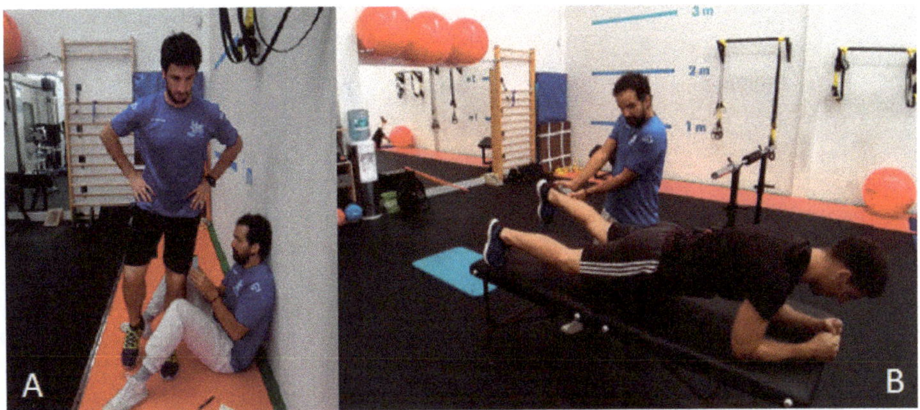

Figure 1. New proposed core functional strength field tests. (**A**) Closed Kinetic Chain Test (CCT). (**B**) Prone Plank Test (PPT).

Study II: (Longitudinal study)

Once the testing procedures were shown to be reliable, a prospective longitudinal study was performed aiming to analyze the core functional strength values in relation to the implementation of a core neuromuscular training program during the competitive season of a professional football team squad (Spanish league 2).

2.2. Participants

Study I: (Cross-sectional validation study)

Seven recreationally active males (6–7 on the Tegner scale) (mean ± standard deviation 19.23 ± 2.43 years old, 181.0 ± 6.23 cm, 75.42 ± 7.25 kg) were recruited for the between-session reliability study. Then, the sample was increased by adding 15 male junior football players competing in Spain's National Junior League (mean ± standard deviation 17.87 ± 1.13 years old, 181.20 ± 6.84 cm, 74.69 ± 6.63 kg). All participants were free from any previous spine or lower leg injury at the time of examination. Participants were instructed not to perform any kind of specific core stability training during the week between the two testing sessions.

Study II: (Longitudinal intervention study)

In total, 20 male professional football players competing in Spain's National Second League (mean ± standard deviation age 27.7 ± 5.1 years old, height 180.9 ± 6.9 cm, weight 76.5 ± 5.4 kg) were included in the longitudinal study.

2.3. Procedures

Two hand-held dynamometers (HOGGAN Health microFET3 Combo Manual Muscle Tester & Digital Inclinometer, United States) were used for data recording. Each tester employed the same device during the entire evaluation process.

Closed Kinetic Chain test (CCT)

The participant was placed in a standing position with the trunk positioned at 50° of flexion, knee positioned at 20° of flexion (0° of abduction/rotation of the knee) and with both hands placed at the iliac crest level. The tester was positioned ipsilateral to the testing leg and the HHD was placed just superior to the lateral femoral condyle. Then, the participant exerted a maximal isometric contraction toward hip abduction and external rotation, "squeezing their glutes" against the HHD which was firmly fixed by the examiner. The applied force had to be maintained for 3 s and 3 attempts were registered for each leg. (Figure 1A). No compensation at the trunk level was allowed during the execution of the test.

Prone Plank Test (PPT)

The participant was placed in a prone position with the ankles positioned at 0° of ankle dorsal flexion. They were instructed to hold their pelvis in a parallel position with respect to the stretcher and to move the testing leg into 20° of hip extension and abduction, with the knee in an extended position. The tester was positioned ipsilateral to the testing leg, and the dynamometer was placed superior to the external malleolus. Then, the participants exerted a maximal isometric contraction toward hip extension and hip abduction against the hand-held dynamometer (HHD) which was firmly fixed by the examiner. Participants had to maintain the generating force for 3 s and 3 attempts were registered for each leg. To ensure appropriate direction of force applied by the participant, they were asked to perform the task with the following instructions: "imagine you had a watch at the bottom of your feet at which you should have to tag the five or ten o'clock hours on it", when testing the right and left legs respectively (Figure 1B). No pelvis compensation was allowed during the execution of the test.

Study I: (Cross-sectional study)

Two test sessions were performed 1 week apart. No physical activity program focusing on the core muscles between the testing sessions was allowed for the participants.

All participants were evaluated by two testers (S.E. and E.B.) following the same testing protocol. Both testers were physical therapists and they had not been familiarized with the procedure of the new proposed core strength tests. Before each testing session, both testers were instructed about the testing procedure and were familiarized with the dynamometer by an expert instructor (I.S). One familiarization trial was followed by three testing measurements for each leg and test type. The two highest recordings were gathered for statistical analysis. If the difference between the trials was higher than 10%, another trial was performed.

Study II: (longitudinal study)

A 42 weeks standardized core stability training program was performed, divided into three different stages with progressively increasing stabilization demands (Figure 2). The difficulty was increased after every 14 weeks. The players performed 28 core program training sessions at each stage and were evaluated at the beginning and at the end of the whole standardized exercise program.

All participants were tested by the same evaluator (I.S) following the same testing protocol described in detail in the validation study. At the follow-up evaluation, 7 players missed the testing session due to either injury or national team recruitment for international call ups, hence the test–retest comparison was done with a sample of 13 football players.

The compliance with the core training program was recorded at every core training session by the athletic trainer of the football team (J.S), using a data log model generated by the authors.

All participants provided informed consent before participation. The football team's medical staff's permission was requested in study II. The study was approved by the ethic committee of the Public University of Navarra and was performed in accordance with the Declaration of Helsinki for longitudinal studies with human beings.

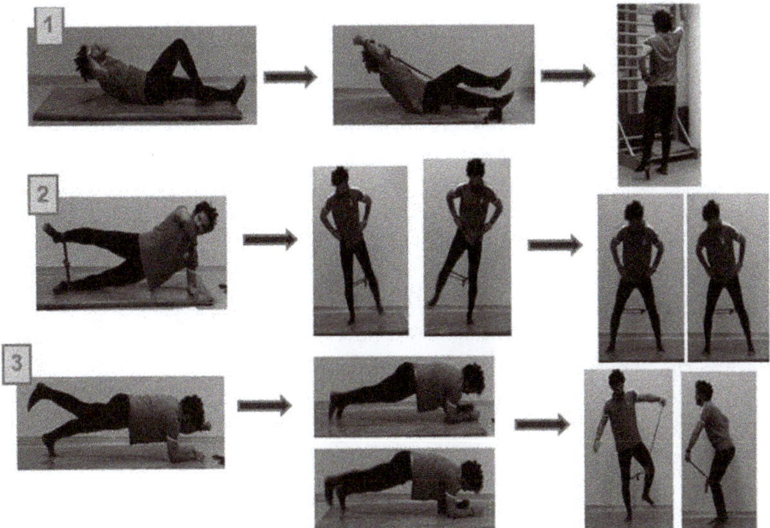

Figure 2. Core stability training program divided into 3 different training phases. The first column of the figure represents the 3 core exercises. The second and third columns exemplify the progression of each exercise, adding motor control difficulty, in ascending order.

2.4. Statistical Analysis

The SPSS statistical package (version 17.0.0.236; SPSS Inc. Chicago, IL, USA) was used to perform all the statistical analyses. All analyses assumed statistical significance at $p \leq 0.05$.

Study I: (Cross-sectional study)

A total of 22 participants were evaluated for the intrasession, and intra- and inter-rater reliability analysis. Furthermore, a subset of seven subjects was randomly selected from the initial cohort and they repeated the whole examination at seven days for the test–retest reliability calculations.

Descriptive statistics, including mean peak torque values with standard deviation, coefficient of variation (%) (expressed as the SD of the mean, multiplied by 100% and expressed as a percentage) and confidence intervals (95%), were calculated. The peak isometric force values obtained from each evaluator during the PPT and CCT were gathered for intra-rater (trial 1, trial 2), inter-rater (male tester, female tester) and intra-rater intersession (assessment day 1/day 2). Afterwards, a Paired *t*-test comparison and the corresponding non parametrical Wilcoxon Signed-rank test were run to primarily test whether the two sets of scores were significantly different from each other, before their test–retest reliability was analyzed [18].

The data was graphically examined using Bland and Altman [19] plots in, which the difference between test sessions was plotted against the mean of the two test sessions. Mean differences between test sessions with 95% CIs were calculated to determine whether there was any systematic bias. The relative reliability of the data was determined using a two-way random effect model intraclass correlation coefficient (2, k). Before calculating absolute reliability, Bland and Altman plots were generated to examine the correlation (R^2) between the absolute differences and the mean values for each test, in order to detect the presence or absence of heteroscedasticity in the data [19]. If the random error increases as the measured values increase, the data is considered heteroscedastic. In contrast, when there is no relation between the error and the size of the measured value, the data is described as homoscedastic. With homoscedastic errors, providing that they are also normally distributed, the raw data can be analyzed with conventional parametric analyses, but heteroscedastic data should be transformed logarithmically before analysis, or investigated via an analysis based on ratio statistics. This comparison method is widely used to verify the reliability and reproducibility of the measurement, rather than the correlation coefficient (r) and regression techniques, which are inadequate and can be very misleading when assessing agreement, as they evaluate only the linear association of two sets of observations. With this graph, evaluation of the magnitude of the disagreement, identification of outliers and the observation of any bias are easily achieved [20]. Due to the simplicity and potential for the identification of pairs of observations whose differences reach beyond clinical tolerances, the Bland and Altman approach is currently the preferred approach to evaluate agreement between two measurements [19].

If R^2 was between 0 and 0.1, the data were considered homoscedastic; consequently, the absolute reliability was assessed by determining the SEM, which was calculated using the square root of the mean square error from the analysis of variance model, as recommended by Atkinson and Nevill [21]. If R^2 was greater than 0.1, the data was considered heteroscedastic; consequently, absolute reliability was determined by calculating the CV. When applicable, the CV was calculated for each subject by dividing the standard deviation of the two testing sessions by the mean of the two testing sessions ×100 (for intersession CV calculation), and the standard deviation of two testers by the mean of the two testers ×100 (for intrasession CV calculation).

Both the intra-rater (rater 1 and 2, between 2 days) and the inter-rater reliability (between rater 1 and 2, the same testing day) peak isometric force values were assessed using intraclass correlation coefficients (ICC) with 95% confidence intervals (CI) in order to assess relative reliability by utilizing a two-way random effects model with single measure reliability (ICC (2,1)). The mean value from each participant was used for the analysis. Values < 0.50 represented poor reliability, values > 0.50 and < 0.75 indicated fair to good reliability, and values > 0.75 marked excellent reliability [20]. Given that ICC values may be influenced by the inter-subject variability of scores, the standard error of measurement (SEM) was reported in conjunction with the ICCs. A large ICC may represent poor trial-to-trial consistency if the inter-participant variability is too high [18], in contrast to SEM, which is not affected by inter-subject variability [18]. SEM was calculated using the equation $SEM = SD \times (\sqrt{1 - ICC})$, where SD corresponds to the sample standard deviation [17,20–22] for intra- and inter-rater reliability analysis. In the case of the intersession reliability analysis, the SD of the T2 data set class was utilized, assuming that between the two assessment sessions, T2 generated lower SD [21]. The error in an individual's score at one point in time was estimated by multiplying the SEM by the z value for the 90% confidence level (z value = 1.65) [23–29]. This value was then multiplied by the square root of 2 (to account for the measurement error in two test sessions) in order to estimate the minimum detectable change (MDC) at the 90% confidence level.

Study II: (Longitudinal study)

A preliminary effect size calculation was performed to determine the adequate sample size required to account for a random error of 0.8. A total of 27 participants would have been necessary, but finally 20 participants were recruited (an entire second Spanish league football club). After that, only 13 players were included in the statistical analysis, because 7 players did not complete the testing session after the core training program due to either injury or national team recruitment for international match play. The compliance of the core training program was 82%, corresponding to 69 core training sessions accomplished out of the 84 total sessions established at the beginning of the intervention.

After checking the assumptions of parametric statistics, the mean comparisons before and after the implementation of the core training program were analyzed using a Paired Sample *t*-test. The estimation of MDC at a 90% confidence level was also calculated to determine whether the observed change in score between the pre- and post-testing evaluations was truly a change in patient status, or was merely associated with an error (noise) in the measurement process [17].

3. Results

Study I: (Cross-sectional study)

Bland and Altman plots for intrasession intratester and intratester intersession reliability (Figure 3) demonstrated no systematic biases between trials and testing days. In relation to intrasession and intertester reliability, the PPT test values were grouped together. Alternatively, in the case of the CCT test values, a systematic bias seemed to occur, given the fact that the higher the force values reported from the participants, the greater the differences encountered between testers.

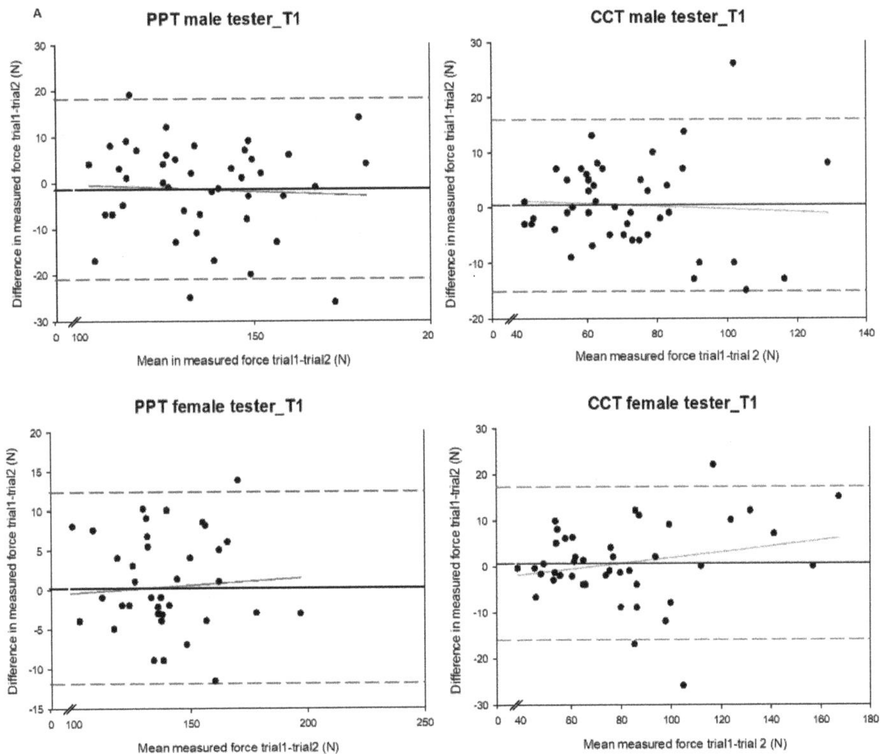

Figure 3. *Cont.*

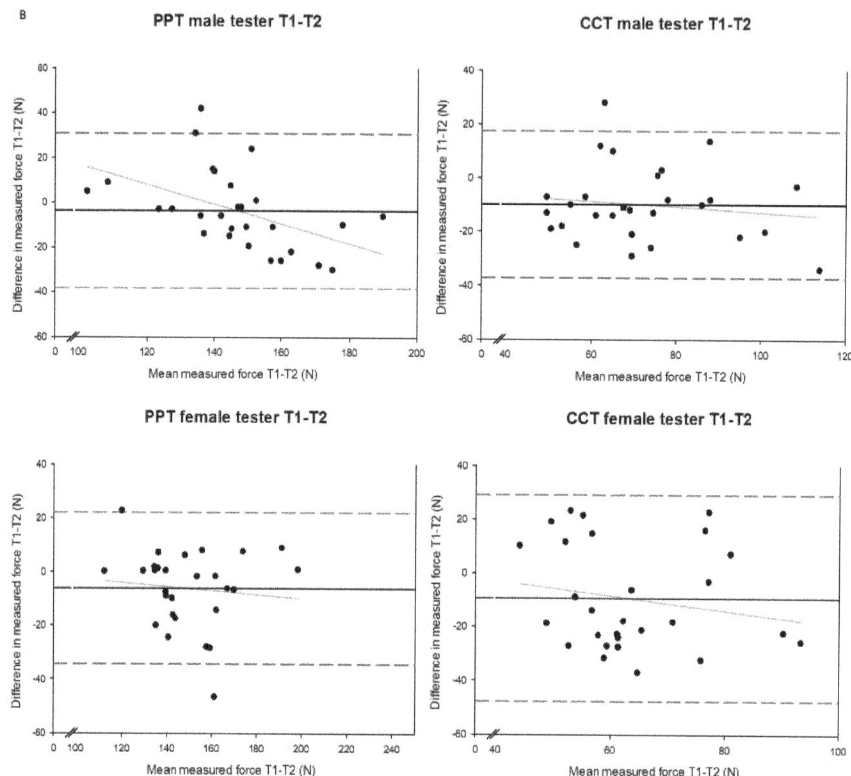

Figure 3. Bland and Altman plots representing mean differences and 95% limits of agreement. (**A**) Between trial 1 and trial 2 for both testers in the first testing session (T1). (**B**) Between first (T1) and second (T2) testing sessions for both testers.

Descriptive data from both testers for the intrasession analysis of the first testing session (T1) are presented in Table 1. Looking at between-trial variability, the mean CV (%) values for both core strength tests for the male tester (PPT = 4.11%; CCT = 5.79%) and the female tester (PPT = 2.73%; CCT = 5.05%) were below 10%, indicating low test–retest variation[22].

Table 1. Descriptive data PPT and CCC test (T1). N = Newton; Δ = mean difference; CV = coefficient of variation; CI = confidence intervals.

	PRONE PLANK TEST					CLOSED KINETIC CHAIN TEST				
	Mean Peak Torque N Trial 1 (±SD)	Mean Peak Torque N Trial 2 (±SD)	Δ (N)	CV (%)	95% CI	Mean Peak Torque N Trial 1 (±SD)	Mean Peak Torque N Trial 2 (±SD)	Δ (N)	CV (%)	95% CI
Tester Male	133.91 (±20.89)	135.31 (±21.49)	+1.4	4.1	128.52–140.69	71.56 (±16.67)	71.14 (±20.16)	−0.42	+5.8	65.58–77.12
Tester Female	133.87 (±26.86)	133.61 (±26.36)	−0.26	2.7	125.93–141.55	80.50 (±31.74)	79.84 (±29.86)	−0.66	+5.0	71.15–89.19

Regarding the intratester and intrasession reliability analysis, the data showed excellent reliability for both the core strength tests for the male tester (PPT = 0.94; CCT = 0.96) and the female tester (PPT = 0.99; CCT = 0.81) (Table 2).

Table 2. Reliability results (N = 22). N = Newton; SD = standard deviation; ICC = intraclass correlation coefficient; SEM = standard error of measurement; MDC_{90} = minimal detectable change 90% confidence level.

	PRONE PLANK TEST				CLOSED KINETIC CHAIN TEST			
	Mean Peak Torque N ($\pm$SD)	ICC (95% CI)	SEM (N)	MDC90 (N)	Mean Peak Torque N ($\pm$SD)	ICC (95% CI)	SEM (N)	MDC90 (N)
Tester Male	134.61 ($\pm$20.59)	0.94 (0.89–0.97)	5.12	11.95	71.35 ($\pm$19.52)	0.96 (0.93–0.98)	4.0	9.4
Tester Female	133.74 ($\pm$26.43)	0.99	3.02	7.04	80.17 ($\pm$30.52)	0.81 (0.68–0.89)	13.5	31.5

In relation to intratester and intersession reliability analysis, the data exhibited excellent reliability for both core strength tests for the male tester (PPT = 0.77; CCT = 0.94) and the female tester for PPT (ICC = 0.84), but poor reliability for the female tester in the case of the CCT (ICC = 0.38). These results are shown in Table 3. The intertester and intrasession reliability analysis is represented in Table 4. For the PPT test, the data exhibit fair to good reliability (ICC = 0.62) and poor reliability for the CCT test (ICC = 0.47). The SEMs and MDC_{90} for each reliability analysis are expressed in conjunction with the ICCs in Tables 2–4.

Table 3. Intratester intersession reliability results (T1) for both tester (N = 22). N = Newton; SD = standard deviation; ICC = intraclass correlation coefficient; SEM = standard error of measurement; MDC_{90} = minimal detectable change 90% confidence level.

	Mean Peak Torque N T1 ($\pm$SD)		Mean Peak Torque N T2 ($\pm$SD)		Δ T2-T1 (N)		ICC (95% CI)		SEM (N)		MDC_{90} (N)	
	Male Tester	Female Tester	Male Tester	Female Tester	Male Tester	Female Tester	Male Tester	Female Tester	Male Tester	Female Tester	Male Tester	Female Tester
Prone plank test	144.20 ($\pm$17.01)	145.44 ($\pm$21.48)	147.79 ($\pm$25.67)	150.34 ($\pm$21.00)	+3.59	+4.9	0.77 (0.28–0.93)	0.94 (0.79–0.98)	12.4	5.4	28.9	16.69
Closed kinetic chain test	66.13 ($\pm$18.05)	58.36 ($\pm$14.68)	73.89 ($\pm$17.83)	67.90 ($\pm$17.03)	−7.77	+9.54	0.84 (0.44–0.95)	0.38 (−0.59–0.78)	7.2	13.5	12.5	31.39

Table 4. Intertester intrasession reliability results (T1) for both test (N = 22). N = Newton; SD = standard deviation; ICC = intraclass correlation coefficient; SEM = standard error of measurement; MDC_{90} = minimal detectable change 90% confidence level.

	Rater Male Mean Peak Torque N ($\pm$SD)	Rater Female Mean Peak Torque N ($\pm$SD)	ICC (95% CI)	SEM (N)	MDC_{90} (N)
Prone plank test	134.61 ($\pm$20.59)	71.35 ($\pm$19.52)	0.62 (0.30–0.8)	14.47	33.76
Closed kinetic chain test	133.74 ($\pm$26.43)	80.17 ($\pm$30.52)	0.47 (0.06–0.70)	18.8	43.88

Study II: (Longitudinal study)

There was a statistically significant increase with respect to peak muscle force exertion between the two testing sessions, for both the PPT ($p < 0.00$) and CCT ($p < 0.00$) (Figure 4). Furthermore, large effect size values (Cohen's D), of 0.9 and 1.0, were reported for PPT and CCT, respectively (Table 5).

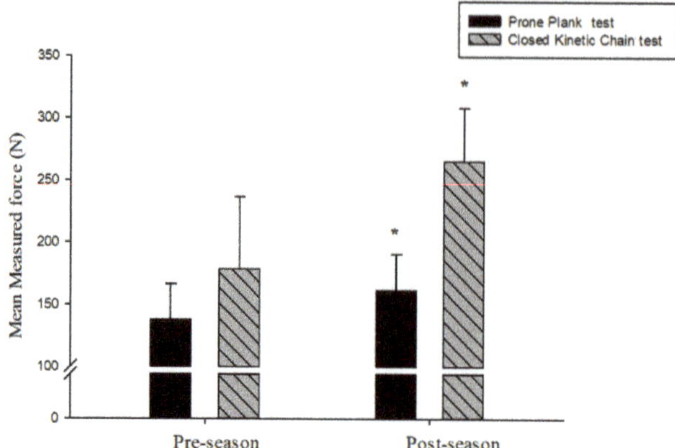

Figure 4. Pre-/post HHD values for PPT and CCT tests. $p < 0.05$. (*) denotes statistically significance between pre to post evaluations.

Table 5. Compared means for pre-season (T1) and post-season (T2) values for the two core strength test (N = 13). N = Newton; SD = standard deviation; Cohen's d = effect size; MDC$_{90}$ = minimal detectable change 90% confidence level.

	Mean Peak Torque N (±SD)		Δ T2-T1 (%)	D	*t*-Test (*p*)	MDC$_{90}$ (N)
	Pre-Season (T1)	Post-Season (T2)				
Prone plank test	137.87 (±28.13)	160.73 (±27.85)	+22.86	0.9	0.00	47.71
Closed kinetic chain test	184.64 (±56.10)	260.92 (±43.61)	+76.28	1.0	0.00	93.29

4. Discussion

Study I: (validation study)

The purpose of the validation study was to determine the validity and reliability of two new proposed core strength field tests, using an HHD to evaluate the ability to produce force in challenging positions, representing core stability. The observed ICCs in this study suggest the excellent test–retest reliability of the Prone Plank Test (PPT) and the Closed Kinetic Chain Test (CCT) (see Tables 3 and 4). The ICC is a measure of relative reliability indicating a measure's ability to discriminate between existing differences among two sets of data. It is calculated by dividing the between-patient variance by the total variance [17] (between-patient and within-patient variance). The present findings therefore suggest that the PPT and CCT demonstrated appropriate reliability for their use in the training field setting as a core strength functional evaluation.

The SEM provides an absolute measure of reliability with regard to the original test units, and can be used to estimate the error associated with an individual patient's test score [17,26]. The smaller the SEM value, the more reliable the measurements [17,20]. In our study, the SEM ranged from 4.01 N to 13.50 N for intratester intrasession and intersession reliability analyses. The results found in our study correlate with a similar study accomplished by De Blaiser et al. [10], in relation to intratester and intrasession reliability, who evaluated maximal isometric trunk flexion and extension strength tests using HHD on a healthy athletic population. They reported excellent intratester and intertester reliability values (ICC = 0.76–0.93) for all tests, except for the intratester reliability of the maximal isometric trunk flexion test starting at 0°, which was good (ICC = 0.67). The ICCs for the intertester and intrasession analysis in our study were lower than what they had found, probably because the testers from their study had similar physical characteristics, in order to counteract the participant's force exertion. Another study conducted by Moreland et al. [30] determined the intertester reliability of six tests of trunk muscle strength evaluation using practical measurement methods which could be applied in most clinical or work settings. For abdominal and back extensor isometric force tests using HHD, they reported poor (0.24 and 0.25, respectively) ICC values. These results contrast with those reported in our study, maybe because the rater's ability to apply force could have been limited, as they argued. In a more recent study conducted by Recio et al. [7], a flexion–extension isokinetic protocol was used to asses trunk muscle strength and endurance in a healthy and physically active population (1–3 h of moderate physical activity in recreational sports, 1–3 days a week). The ICC values obtained for the strength variables in this study ranged from 0.57 to 0.77 in extension, and from 0.62 to 0.84 in flexion, with low SEM values. Another study designed by Essendrop et al. [31] evaluated the intratester reliability of isometric strength tests by using a HHD device in a workplace environment, and found excellent ICCs for back flexion and extension tests (0.97–0.93, respectively).

Two distinct methodologies have been traditionally used in sport science for strength evaluation: isokinetic testing or isometric HHD [19]. Isokinetic dynamometers are computerized machines able to provide different variables of muscle strength (i.e., peak force, endurance, power, optimum angle of torque production, etc.). Isokinetic dynamometry is considered the "gold standard" for trunk muscle strength evaluation, mainly because it allows a controlled and accurate assessment of a large number of muscle force parameters, but the high technology requirements, as well as the time consumption, of these isokinetic protocols, along with their substantial economic cost, could limit the generalization of its use.

Alternatively, hand-held dynamometry (HHD) strength evaluation has been shown to be a reliable, objective way to obtain strength measurements [23–26]. HHD is a tool that can be placed between the practitioner's hand and the tested body part of the athlete. Stark et al. [22] recently demonstrated, in a systematic review, the moderate-to-good reliability and validity scores between the HHD vs. isokinetic testing procedures. Considering the cost of isokinetic testing devices and their logistical limitations for routine clinical testing purposes, the HHD could be considered a practical standard for muscle strength assessment in the training field or clinical setting. These results support those from our study, and enlarge the scientific verification of the use of HHD-based core strength evaluations in the field training arena, or the clinical setting itself, enabling clinicians to routinely perform quantitative and objective core body region strength evaluations on a science-based foundation. In the authors opinion, the employment of previously analyzed and validated tests, with adequate reliability and reproducibility, guarantees the internal methodological quality standards that this kind of in-the-field performed test have to demonstrate in order to ensure the appropriate extrapolation of these evaluation methodologies. Moreover, a robust validation process can also enable different researchers in the development of these routines and result interpretations.

Study II: (longitudinal study)

The purpose of the study was to determine whether these preliminary validated tests (PPT and CCT) would be able to detect changes in the force application capacities of professional football players after the implementation of a motor control-based core stability training program.

We found a significant improvement in the measured core strength values (PPT and CCT tests) after the core stability training program was implemented. The original result obtained from this longitudinal study was that participants were able to produce higher strength values in spite of the fact that the progression criteria were core neuromuscular control challenge-based, instead of being based on traditional resistance progression criteria. In fact, neuromuscular control training for athletes has traditionally been focused on resistance exercises for trunk musculature. However, these types of core training programs could not only lead to excessive applications of mechanical loads to the spine, but they can also fail to satisfy the biomechanical stabilization requirements necessary to counteract the acting internal and external forces during athletic maneuvers [30,31]. Even so, previous studies using non-elite or sub-elite instead of elite subjects have clearly demonstrated the effectiveness of core stabilization programs in improving core strength when compared with no or regular training only [32,33]. Regarding the literature available, it is difficult to find longitudinal studies recruiting elite professional football players due to the inability to have a control group and the high level competitive demand that often challenges this type of scientific effort.

A weak core body region has been linked with alterations in the transfer of energy throughout the whole kinetic chain, potentially resulting in reduced sport performance and increased injury risk [2,15,34]. As a result, the implementation of a core strength training program has become popular in sport and clinical settings as a means to improve performance and reduce injury risk. Several studies have examined the effects of improving core-related strength following a core training program, finding a wide variety of results. Similarly, few studies have analyzed the possible relationship between core strength and sport-performance, concluding that if founded, the correlations were weak to moderate [31,34–36]. One of the main sources that could explain the inconsistency of these results could involve the methodology used to measure the core strength. Most of the studies followed McGill testing protocol-derived methods. These studies, despite demonstrating excellent test–retest reliability, seem to be more related to trunk muscle endurance capacity, rather than being focused on the more functional and faster movements demands that most multi-directional sports seem to require [35].

Furthermore, in the present study, we found that those players displaying a limb symmetry index greater than 15% (described in the literature as the maximum physiological difference) displayed a trend toward the equalization of the observed deficit at the end of the season. In contrast, players reporting more symmetric limb-to-limb values at baseline significatively improved their performance in the tests at the end of the season. Whether this fact could be linked to both injury risk reduction and/or athletic performance enhancement should be appropriately addressed in future studies.

The present study had several limitations. The first is associated with the use of the HHD, which has been demonstrated to reproduce some tester bias as the evaluator has to produce sufficient force to ensure the actual isometric contraction of the participant. As such, inadequate force application by the evaluator may lead to decreased intertester reliability [23,24]. To avoid or diminish this bias, both testing sessions (test 1 and test 2) were conducted by the same tester in the longitudinal study [25]. Another limitation of HHD is that the force recorded by male and female testers is different. The female tester recorded lower values, as the subject could produce higher forces, and the force exerted by the same participant was different in the reports of different examiners [23]. This finding can be verified in the Bland and Altman plots. In relation to the longitudinal study, the a priori effect size calculation showed the necessity of obtaining 27 players for a random error of 0.8. This fact highlighted the fact that the study was slightly underpowered. Furthermore, seven dropouts were registered in the post-intervention evaluations. However, the statistically significant improvements, as well as the large to very large effect sizes achieved from pre- to post-evaluation, may serve as a starting point for future adequately powered studies targeting this issue. Another limitation of this study could be that there was not a control group, but the lack of a control group is common in interventions among professional players, because leaving a player out of an intervention that could be beneficial is not recommended due to potential ethical objections.

Despite these limitations, in the authors' opinion the HHD is easy to use, portable, inexpensive, and requires little training for proficient application. The present validation study revealed that the tester's experience has little or no bearing on intratester reliability measures. However, greater participant strength values could produce weaker intertester reliability measures [26,37]. To avoid this error, one solution could be to couple the HHD device to a rigid customized device. In this manner, the subject would exert the force against a structure that it is not possible to move, ensuring an actual isometric contraction [38].

5. Practical Applications

The Prone Plank Test (PPT) and Closed Kinetic Chain Test (CCT) both demonstrated good test–retest reliability and acceptable error measurement. This kind of easy-to-implement core region functional test could provide both athletic trainers and clinicians with a new and reliable tool for carrying out their core body region training routines in a more objective and standardized way.

The implementation of a core stability training program with a professional football team during the competitive season resulted in a significant core muscle force improvement, achieved with the PPT and CCT. This finding demonstrated that a progressive standardized core neuromuscular training program could be satisfactorily implemented in a professional football team environment, where competition pressure and results sometimes compromise the adequate implementation of an ideal training program.

Author Contributions: Data curation, S.E., E.B., I.S. (Iosu Sesma) and I.S. (Igor Setuain); Formal analysis, S.E., E.B. and I.S. (Igor Setuain); Investigation, M.I., E.B., J.G.A., I.S. (Iosu Sesma), I.S. (Iñigo Sarriegi) and I.S. (Igor Setuain); Methodology, S.E., M.I., J.G.A., I.S. (Iosu Sesma), I.S. (Iñigo Sarriegi) and I.S. (Igor Setuain); Project administration, M.I. and I.S. (Igor Setuain); Resources, M.I. and I.S. (Igor Setuain); Supervision, M.I., J.G.A., I.S. (Iñigo Sarriegi) and I.S. (Igor Setuain); Validation, M.I.; Writing—original draft, S.E. and E.B.; Writing—review & editing, M.I., I.S. (Iñigo Sarriegi) and I.S. (Igor Setuain). All authors have read and agreed to the published version of the manuscript.

Funding: This research received no external funding.

Acknowledgments: The authors would like to express their great appreciation to the participants, as well as to the football club medical and technical stuff and to the players for their predisposition in taking part in the study.

Conflicts of Interest: The authors declare no conflict of interest.

References

1. Akuthota, V.; Ferreiro, A.; Moore, T.; Fredericson, M. Core Stability Exercise Principles. *Curr. Sports Med. Rep.* **2008**, *7*, 39–44. [CrossRef] [PubMed]
2. Kibler, W.B.; Press, J.; Sciascia, A. The role of core stability in athletic function. *Sports Med.* **2006**, *36*, 189–198. [CrossRef] [PubMed]
3. Mendiguchia, J.; Ford, K.R.; Quatman, C.E.; Alentorn-Geli, E.; Hewett, T.E. Sex differences in proximal control of the knee joint. *Sports Med.* **2011**, *41*, 541–557. [CrossRef] [PubMed]
4. Hodges, P.W.; A Richardson, C. Contraction of the Abdominal Muscles Associated With Movement of the Lower Limb. *Phys. Ther.* **1997**, *77*, 132–142. [CrossRef]
5. Ohnson, C.D.; Whitehead, P.N.; Pletcher, E.R.; Faherty, M.S.; Lovalekar, M.; Eagle, S.R.; Keenan, K.A. The Relationship of Core Strength and Activation and Performance on Three Functional Movement Screens. *J. Strength Cond. Res.* **2018**, *32*, 1166–1173. [CrossRef]
6. De Blaiser, C.; De Ridder, R.; Willems, T.; Danneels, L.; Roosen, P. Reliability and validity of trunk flexor and trunk extensor strength measurements using handheld dynamometry in a healthy athletic population. *Phys. Ther. Sport* **2018**, *34*, 180–186. [CrossRef]
7. Juan-Recio, C.; López-Plaza, D.; Murillo, D.B.; García-Vaquero, M.P.; Vera-Garcia, F.J. Reliability assessment and correlation analysis of 3 protocols to measure trunk muscle strength and endurance. *J. Sports Sci.* **2017**, *14*, 1–8. [CrossRef]
8. Reeves, N.P.; Popovich, J.M.; Priess, M.C.; Cholewicki, J.; Choi, J.; Radcliffe, C.J. Reliability of assessing trunk motor control using position and force tracking and stabilization tasks. *J. Biomech.* **2013**, *47*, 44–49. [CrossRef]
9. Barbado, D.; Barbado, L.C.; Elvira, J.L.L.; Van Dieën, J.H.; Vera-Garcia, F.J. Sports-related testing protocols are required to reveal trunk stability adaptations in high-level athletes. *Gait Posture* **2016**, *49*, 90–96. [CrossRef]
10. De Blaiser, C.; De Ridder, R.; Willems, T.; Danneels, L.; Bossche, L.V.; Palmans, T.; Roosen, P. Evaluating abdominal core muscle fatigue: Assessment of the validity and reliability of the prone bridging test. *Scand. J. Med. Sci. Sports* **2017**, *28*, 391–399. [CrossRef] [PubMed]
11. Sorensen, F.B. Physical measurements as risk indicators for low-back trouble over a one-year period. *Spine* **1984**, *9*, 106–119.
12. Krause, D.A.; Youdas, J.W.; Hollman, J.H.; Smith, J. Abdominal Muscle Performance as Measured by the Double Leg-Lowering Test. *Arch. Phys. Med. Rehabilit.* **2005**, *86*, 1345–1348. [CrossRef] [PubMed]

13. McGill, S.M. Low Back Stability: From Formal Description to Issues for Performance and Rehabilitation. *Exerc. Sport Sci. Rev.* **2001**, *29*, 26–31. [CrossRef]

14. Cowley, P.M.; Swensen, T.C. Development and Reliability of Two Core Stability Field Tests. *J. Strength Cond. Res.* **2008**, *22*, 619–624. [CrossRef]

15. Peco, N.; Vera-garcia, F.J. Flexion-Rotation Trunk Test to Assess Abdominal Muscle Endurance. *J. Strength Cond. Res.* **2013**, *27*, 1602–1608.

16. Bliven, K.C.H.; Anderson, B.E. Core Stability Training for Injury Prevention. *Sports Health* **2013**, *5*, 514–522. [CrossRef]

17. Leetun, D.T.; Ireland, M.L.; Willson, J.D.; Ballantyne, B.T.; Davis, I.M. Core Stability Measures as Risk Factors for Lower Extremity Injury in Athletes. *Med. Sci. Sports Exerc.* **2004**, *36*, 926–934. [CrossRef]

18. Hernaez, R. Reliability and agreement studies: A guide for clinical investigators. *Gut* **2015**, *64*, 1018–1027. [CrossRef]

19. Bland, J.M.; Altman, D.G. Measuring agreement in method comparison studies. *Stat. Methods Med. Res.* **1999**, *8*, 135–160. [CrossRef]

20. Weir, J.P. Quantifying test-retest reliability using the intraclass correlation coefficient and the SEM. *J. Strength Cond. Res.* **2005**, *19*, 231–240. [PubMed]

21. Atkinson, G.; Nevill, A. Statistical Methods for Assssing Measurement Error (Reliability) in Variables Relevant to Sports Medicine. *Sports Med.* **1998**, *26*, 217–238. [CrossRef] [PubMed]

22. Stark, T.; Walker, B.F.; Phillips, J.K.; Fejer, R.; Beck, R. Hand-held Dynamometry Correlation With the Gold Standard Isokinetic Dynamometry: A Systematic Review. *PM&R* **2011**, *3*, 472–479.

23. Denegar, C.R.; Ball, D.W. Assessing Reliability and Precision of Measurement: An Introduction to Intraclass Correlation and Standard Error of Measurement. *J. Sport Rehabilit.* **1993**, *2*, 35–42. [CrossRef]

24. De Vet, H.C.; Terwee, C.B.; Knol, D.L.; Bouter, L.M. When to use agreement versus reliability measures. *J. Clin. Epidemiol.* **2006**, *59*, 1033–1039. [CrossRef]

25. Kelln, B.M.; McKeon, P.; Gontkof, L.M.; Hertel, J. Hand-held dynamometry: Reliability of lower extremity muscle testing in healthy, physically active, young adults. *J. Sport Rehabilit.* **2008**, *17*, 160–170. [CrossRef]

26. Lu, T.-W.; Hsu, H.; Chang, L.; Chen, H.-L. Enhancing the examiner's resisting force improves the reliability of manual muscle strength measurements: Comparison of a new device with hand-held dynamometry. *Acta Derm. Venereol.* **2007**, *39*, 679–684. [CrossRef]

27. Celik, D.; Dirican, A.; Baltaci, G. Intrarater Reliability of Assessing Strength of the Shoulder and Scapular Muscles. *J. Sport Rehabilit.* **2017**, *3*, 1–5. [CrossRef]

28. Bohannon, R.W.; Andrews, A.W. Interrater Reliability of Hand-Held Dynamometry. *Phys. Ther.* **1987**, *67*, 931–933. [CrossRef]

29. Stratford, P.W.; Goldsmith, C.H. Use of the standard error as a reliability index of interest: An applied example using elbow flexor strength data. *Phys. Ther.* **1997**, *77*, 745–750. [CrossRef]

30. Moreland, J.; Finch, E.; Stratford, P.; Balsor, B.; Gill, C. Interrater Reliability of Six Tests of Trunk Muscle Function and Endurance. *J. Orthop. Sports Phys. Ther.* **1997**, *26*, 200–208. [CrossRef] [PubMed]

31. Essendrop, M.; Schibye, B.; Hansen, K. Reliability of isometric muscle strength tests for the trunk, hands and shoulders. *Int. J. Ind. Ergon.* **2001**, *28*, 379–387. [CrossRef]

32. Zazulak, B.T.; Hewett, T.E.; Reeves, N.P.; Goldberg, B.; Cholewicki, J. Deficits in Neuromuscular Control of the Trunk Predict Knee Injury Risk. *Am. J. Sports Med.* **2007**, *35*, 1123–1130. [CrossRef] [PubMed]

33. Prieske, O.; Muehlbauer, T.; Borde, R.; Gube, M.; Bruhn, S.; Behm, D.G.; Granacher, U. Neuromuscular and athletic performance following core strength training in elite youth soccer: Role of instability. *Scand. J. Med. Sci. Sports* **2015**, *26*, 48–56. [CrossRef] [PubMed]

34. Sharma, A.; Geovinson, S.G.; Sandhu, J.S. Effects of a nine-week core strengthening exercise program on vertical jump performances and static balance in volleyball players with trunk instability. *J. Sports Med. Phys. Fit.* **2012**, *52*, 606–615.

35. Hoshikawa, Y.; Iida, T.; Muramatsu, M.; Ii, N.; Nakajima, Y.; Chumank, K.; Kanehisa, H. Effects of Stabilization Training on Trunk Muscularity and Physical Performances in Youth Soccer Players. *J. Strength Cond. Res.* **2013**, *27*, 3142–3149. [CrossRef]

36. Nesser, T.W.; Huxel, K.C.; Tincher, J.L.; Okada, T. The Relationship Between Core Stability and Performance in Division I Football Players. *J. Strength Cond. Res.* **2008**, *22*, 1750–1754. [CrossRef]

37. Roth, R.; Donath, L.; Zahner, L.; Faude, O. Muscle Activation and Performance During Trunk Strength Testing in High-Level Female and Male Football Players. *J. Appl. Biomech.* **2016**, *32*, 241–247. [CrossRef]
38. Hölmich, P.; Larsen, K.; Krogsgaard, K.; Gluud, C. Exercise program for prevention of groin pain in football players: A cluster-randomized trial. *Scand. J. Med. Sci. Sports* **2009**, *20*, 814–821. [CrossRef]

Article

Impact of Quadriceps/Hamstrings Torque Ratio on Three-Dimensional Pelvic Posture and Clinical Pubic Symphysis Pain-Preliminary Results in Healthy Young Male Athletes

Oliver Ludwig [1,*], Jens Kelm [2,3] and Sascha Hopp [4]

[1] Fachgebiet Sportwissenschaft, Technische Universität Kaiserslautern, 67663 Kaiserslautern, Germany
[2] Chirurgisch-Orthopädisches Zentrum Illingen, 66557 Illingen, Germany; kelm@chirurgie-illingen.de
[3] Klinik für Orthopädie und Orthopädische Chirurgie, Universitätsklinikum des Saarlandes, 66424 Homburg, Germany
[4] Lutrina Klinik, 67663 Kaiserslautern, Germany; sascha.hopp@uks.eu
* Correspondence: oliver.ludwig@sowi.uni-kl.de; Tel.: +49-631-205-5423

Received: 27 June 2020; Accepted: 27 July 2020; Published: 29 July 2020

Featured Application: The crossed torque ratio analyzed in this study as a possible pathogenic mechanism for symphyseal pain could be used as a preventive guiding criterion in sports prevention and sports medicine diagnostics.

Abstract: Pain in the pubic symphysis is of significance, especially in high-performance sports. Pelvic torsion, possibly caused by muscular imbalances, is discussed as a pathogenic mechanism. This study examined a possible interrelationship between the maximum torques of quadriceps femoris and hamstrings and the spatial positioning of the hemi-pelvises, as well as the tenderness to palpation of the pubic symphysis. The three-dimensional pelvic contour of 26 pain free adolescents (age 16.0 ± 0.8 years, weight 66.3 ± 9.9 kg, height 176.2 ± 6.0 cm) was registered by means of an 3D optical system and the torsion of both hemi-pelvises against each other was calculated based on a simplified geometrical model. Tenderness on palpation of the pubic symphysis was assessed by means of a visual analogue scale, and isometric torques of knee extensors and flexors were measured for both legs. The torque ratio between knee extensors and flexors was calculated for both sides, as was the crossed torque ratio between the two legs. On the basis of a MANOVA, possible significant differences in torques and torque ratios between subgroups with lower and higher pelvic torsion were analyzed. The crossed torque ratio (F = 19.55, $p < 0.001$, partial $\eta^2 = 0.453$) and the tenderness to palpation of the pubic symphysis (F = 10.72, $p = 0.003$, partial $\eta^2 = 0.309$) were significantly higher in the subgroup with higher pelvic torsion. The results indicate the crossed torque ratio of knee flexors and extensors as a potential biomechanical-pathogenic mechanism to be considered in the primary prevention and diagnosis of symphyseal pain.

Keywords: pubic pain; pubic symphysis; symphyseal pain; force ratio; hamstrings; rectus femoris; pelvic torsion; pelvic position; hemi pelvises; athletic pubalgia

---

1. Introduction

The pelvis is a complex ring structure in which both hemi-pelvises are connected dorsally via the sacrum and ventrally via the pubic symphysis. The pubic symphysis is a non-synovial, amphiar-throtic joint with the opposing pubic bone surfaces covered by a thin, hyaline layer of cartilage and with the discus interpubicus, a fibrocartilaginous disc, firmly anchored in between them. Circumfer-ential ligaments provide the pubic symphysis with a passive, mechanical stability [1]. The pubic symphysis is subjected to tensile strain when standing (bipedal stand), to pressure while in a supine position, and to shear while walking (lifting the swinging leg) [2]. Injuries and damage to the pubic symphysis are of the utmost significance, especially in high-performance and professional sports, and can lead to prolonged idle times of up to 2 years. Predisposed sports include soccer, rugby, and American football, that is, sports with quick changes of direction or sprint and interval strain [3]. Regardless of the sports practiced, however, athletes with biomechanical deficits, such as different leg lengths, pelvic instability, or restricted flexibility in the hip joints, can also be affected. For professional athletes, the incidence of chronic athletic pubalgia is estimated to range between 0.5 and 6.2% [4]. The highest values (up to 18%) are found in ice hockey and soccer players [5]. More detailed numbers, particularly pertaining to young athletes, cannot be derived from international literature owing to the inhomogeneity of studies.

If both hemi-pelvises move in opposite directions in the sagittal plane, minimal torsions in a millimeter range occur in the joints involved. Such a pelvic torsion triggers shearing forces in the sacroiliac joints and the pubic symphysis. Repetitive mechanical strain can lead to symphyseal instability and consequently cause unphysiologically high shearing forces in the pubic symphysis. This is considered the etiological factor that most frequently leads to complaints [6,7].

For prevention purposes, it is thus important to identify at an early stage potential factors that may lead to unfavorable strain of the pubic symphysis. Potential causes being discussed include reduced muscular flexibility, an existing deformity (e.g., femoroacetabular impingement) [8,9], or muscular imbalances [10,11]. Muscular, antagonist concatenations with insertion directly at the pubic symphysis (esp. rectus abdominis, adductors) or with an immediate effect on the pubic symphysis (especially psoas/gluteus, rectus femoris/hamstrings) play a significant role in the mechanical strain of the pubic symphysis. Early on, Meyers et al. postulated that imbalances between rectus abdominis and adductor muscles would result in secondary symptomatic pubic symphysis instability and pain [12]. Tyler et al. showed that, in ice hockey players, injuries of the adductor muscles and tendons close to the pubic symphysis were 17 times more likely to occur when the strength of the adductors was 80% less than that of the abductor muscles [13].

The influence of muscular activity on the pelvic position is particularly interesting because it might be a possible approach to derive both preventive and curative intervention methods. Also of particular interest are those muscle groups that are able to influence the anteversion of the hemi-pelvises, such as rectus femoris (as part of the quadriceps femoris muscle) and hamstrings [14]. With extended knees, both muscle groups act as antagonists to the pelvic anteversion or retroversion, respectively.

This study thus aims to identify any potential interconnection between the maximum torques of quadriceps and hamstrings and the spatial positioning of the hemi-pelvises, as well as the tenderness to palpation of the pubic symphysis in healthy young athletes.

This study aimed to verify the following:

(i) whether imbalances of the force ratio of quadriceps and hamstrings when comparing sides have an impact on the three-dimensional positioning of the hemi-pelvises in habitual upright posture;

(ii) whether changes to the pain on palpation occur in the pubic symphysis depending on the three-dimensional positioning of the hemi-pelvises in habitual upright posture.

2. Materials and Methods

2.1. Test Subjects

Thirty-seven adolescents participated in a routine posture examination within the framework of an interdisciplinary research project (Kid-Check). At the time of testing, they were asymptomatic and without any obvious orthopedic, neurological, or internal diagnoses. Exclusion criteria were anatomical or functional leg length discrepancy or functional limitations in the sacroiliac joints. The study includes data from 26 male young athletes who fulfilled the inclusion criteria (see Table 1 for anthropometric data). Of the 26 athletes, 18 were soccer players, 4 track and field athletes, 3 handball players, 1 biathlete, 1 fitness athlete (machine-assisted resistance training), and 1 fencer. The adolescents and their parents were informed of the intention and procedure of the study and gave their written informed consent in accordance with the Declaration of Helsinki. The university's ethics commission approved the study (Ref. Nr. 15-03/05).

Table 1. Anthropometric data of total group and subgroups.

Total Group (N = 26)			Group with Low Pelvic Torsion (N = 13)			Group with High Pelvic Torsion (N = 13)		
Age [years]	Weight [kg]	Height [cm]	Age [years]	Weight [kg]	Height [cm]	Age [years]	Weight [kg]	Height [cm]
16.0 ± 0.8	66.3 ± 9.9	176.2 ± 6.0	16.0 ± 1.0	66.0 ± 6.2	176.3 ± 6.1	16.0 ± 0.6	66.6 ± 12.8	176.0 ± 6.1

2.2. Manual Testing

Before the actual measurement, the sacroiliac joints were tested for dysfunction by means of a test cluster comprising a sacral thrust test, distraction test, and Gaenslen's test [15]. Applying a visual metric analog scale from 0 (no pain) to 10 (maximum pain), the test persons assessed the tenderness to palpation of the pubic symphysis [16]. As it is known that a leg length discrepancy influences pelvic torsion [17,18], it was checked in advance by photometric comparison of the positions of the posterior superior iliac spines (PSIS) and the anterior superior iliac spines (ASIS). Furthermore, 3 of the 37 interested subjects were excluded during posture analysis because of a scoliotic malposition of the spine owing to a potential correlation with malpositioning of the hemi-pelvises [19].

2.3. Three-Dimensional Posture Measurement

To determine the three-dimensional pelvis contour, marker spheres (diameter 8 mm) were applied to anatomical landmarks on the shaved skin in the pelvic area [20] by means of double-sided adhesive pads (Figure 1). All palpations were performed by the same experienced examiner. The test persons were standing in habitual posture, arms crossed in front of the chest. Three-dimensional posture scans were performed dorsally and ventrally using the Rothballer 3D Scanner (Rothballer electronic systems, Weiden, Germany). This system generated a three-dimensional image of the pelvis in the shape of a mathematical point cloud. System reliability and validity for postural measurements have been confirmed by other studies (comparable inter-trial reliability to 3D motion analysis systems: intraclass correlation coefficient difference = 0.06 ± 0.05, range 0.00–0.16; very good validity: Pearson's r = 0.96 ± 0.04; range 0.84–0.99 [21,22]).

For the evaluation, the position of the marker spheres was determined individually using the analysis software, and their spatial coordinates were extracted. The marker spheres on the trochanters were visible on both 3D images and served as reference points for the spatial coordinates of all points. Figure 2 shows the simplified geometries of both hemi-pelvises, calculated trigonometrically from the coordinates. The torsion of both triangles against each other was calculated as an angle using MATLAB (R230a, MathWorks Inc., Natick, MA, USA) and Microsoft Excel (Microsoft, Redmond, WA, USA); this angle was then used as the measured value for pelvic torsion. In addition, pelvic anteversion was calculated as the angle of the connection line between pubic symphysis and S1 and the horizontal.

2.4. Error Analysis

The accuracy of the determination of the forward tilt of the pelvis and thus pelvic torsion depends on the accuracy of marker spheres' placement on the skin. In a preliminary study, the retest reliability of marker positioning was determined by the same investigator on twelve subjects from the test group at intervals of one day. Cronbach's alpha was 0.998 and confirmed a very good reliability. The mean deviation of the marker spheres was 1.4 ± 0.9 mm. Therefore, an assumed inaccuracy of ±1 mm in marker positioning and an average pelvis depth of 200 mm lead to a calculated error of <1.5°, which ensures sufficient distinguishability of the groups when looking at the group mean values (Table 2), and thus an acceptable inaccuracy.

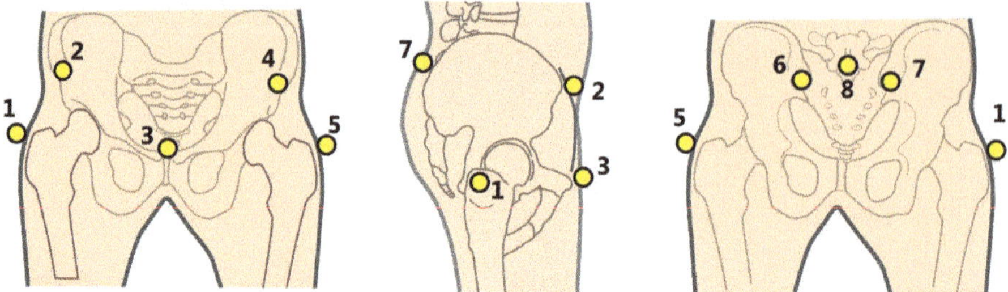

Figure 1. Position of the marker spheres. 1: Trochanter major right, 2: spina iliaca anterior superior (ASIS) right, 3: symphysis, 4: ASIS left, 5: trochanter major left, 6: spina iliaca posterior superior (PSIS) left, 7: PSIS right, 8: S1.

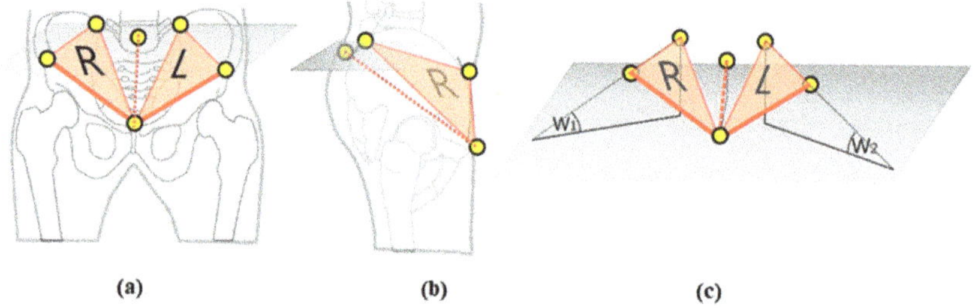

(a)   (b)   (c)

Figure 2. Simplified geometrical model: representation of the left and right hemi-pelvises by triangles between ASIS, PSIS, and symphysis. (**a**) Frontal view; (**b**) view from the right side; (**c**) angles W1 and W2 between the horizontal (transversal) plane and the triangles are the basis for the calculation of the pelvic torsion. Dotted lines: pelvic anteversion.

2.5. Isometric Torque Measurement

The maximum isometric torques for knee flexion and extension were measured in a sitting position using an isometry measurement device (Easytorque, Fa. Tonus, Zemmer, Germany). The knee angle was set to 80 degrees for measuring knee flexion and to 90 degrees for knee extension (Figure 3). The thighs were fixated with pads and the calves pressed isometrically against the pads, and thus against the force transducers. The test persons had to build up strength for five seconds and hold it for five more. For familiarization purposes, a trial was performed for each measurement. Both sides were tested alternatingly. At least three isometric measurements were performed for each side. Whenever the third value turned out to be the maximum value, further tests were performed until the last measurement value was smaller than the penultimate value. Only the maximum value achieved was included in the subsequent calculations. The reproducibility of this type of isometric measurement has been shown in other studies (intraclass correlation coefficient 0.97–0.99, Pearson's correlation coefficient 0.95 [23,24].

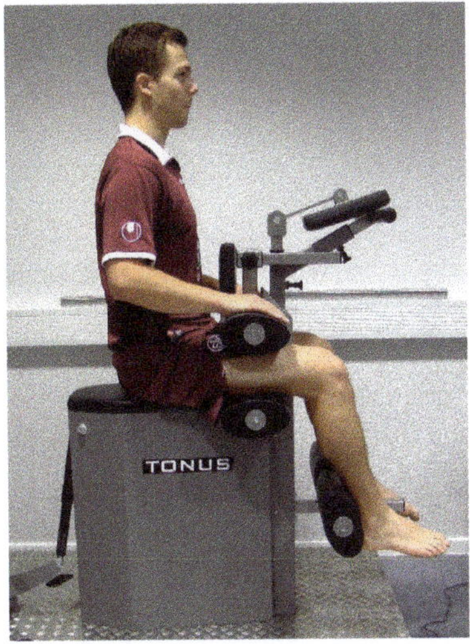

Figure 3. Measurement of the isometric torques of the hamstrings (pelvic fixation not shown).

2.6. Statistics

On the basis of the 3D model, the side with the strongest pelvic rotation was determined. The ipsilateral torque ratio,

$$\text{IPSI Ex/Flex} = (\text{Ex}_\text{IPSI}/\text{Flex}_\text{IPSI}) \tag{1}$$

(with Ex_IPSI = torque of the extension movement of the ipsilateral side and Flex_IPSI = torque of the flexion movement of the ipsilateral side), between knee extensors and flexors for this side was calculated, as well as the contralateral torque ratio,

$$\text{CON Ex/Flex} = (\text{Ex}_\text{CON}/\text{Flex}_\text{CON}) \tag{2}$$

(with Ex_CON = torque of the extension movement of the contralateral side and Flex_CON = torque of the flexion movement of the contralateral side), and the crossed torque ratio,

$$\text{CTR} = (\text{Ex}_\text{IPSI} * \text{Flex}_\text{CON})/(\text{Flex}_\text{IPSI} * \text{Ex}_\text{CON}). \tag{3}$$

Afterwards, the test persons were divided into two groups based on the pelvic torsion median (2.15°): one group with lower pelvic torsion (pelvic torsion $\leq$ 2.15°; n = 13) and one with higher pelvic torsion (>2.15°; n = 13).

IBM SPSS Statistics 23 was used for calculating the statistical tests. The significance level was set to 0.05. The sample size could not be determined a priori, but was given by the number of subjects participating in the interdisciplinary research project (Kid-Check) at that time.

Testing for multicollinearity was done by calculating Pearson's correlation coefficient r for all com bi nations of variables. After checking the starting conditions, a MANOVA with post hoc univariate ANOVAs was performed between the two groups for the variables IPSI Ex/Flex, CON Ex/Flex, and CTR. Partial η^2 was calculated as effect size, with $\eta^2 < 0.06$ indicating a small effect, values between 0.06 and 0.14 a medium effect, and $\eta^2 > 0.14$ a large effect.

The variable 'tenderness on palpation of the pubic symphysis' (TPS) was not included in the MANOVA, because, according to our model, TPS would be the result and not the cause of pelvic torsion. Therefore, differences between the groups for tenderness on palpation were calculated separately using a one-way ANOVA.

Linear regression models were calculated for pelvic torsion, crossed torque ratio, and tenderness on palpation (bivariate regressions with pelvic torsion = dependent variable, CTR = independent variable, and pelvic torsion = independent variable, TPS = dependent variable), and R^2 was determined.

3. Results

Correlations between the absolute torque values were high (Pearson's correlation coeffi cient: IPSI Flex–CON Flex: r = 0.905; IPSI Ex–CON Ex: r = 0.871), and multicollinearity was thus present. Therefore, only the ratios IPSI Ex/Flex, CON Ex/Flex, and CTR were included in the following MANOVA. These variables were normally distributed in both groups, as assessed by the Shapiro–Wilk test (IPSI Ex/Flex: $p = 0.224$; CON Ex/Flex: $p = 0.321$; CTR: $p = 0.876$). There were no outliers in the data. No multivariate outliers were found, as assessed by the Mahalanobis distance MD (cut-off (3) = 16.27, MDmax = 7.78, $p > 0.001$). Linearity between the dependent variables was verified by visual inspection of the scatter plots.

The homogeneity of the error variances was checked with Levene's test and showed no significance for the dependent variables (IPSI Ex/Flex: $p = 0.517$; CON Ex/Flex: $p = 0.881$; CTR: $p = 0.779$). Box's test confirmed the homogeneity of covariances ($p > 0.001$).

The one-way MANOVA showed a statistically significant difference between the groups on the combined dependent variables, with F(3, 22) = 6.640, $p = 0.002$, partial $\eta^2 = 0.475$, Wilk's $\Lambda = 0.525$.

Post-hoc univariate ANOVAs showed no statistically significant difference between the groups for CON Ex/Flex, with F(1, 24) = 0.006, $p = 0.941$, and partial $\eta^2 < 0.001$. However, a statistically significant difference between the two groups was found for IPSI Ex/Flex (F(1, 24) = 5.864, $p = 0.023$, partial $\eta^2 = 0.196$) and for CTR (F(1, 24) = 19.885, $p < 0.001$, partial $\eta^2 = 0.453$).

The group with the higher pelvic torsion angle exhibited a higher (but statistically not significant) torque in the extensors on the side of the anteverted hemi-pelvis (IPSI Ex), and consequently a statistically significant higher torque ratio between extensors and flexors (Table 2). A contralateral difference between the flexor and extensor torque in both subgroups was not found. The torque ratio on the contralateral side (CON Ex/Flex) was also not significantly different. However, the crossed torque ratio at $p < 0.001$ was significantly higher in the group with higher pelvic torsion (Table 2). The effect size was partial $\eta^2 = 0.453$, which represents a large effect.

Table 2. Parameters of the subgroups with low and with high pelvic torsion. Last row: results of the ANOVAs.

Group	Pelvic Torsion [°]	IPSI Flex [Nm]	IPSI Ex [Nm]	IPSI Ex/Flex [−]	CON Flex [Nm]	CON Ex [Nm]	CON Ex/Flex [−]	CTR [−]	TPS [VAS 0–10]
Low pelvic torsion	0.88 ± 0.57	159.54 ± 58.23	234.77 ± 107.94	1.46 ± 0.39	151.23 ± 52.80	227.23 ± 96.68	1.48 ± 0.33	0.99 ± 0.16	1.57 ±1.49
High pelvic torsion	3.74 ± 1.00	174.08 ± 44.63	340.8 ± 71.8	1.83 ± 0.40	177.62 ± 60.45	256.08 ± 86.96	1.47 ± 0.35	1.26 ± 0.15	3.55 ± 1.58
ANOVA: F(1,24) =				5.864			0.006	19.885	10.72
$p =$				0.023 *			0.941	<0.001 **	0.0032 *

IPSI: ipsilateral side with increased pelvic tilt, CON: contralateral, Ex: extension torque, Flex: flexion torque, CTR: crossed torque ratio, TPS: tenderness to palpation of the pubic symphysis on the visual analog scale (VAS), Nm = Newtonmeter, bold values: * $p < 0.05$, ** $p < 0.001$.

In the linear regression models, the crossed torque ratio (CTR) showed the strongest correlation with the pelvic torsion angle with a correlation coefficient of 0.668 (Table 3).

At the same time, the subgroup with higher pelvic torsion reported statistically significant higher values for tenderness to palpation (Table 2, partial $\eta^2 = 0.309$, large effect). The linear regression models showed that 39% of the variance of tenderness to palpation can be explained by the pelvic torsion angle and 44% of the variance of the pelvic torsion angle can be explained by the crossed torque ratio (Figure 4).

Table 3. Correlations (Pearson's r) between the torque parameters and the pelvic torsion angle.

Parameter	r
IPSI Flex	0.218
IPSI Ex	0.476
IPSI Ex/Flex	0.472
CON Flex	0.262
Con Ex	0.203
CON Ex/Flex	<0.001
CTR (crossed torque ratio)	0.668

IPSI = ipsilateral, CON = contralateral, Ex = extension, Flex = flexion.

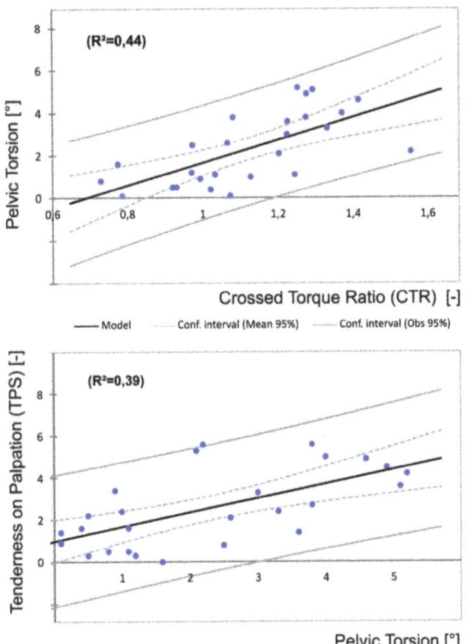

Figure 4. Regression of crossed torque ratio and pelvic torsion angle (**above**) and of pelvic torsion angle and tenderness on palpation (**below**).

4. Discussion

This study aimed to verify the existence of interconnections between contralateral imbalances of the torque ratios of quadriceps and hamstrings and the three-dimensional positioning of the hemi-pelvises. As a matter of fact, our results support this assumption because a significant difference of the calculated crossed torque ratio was found between the two groups with a lower and higher pelvic torsion angle.

The biomechanics of the pelvic bones and their joints is complex and has come into scientific focus over the past few years. Very low amplitude movements occur in the sacroiliac joints and the pubic symphysis, which could be the cause for complaints [1,25]. If we assume, in a simplified biomechanical model, that both hemi-pelvises are slightly twistable against each other ('inflare' [26,27]), three different cases can arise:

1. The torque ratios are balanced both anterior–posterior and left–right (Figure 5a). Both hemi-pelvises are not tilted toward each other.

2. The absolute anterior and posterior torques are left–right different, but show the same ratio, with the torque of the knee extensors (more precisely, the torque of the rectus femoris as part of the

quadriceps femoris) being predominant. This can lead to an anteversion of the pelvis, which affects both hemi-pelvises equally (Figure 5b). Shearing forces would not develop in the pubic symphysis. The influence of the relevant muscle groups on the positioning of the pelvis has been analyzed in other studies [14].

3. The anterior–posterior torque ratios differ. If this is the case, the pelvis tilts forward on the side with the higher quadriceps–hamstrings ratio. The pubic symphysis may experience shearing forces (Figure 5c).

The crossed torque ratio had the most expressed influence on the pelvic torsion angle. This is interesting because this harmonizes with our model-based assumption, according to which differing agonist/antagonist torque ratios between the left and right side are more decisive for torsion than the absolute values of the agonist–antagonist ratios. Differences in the agonist–antagonist torque ratios would, if they were identical left and right, allow the pelvis to antevert, but they would not necessarily have an effect on the torsion of the hemi-pelvises. The absolute torque of the muscles involved, however, do play a role, as shown by the correlation coefficient of the quadriceps at the side of the anteverted hemi-pelvis (quadriceps torque IPSI Ex, Table 3). The crossed torque ratio (Equation (3)) seems to have a more expressed influence, though. Nevertheless, our results do not allow a direct derivation of a causal connection. This means that we cannot identify the muscular torque ratios measured as a reason for the three-dimensional torsion of the hemi-pelvises and the increased tenderness to palpation in the pubic symphysis. However, the findings still fit in the pelvis movement model described.

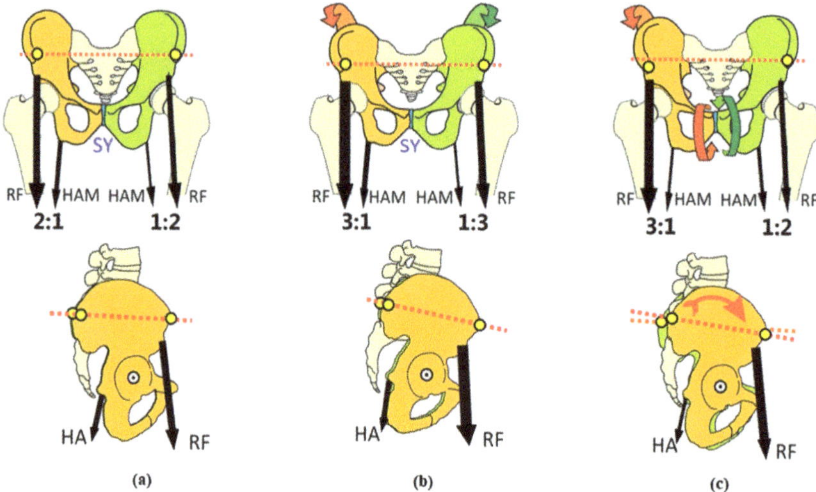

Figure 5. Model of the development of pelvic torsion owing to muscular torque ratios between rectus femoris (RF) and hamstrings (HA). Arrow thickness represents traction force. (**a**) Different torque ratios right and left, but identical anterior–posterior torque ratio (2:1): pelvis neutral. (**b**) Increased anterior–posterior torque ratio (3:1): anterior pelvic tilt. (**c**) Different torque ratios left and right: unilateral pelvic tilt and pelvic torsion. SY = pubic symphysis.

Only a few studies focus on the three-dimensional movement of the hemi-pelvises against each other and the sacrum. Becker et al. describe physiological, rotatory movements of one degree in the pubic symphysis [28]. Strain caused by standing on one leg results in relative vertical shearing stress and tensile stress on the stabilizing sacroiliac joints, as well as in vertical shearing and pressure strain on the pubic symphysis [25,29]. In vivo measurements of healthy test persons with metal wires implanted in the pubic bone ramus (ramus ossis superior pubis) showed translation values of <2 mm and <3° rotation at the pubic symphysis when subjected to various strain-causing maneuvers [30].

In a radiological study with asymptomatic test persons who were standing alternatingly on the left and right leg, Garras et al. were able to prove that physiological movements up to 5 mm can occur at the pubic symphysis (functional instability) and that higher values are to be interpreted as a sign of "macro" instability [2]. Macro instabilities do not necessarily cause direct problems. They may, however, represent a significantly increased risk factor for the development of groin pain/injuries. Golden and colleagues define a pelvis as stable if a vertical rotational symphyseal displacement of less than 11 mm occurs [31]. These data refer to traumatic pelvis injuries only, therefore, they cannot be generalized or even transferred to an athletic population. It is known that routine movements like walking already lead to slight physiologic movements that cause shearing forces in the pubic symphysis during the one-leg stance phase. This may explain why our test persons with higher pelvic torsion exhibited a significantly increased tenderness to palpation of the pubic symphysis. Even if they did not complain of pain in the pubic symphysis during everyday (sports) activities, we assume that strain in the area of the articular capsule caused by shearing forces may be a potential factor that might increase tenderness to palpation and sensitivity. Therefore, it may make sense to keep this in mind as a potential precursor for later pathologies and perform further studies on this topic.

Some studies could show that the muscular balance between quadriceps and hamstrings depends on the type of sports performed. For example, Iga et al. found significant differences in the hamstrings–quadriceps ratio between the dominant and non-dominant leg of adolescent soccer players [32]. Gür et al. were able to show that these changes are adaptations to soccer-specific requirements [33]. This torsion of the hemi-pelvises may play an important role as a potentially pathogen mechanism in athletes, especially those with a heavy muscular imbalance between the dominant and non-dominant leg. This information is particularly relevant to our study because most of our participants were indeed soccer players.

Key aspects pertaining to the prevention and rehabilitation of groin/pubic bone problems could be derived if the strength ratios of quadriceps and hamstrings were analyzed and improved, in the case of contralateral imbalances. Iga et al. emphasize that imbalances of the leg muscles might be improved by including resistance exercises in the habitual training routines of youth soccer players [32]. To date, therapeutic focus has mostly been placed on strengthening adductors and rectus abdominis. In a prospectively randomized study, Holmich et al. were able to show that a group achieved significantly better results through active adductor training than through physical therapy treatment [34].

The results of our study point to a potential biomechanical-pathogen mechanism that should be taken into account in primary prevention and diagnostics of groin pain.

Limitations

This study analyzed the potential influence factor of only two major muscle groups on the three-dimensional pelvis orientation. Many other muscles and their torque ratios also have an influence on the orientation of the pelvis and that of the hemi-pelvises against each other. This includes mainly the gluteus maximus, the adductor muscles, iliopsoas muscle, and the abdominal and pelvic floor muscles. Our study is thus based only on a simplified model.

In daily athletic routine, not only the absolute isometric strength plays a key role, but also the neuromuscular responsivity of the muscles when accelerating or changing directions, which was not investigated in our study. For example, Cowan et al. found that the feed-forward activation of transversus abdominis during a straight leg raise movement was delayed in patients with persistent groin pain [35]. Owing to the potential stabilizing role of the transversely oriented abdominal muscles for the anterior pelvic ring, a late reaction of these muscles can have consequences for the stability of the pelvis.

We also need to emphasize that we only examined healthy male test persons. Therefore, the results cannot be applied directly to patients or to female athletes. However, we still need to state that a notable pain on palpation of the pubic symphysis was often observed in asymptomatic test persons.

5. Conclusions

As interconnections exist between pelvic torsion and the quadriceps/hamstrings torque ratio on one hand and pain on palpation on the other, a focus needs to be placed on the torque ratio of quadriceps and hamstrings in preventive interventions, particularly in the case of problems with pubic symphysis and groin. This applies specially to types of sports that are predisposed to muscular imbalances in these muscle areas.

Author Contributions: Conceptualization and methodology, O.L., J.K., and S.H.; formal analysis, O.L.; writing—original draft preparation, O.L., J.K., and S.H.; visualization, O.L.; supervision, J.K. and S.H. All authors have read and agreed to the published version of the manuscript.

Funding: This research received no external funding.

Acknowledgments: The authors wish to thank Monika Schutz for translating the manuscript.

Conflicts of Interest: The authors declare no conflict of interest.

References

1. Gamble, J.G.; Simmons, S.C.; Freedman, M. The symphysis pubis. Anatomic and pathologic considerations. *Clin. Orthop. Relat. Res.* **1986**, *203*, 261–272. [CrossRef]
2. Garras, D.N.; Carothers, J.T.; Olson, S.A. Single-leg-stance (flamingo) radiographs to assess pelvic instability: How much motion is normal? *JBJS* **2008**, *90*, 2114–2118. [CrossRef]
3. Pizzari, T.; Coburn, P.T.; Crow, J.F. Prevention and management of osteitis pubis in the australian football league: A qualitative analysis. *Phys. Ther. Sport* **2008**, *9*, 117–125. [CrossRef]
4. Campanelli, G. *Pubic Inguinal Pain Syndrome: The So-Called Sports Hernia*; Springer: Berlin/Heidelberg, Germany, 2010.
5. Via, A.G.; Frizziero, A.; Finotti, P.; Oliva, F.; Randelli, F.; Maffulli, N. Management of osteitis pubis in athletes: Rehabilitation and return to training—A review of the most recent literature. *Open Access J. Sports Med.* **2018**, *10*, 1–10.
6. Melegati, G.; Elli, S. Osteitis Pubis. In *Groin Pain Syndrome*; Zini, R., Volpi, P., Bisciotti, G., Eds.; Springer: Cham, Switzerland, 2017; pp. 135–140.
7. Garvey, J.; Read, J.; Turner, A. Sportsman hernia: What can we do? *Hernia* **2010**, *14*, 17–25. [CrossRef]
8. Fraitzl, C.; Kappe, T.; Reichel, H. Femoroacetabular impingement-a frequent cause for groin pain in the athlete. *Dtsch. Z. Sportmed.* **2010**, *61*, 292–298.
9. Larson, C.M.; Pierce, B.R.; Giveans, M.R. Treatment of athletes with symptomatic intra-articular hip pathology and athletic pubalgia/sports hernia: A case series. *Arthrosc. J. Arthrosc. Relat. Surg.* **2011**, *27*, 768–775. [CrossRef]
10. Cunningham, P.M.; Brennan, D.; O'Connell, M.; MacMahon, P.; O'Neill, P.; Eustace, S. Patterns of bone and soft-tissue injury at the symphysis pubis in soccer players: Observations at mri. *Am. J. Roentgenol.* **2007**, *188*, W291–W296. [CrossRef]
11. Hiti, C.J.; Stevens, K.J.; Jamati, M.K.; Garza, D.; Matheson, G.O. Athletic osteitis pubis. *Sports Med.* **2011**, *41*, 361–376. [CrossRef]
12. Meyers, W.C.; McKechnie, A.; Philippon, M.J.; Horner, M.A.; Zoga, A.C.; Devon, O.N. Experience with "sports hernia" spanning two decades. *Ann. Surg.* **2008**, *248*, 656–665. [CrossRef]
13. Tyler, T.F.; Nicholas, S.J.; Campbell, R.J.; McHugh, M.P. The association of hip strength and flexibility with the incidence of adductor muscle strains in professional ice hockey players. *Am. J. Sports Med.* **2001**, *29*, 124–128. [CrossRef]
14. Buchtelová, E.; Tichy, M.; Vaniková, K. Influence of muscular imbalances on pelvic position and lumbar lordosis: A theoretical basis. *J. Nurs. Soc. Stud. Public Health Rehabil.* **2013**, *1–2*, 25–36.
15. Grieve, E. Diagnostic tests for mechanical dysfunction of the sacroiliac joints. *J. Man. Manip. Ther.* **2001**, *9*, 198–206. [CrossRef]
16. Harms-Ringdahl, K.; Carlsson, A.M.; Ekholm, J.; Raustorp, A.; Svensson, T.; Toresson, H.-G. Pain assessment with different intensity scales in response to loading of joint structures. *Pain* **1986**, *27*, 401–411. [CrossRef]

17. Rhodes, D.W.; Bishop, P.A. The relationship between pelvic torsion and anatomical leg length inequality: A review of the literature. *J. Chiropr. Med.* **2010**, *9*, 95–96. [CrossRef]

18. Betsch, M.; Wild, M.; Große, B.; Rapp, W.; Horstmann, T. The effect of simulating leg length inequality on spinal posture and pelvic position: A dynamic rasterstereographic analysis. *Eur. Spine J.* **2012**, *21*, 691–697. [CrossRef]

19. Betsch, M.; Rapp, W.; Przibylla, A.; Jungbluth, P.; Hakimi, M.; Schneppendahl, J.; Thelen, S.; Wild, M. Determination of the amount of leg length inequality that alters spinal posture in healthy subjects using rasterstereography. *Eur. Spine J.* **2013**, *22*, 1354–1361. [CrossRef]

20. van Sint Jan, S. *Color Atlas of Skeletal Landmark Definitions E-Book: Guidelines for Reproducible Manual and Virtual Palpations*; Elsevier Health Sciences: Edinburgh, UK, 2007.

21. Clark, R.A.; Pua, Y.H.; Oliveira, C.C.; Bower, K.J.; Thilarajah, S.; McGaw, R.; Hasanki, K.; Mentiplay, B.F. Reliability and concurrent validity of the microsoft xbox one kinect for assessment of standing balance and postural control. *Gait Posture* **2015**, *42*, 210–213. [CrossRef]

22. Clark, R.A.; Pua, Y.H.; Fortin, K.; Ritchie, C.; Webster, K.E.; Denehy, L.; Bryant, A.L. Validity of the Microsoft Kinect for assessment of postural control. *Gait Posture* **2012**, *36*, 372–377. [CrossRef]

23. Katoh, M.; Yamasaki, H. Comparison of reliability of isometric leg muscle strength measurements made using a hand-held dynamometer with and without a restraining belt. *J. Phys. Ther. Sci.* **2009**, *21*, 37–42. [CrossRef]

24. Kim, W.K.; Kim, D.K.; Seo, K.M.; Kang, S.H. Reliability and validity of isometric knee extensor strength test with hand-held dynamometer depending on its fixation: A pilot study. *Ann. Rehabil. Med.* **2014**, *38*, 84–93. [CrossRef] [PubMed]

25. Icke, C.; Koebke, J. Normal stress pattern of the pubic symphysis. *Anat. Cell Biol.* **2014**, *47*, 40–43. [CrossRef] [PubMed]

26. Schamberger, W. *The Malalignment Syndrome: Diagnosis and Treatment of Common Pelvic and back Pain*; Elsevier Health Sciences: Edinburgh, UK, 2012.

27. Timgren, J. Role of pelvic asymmetry in skeletal posture. *PRM+* **2018**, *1*, 26–29.

28. Becker, I.; Woodley, S.J.; Stringer, M.D. The adult human pubic symphysis: A systematic review. *J. Anat.* **2010**, *217*, 475–487. [CrossRef] [PubMed]

29. Hagen, R. Pelvic girdle relaxation from an orthopaedic point of view. *Acta Orthop. Scand.* **1974**, *45*, 550–563. [CrossRef]

30. Walheim, G.G.; Selvik, G. Mobility of the pubic symphysis: In vivo measurements with an electromechanic method and a roentgen stereophotogrammetric method. *Clin. Orthop. Relat. Res.* **1984**, *191*, 129–135. [CrossRef]

31. Golden, R.D.; Kim, H.; Watson, J.D.; Oliphant, B.W.; Doro, C.; Hsieh, A.H.; Osgood, G.M.; O'Toole, R.V. How much vertical displacement of the symphysis indicates instability after pelvic injury? *J. Trauma Acute Care Surg.* **2013**, *74*, 585–589. [CrossRef]

32. Iga, J.; George, K.; Lees, A.; Reilly, T. Cross-sectional investigation of indices of isokinetic leg strength in youth soccer players and untrained individuals. *Scand. J. Med. Sci. Sports* **2009**, *19*, 714–719. [CrossRef]

33. Gur, H.; Akova, B.; Punduk, Z.; Kucukoglu, S. Effects of age on the reciprocal peak torque ratios during knee muscle contractions in elite soccer players. *Scand. J. Med. Sci. Sports* **1999**, *9*, 81–87. [CrossRef]

34. Holmich, P.; Uhrskou, P.; Ulnits, L.; Kanstrup, I.L.; Nielsen, M.B.; Bjerg, A.M.; Krogsgaard, K. Effectiveness of active physical training as treatment for long-standing adductor-related groin pain in athletes: Randomised trial. *Lancet* **1999**, *353*, 439–443. [CrossRef]

35. Cowan, S.M.; Schache, A.G.; Brukner, P.; Bennell, K.L.; Hodges, P.W.; Coburn, P.; Crossley, K.M. Delayed onset of transversus abdominus in long-standing groin pain. *Med. Sci. Sports Exerc.* **2004**, *36*, 2040–2045. [CrossRef] [PubMed]

Article

The Acute Effect of Match-Play on Hip Isometric Strength and Flexibility in Female Field Hockey Players

Violeta Sánchez-Migallón [1], Alvaro López-Samanes [1,*], Pablo Terrón-Manrique [1], Esther Morencos [2], Vicente Fernández-Ruiz [1], Archit Navandar [3] and Victor Moreno-Pérez [4]

[1] Exercise Physiology Group, School of Physiotherapy, Faculty of Health Sciences, Universidad Francisco de Vitoria, Pozuelo de Alarcón, 28049 Madrid, Spain; violeta.smigallon@ufv.es (V.S.-M.); p.terron.prof@ufv.es (P.T.-M.); vicente.fernandez@ufv.es (V.F.-R.)

[2] School of Exercise and Sports Sciences, School of Health Sciences, Universidad Francisco de Vitoria, Pozuelo de Alarcón, 28049 Madrid, Spain; esther.morencos@ufv.es

[3] Faculty of Sport Sciences, Europea de Madrid University, Villaviciosa de Odón, 28670 Madrid, Spain; archit.navandar@universidadeuropea.es

[4] Center for Translational Research in Physiotherapy, Department of Pathology and Surgery, Miguel Hernández University, 03202 Elche, Alicante, Spain; vmoreno@goumh.umh.es

* Correspondence: alvaro.lopez@ufv.es; Tel.: +34-91-709-1400 (ext. 1955)

Received: 22 June 2020; Accepted: 15 July 2020; Published: 17 July 2020

Abstract: The aim of this study was to determine the acute effect of simulated field hockey match-play on isometric knee flexion, adductor (ADD) and abductor (ABD) strength, adductor/abductor (ADD/ABD) strength ratio, countermovement jump height (CMJ), hip flexion and ankle dorsiflexion range of motion (ROM). Thirty competitive female field hockey players (23.0 ± 3.9 years old) participated in the study. Apart from the afore-mentioned variables, external (through GPS) and internal load (through RPE) were measured before (pre-match) and immediately after simulated hockey match-play (post-match) in both limbs. Isometric knee flexion strength (+7.0%, $p = 0.047$) and hip flexion ROM (+4.4%, $p = 0.022$) were higher post-match in the non-dominant limb, while CMJ values reduced (-11.33%, $p = 0.008$) when comparing from pre-match data. In addition, no differences were observed for isometric hip ADD, ABD, ADD/ABD strength ratio, passive hip flexion ROM and ankle dorsiflexion ROM test. A simulated field-hockey match produces an increment in hip isometric strength and hip flexion ROM values in the non-dominant limb and a decrease in jump height capacity. As a result, CMJ assessment should be considered post-match in order to identify players who would require further rest before returning to training.

Keywords: risk factors; performance; team sport; fatigue; groin; hamstring

---

1. Introduction

Field hockey is an intermittent sport where hockey players perform repeated actions such as changes of direction, dribbles, sprints, accelerations, decelerations and body impacts alternating high and moderate with low intensity efforts [1]. During an official field hockey match, consisting of four quarters of fifteen minutes each, hockey players cover around 6000–8000 m [2,3] primarily at low and medium intensities, with high-intensity efforts (>19 km/h^{-1}) making up around 6% of the total playing time [2]. Previous studies have reported an average of 14 to 48 injuries per 1000 h attributed to the high physical demands of this sport [4,5]. Specifically, most of these field hockey injuries have been reported in the lower limbs, especially in the thigh and groin [6], with the hamstring strain injury being the most frequent muscle injury (32%) [4], followed by the groin injury (10%) [6]. Consequently, identification of the risk factors associated with groin and/or hamstring injury occurrence is essential.

In this sense, previous studies in different intermittent sports such as football [7], tennis [8] and ice hockey [9], have identified several modifiable intrinsic risk factors as causing an increased likelihood of developing groin and hamstring injuries. Among them, a weakness in the isometric adductor strength (ADD) [7,10] and lower adduction/abduction strength ratio (ADD/ABD) [9] have been associated with a higher risk of sustained groin injuries, while a lower hamstring strength [11–13], decreased range of hip flexion [14] and ankle dorsiflexion range of motion (ROM) have been associated with hamstring strain injuries [15]. However, some conflicting results have been found in literature regarding these factors [16–18]. Notably, most researches have investigated these risk factors before the commencement of the season or in off-season situations [19]. However, the ability to capture fluctuations in ROM and/or strength profile in-season, specifically in response to match-play, has not been studied [19].

Similar to other intermittent sports, field hockey players reported a higher incidence of injuries during matches compared to training [6] probably due to the higher intensity reported in matches versus training, and the appearance of fatigue [20,21]. It is well known that there is a decrement of lower limbs' power performance after match-play, and this measure is one of the potential factors in injury causation in intermittent sport [22]. Appearance of fatigue is known to reduce sports performance through reduced muscle strength, neuromuscular control and ROM [22]. In this line, the impact of match-play on ADD, ABD, hamstring strength and ROM during hip flexion and ankle dorsiflexion has been studied recently in several sports such as tennis, football and basketball [19,23–25]; however, to the best of the authors' knowledge, there is no information regarding this effect in field hockey. Therefore, the aim of this study was to examine the acute effect of hockey match-play on several risk factors such as isometric knee flexion and hip ADD and ABD strength, ADD/ABD strength ratio, passive hip flexion ROM and ankle dorsiflexion ROM and countermovement jump (CMJ) in elite female hockey players.

2. Materials and Methods

2.1. Subjects

Thirty highly competitive female hockey players (age, 23.0 ± 3.9 years; body mass, 60.0 ± 7.5 kg; height, 1.60 ± 0.09 m; body mass index, 22.0 ± 2.1 kg·m^2; hours per week, 9.4 ± 4.4; playing experience, 14.3 ± 4.9 years) volunteered to participate in this investigation. The players were recruited from two different professional teams. The inclusion criteria were: (a) To be healthy and able to complete a full game of field hockey; (b) to be uninjured and declared match-fit by the medical and coaching staff at the time of the experiment and not to have taken any type of medication to treat pain or musculoskeletal injuries at the time of the study; and (c) to have an absence of late onset muscle pain during the training session [26]. All players were informed of the tests they were to perform and signed the consent form. The experimental procedure of this study was conducted in accordance with the Declaration of Helsinki and the approval by the Ethics Committee of the University Francisco de Vitoria, number 45/2018.

2.2. Experimental Protocol

Following their arrival, female hockey players filled out a questionnaire which included personal information such as body mass, height, medical history, training frequency and experience (practice hours per week and playing experience in field hockey). Testing (i.e., ROM, isometric strength, and countermovement jump) was performed in the clinical area of each field hockey club. Testing was conducted by two sports physiotherapists: a senior physiotherapist with over nineteen years of experience and a junior physiotherapist with two years of experience, to ensure participants' positioning during measurements. Considering recommendations by Wollin, Thorborg and Pizzari [19], the testing order of the players and the selection of the leg tested were randomly chosen prior to the pre-match test. Pre-match testing was performed 60 min prior to match-play, and the post-match re-testing was performed immediately after the match. At the beginning of the pre-match testing, participants carried out a standardized warm-up that consisted of 5 min of jogging at 10 km·h^{-1} and 5 min of static stretches and joint mobility exercises [27]. Subsequently, participants played the simulated field hockey match according to the International Hockey Federation rules on a rectangular surface, 91.40 m long and 55.00 m wide. The external load of the simulated matches were estimated using a global positioning system (GPS) (Wimu ProTM, RealTrack Systems, Spain) placed in specific vests worn by the players, these devices operated at a sampling frequency of 10 Hz and its validity and reliability have been reported previously [28]. In addition, subjective internal load of the game was obtained using the modified RPE scale (i.e., 0–10 points) within 30 min of match termination [29]. The following variables were used to assess the external load during match-play, total distance covered per minute at different velocities ranges during a 60 min match as previously reported [3]. To reduce the interference of uncontrolled variables, all subjects were instructed to maintain their habitual lifestyle and normal dietary intake before and during the study, and refrain from caffeine ingestion 24 h before the experiment [30].

2.3. Isometric Strength of Abductors (ABD) Adductors (ADD) and Knee Flexion

Hip isometric ADD and ABD strength were measured according to the methodology previously reported [31] using a portable handheld dynamometer (Nicholas Manual Muscle Tester; Lafayette Indiana Instruments, Lafayette, IN, USA). Participants were placed in a supine position with their hips in a neutral position and told to stabilize themselves by holding onto the sides of the table. Examiner 1 applied a resistance on a fixed position (ABD: At 5 cm proximal to the lateral malleolus; ADD: At 5 cm proximal to the medial malleolus). The hockey players were instructed to exert a voluntary contraction for a maximum of 5 s against the dynamometer [31]. Two attempts were registered for each contraction of each limb and a 30 s rest period between attempts. Regarding the isometric knee flexion, the strength test was evaluated by placing the subject in the prone position, with 15 degrees of knee flexion and with their hips in a neutral position [32]. Examiner 1 placed the dynamometer on the distal portion of the sural triceps, three centimeters above the bimalleolar line. Examiner 2 clamped the subject's pelvis over the sacrum, to prevent elevation during the test. Examiner 1 requested the participant to flex their knee with the intention of bringing the heel of the foot to the gluteus. Similarly, two repetitions were recorded for each limb with a 30 s rest between attempts. Isometric hip ADD, ABD and knee flexion strength was expressed as the maximal hip and knee torque per kilogram of body weight (Nm·kg^{-1}) using the external lever arm and body weight of each participant. The mean value out of two attempts was recorded and selected for further analysis.

2.4. Ankle Dorsiflexion ROM

Unilateral ankle dorsiflexion ROM was measured with LegMotion System (LegMotion, Check your Motion, Albacete, Spain). The testing was carried out following the methodology previously described by Calatayud et al. [33]; participants were in a standing position on the LegMotion System with the tested foot on the measurement platform and the contralateral foot out of the platform with the toes at the edge of it. Each player performed the test with their hands on the hips and the assigned foot in the middle of the longitudinal line behind the transversal line of the platform. From this position, subjects were instructed to flex the knee forwards, placing it in contact with the metal stick. When the subject was able to maintain heel and knee contact, the metal stick was progressively moved away from the knee, and the following achieved distance was recorded. Two attempts were allowed for each limb (i.e., left and right), with 15 s of passive recovery between trials. The mean value of the two attempts was selected for further analysis.

2.5. Hip Flexion ROM

Passive hip flexion ROM values with the knee extended were evaluated with the Straight Leg Elevation Test (SLET). Participants made two maximum passive attempts for the dominant and non-dominant leg, when the difference between one attempt and another was greater than 5%, a third attempt was made, selecting the mean value of the two attempts whose results were similar for further statistical analysis [34]. A unilevel inclinometer ISOMED (Portland, OR, USA) with a telescopic was used for the measurement. The test ended with one or more of the following criteria: (a) The examiner was unable to continue the joint movement evaluated due to the high resistance developed by the stretched muscle group; (b) The participant reported an important sensation of discomfort; or (c) The examiners noted compensations that could increase the ROM [35]. The inclinometer was placed approximately on the external malleolus and the distal arm was aligned parallel to an imaginary bisecting line of the extremity [35]. The mean value out of two attempts was recorded and selected for further analysis.

2.6. Countermovement Jump (CMJ)

Participants carried out three repetitions of CMJ using a contact mat jump system (Chronojump Boscosystem, Barcelona, Spain) with their arms on hips [36]. They were instructed to jump and land in the same place, with the body in an erect position during the jump until landing. Each participant performed two maximal CMJs interspersed with 45 s of passive recovery. In addition, the mean value out of two attempts was recorded and selected for further analysis.

2.7. Statistical Analysis

Data were calculated as means/standard deviation. The Shapiro–Wilk test was selected to assess the normal distribution. All study variables were compared using a t test (pre- vs. post-match). The statistical significance level was set at $p < 0.05$. Cohen's effect sizes were calculated and presented with their respective 95% confidence intervals (C.I.) based on the following criteria: Trivial effect (0–0.19), small effect (0.20–0.49), medium effect (0.50–0.79) and large effect (0.80 and greater) [37]. All the statistical analyses were completed using the SPSS software version 25 (SPSS Inc., Chicago, IL, USA).

3. Results

3.1. Match-Play Workload

The internal match-play workload was 6.83 ± 0.80 units (RPE). In addition, female hockey players covered a mean distance of 5456.50 ± 699.09 m across different velocity profiles (Table 1).

Table 1. Mean distances and % of total distance covered during the match at different velocity ranges.

Velocity Range	Distance Covered (m)	Total Distance (%)
0.0–6.0 km·h^{-1}	2692.04 ± 614.70	49.36 ± 9.71
6.1–12.0 km·h^{-1}	1781.15 ± 380.97	32.47 ± 4.53
12.1–18.0 km·h^{-1}	851.66 ± 282.16	15.74 ± 5.48
18.1–21.0 km·h^{-1}	107.59 ± 75.87	1.98 ± 1.40
21.1–24.0 km·h^{-1}	21.59 ± 23.66	0.40 ± 0.44
>24.1 km·h^{-1}	2.47 ± 6.19	0.05 ± 0.12

Abbreviations: km·h^{-1} = kilometers/hour; m = meters.

3.2. Isometric Strength and Countermovement Jump

No statistical differences were seen in relative isometric hip ABD strength in the dominant (+2.42%, $p = 0.864$, ES [C.I.] = 0.01 [−0.08, 0.11]) (Figure 1a) and non-dominant limb (−1.46%, $p = 0.834$, ES [C.I.] = −0.02 [−0.12, 0.07]) (Figure 1b); nor in the relative isometric hip ADD strength in the dominant(−2.10%, $p = 0.399$, ES [C.I.] = 0.10 [0.00, 0.19]) (Figure 1c) and non-dominant limb (+3.38%, $p = 0.349$, ES [C.I.] = 0.11 [0.02, 0.21]) (Figure 1d). In addition, no differences were obtained in ADD/ABD strength ratios in dominant (1.14 vs. 1.19, $p = 0.220$, ES [C.I.] = 0.28 [0.19–0.37]) and non-dominant limbs (0.98 vs. 0.96, $p = 0.600$, ES [C.I.] = 0.14 [0.05–0.24]) when comparing them pre and post-match. However, for isometric knee flexion strength, statistical differences were obtained in the non-dominant limb (7.0%, $p = 0.047$; ES [C.I.] = 0.29 [0.20, 0.38]) (Figure 1f) but no differences were reported in the dominant limb (0.1%, $p = 0.983$; ES [C.I.] = 0.11 [0.02, 0.21]) (Figure 1f). Finally, neuromuscular fatigue was measured by a countermovement jump test after match-play (23.0 ± 4.9 vs. 20.5 ± 6.6 cm, $p = 0.008$, ES [C.I.] = 0.44 [0.34, 0.53]).

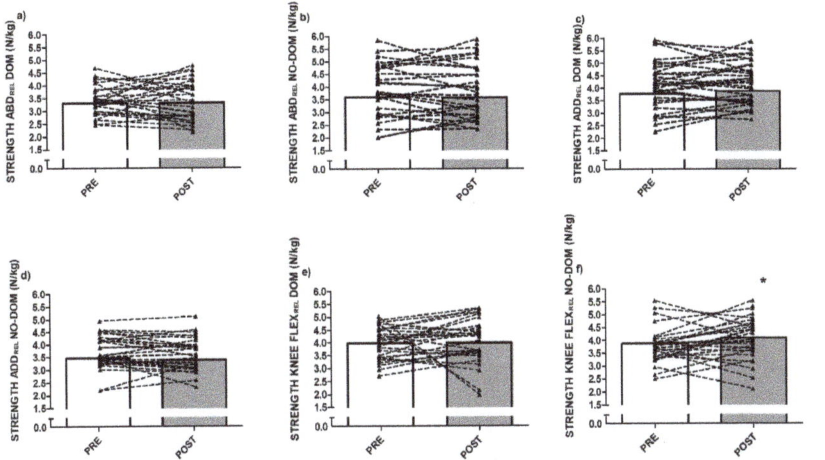

Figure 1. Hip and knee isometric hip abduction (ABD), adduction (ADD) and knee flexion strength values. (**a**) Relative isometric hip ABD strength in the dominant limb; (**b**) relative hip abductor strength in the non-dominant limb; (**c**) relative hip adductor strength in the dominant limb; (**d**) relative hip adductor strength in the non-dominant limb; (**e**) relative knee flexion strength in the dominant limb; (**f**) relative knee flexion strength in the non-dominant limb.

3.3. Hip Flexion and Ankle ROM

A significant increase was found when comparing the ROM in the hip flexion (straight leg elevation raise test) for the non-dominant limb (+4.38% $p = 0.022$). However, no differences were found in the dominant limb (+1.19%, $p = 0.753$) (Table 2). In addition, no differences were obtained for ankle dorsiflexion ROM values after field hockey match in the dominant limb (−3.77%; $p = 0.316$) and non-dominant limb (−2.34%; $p = 0.362$) (Table 2).

Table 2. Hip and ankle range of motion (ROM).

Variables	Pre-Match	Post-Match	*p*-Value	ES [95%CI]
Straight Leg Elevation Test DOM (°)	81.1 ± 12.3	82 0 ± 9.1	0.753	0.09 [0.00–0.18]
Straight Leg Elevation Test NO-DOM (°)	81.4 ± 10.3	84.9 ± 9.9	0.022 *	0.35 [0.26–0.45]
Ankle dorsiflexion DOM (cm)	10.9 ± 2.5	10.4 ± 2.7	0.316	0.16 [0.06–0.25]
Ankle dorsiflexion NO-DOM (cm)	11.1 ± 2.7	10.8 ± 2.7	0.362	0.10 [0.00–0.19]

Abbreviations: DOM = dominant-side; NO-DOM = non-dominant side; ° = degrees; cm = centimeters; * Significant differences compared to the PRE values at $p < 0.05$.

4. Discussion

The aim of this study was to determine the acute effect of hockey match-play on several risk factors such as isometric knee flexion, hip ADD and ABD strength, ADD/ABD strength ratio, passive hip flexion ROM, ankle dorsiflexion ROM and CMJ in competitive female hockey players. To the best of the authors' knowledge, this is the first study that analyzed the acute effects of hockey match-play on several risk factors in female athletes. The main results showed that hockey match-play acutely produced a decrease in CMJ performance, and an increase in isometric knee flexion strength and hip flexion ROM with knee extension in the non-dominant limb. However, no significant differences were found in isometric hip ADD, ABD strength and ADD/ABD strength ratio and ankle dorsiflexion ROM in both limbs.

The CMJ is one of the most important tests used to evaluate the lower-limb muscles fatigue [38,39]. The results in this study showed a significant reduction in levels of performance in CMJ (−11.33%) after match-play which was in agreement with previous studies conducted in other intermittent sports [40,41]. Recent research from Kim and Kipp [42] has shown that the gastrocnemius, soleus and vastus muscles have the largest contribution to vertical center of mass (COM) acceleration during the CMJ, and the soleus and gastrocnemius muscles function closest to their maximal capacities. If one were to look at the distances covered by the players (Table 1), one can observe that most of the distance had been covered at lower intensities. Here, the production of the horizontal force has been attributed to the muscles of the lower limb: namely tibialis anterior, gastrocnemius and soleus [2]. This indicates that a greater fatigue caused by distances ran at these intensities appears to have increased the neuromuscular fatigue associated with the CMJ, leading to a decrease in performance. In addition, albeit speculative, another possible reason for the decrease values in the CMJ test after match-play has been attributed to disruptions within the muscular fibers [41], increasing some markers of muscle damage (e.g., creatine-kinase, myoglobin) after a match in intermittent sports [41,43]. Moreover, previous literature described the inflammatory responses and fibrillar damage to the muscles and showed a decrease in metabolic indices in athletes after playing a match; however, this speculation requires further investigation.

Muscle strength in the lower limbs is essential to produce explosive actions in hockey (e.g., accelerations, changes of direction). The results obtained in the present study showed improvements in the isometric knee flexion strength (+7.0%) immediately post-match in the non-dominant limb. While no previous study has reported knee flexion strength values in field hockey players, these results differ from previous studies conducted in other intermittent sports, which revealed a lower knee flexion strength post-match-play [19,44]. The lack of agreement between studies could be related to the different match-play demands of the participants (e.g., total distance covered, duration of match, etc.). While in the current study hockey players reported an average of 5456.50 ± 699.09 m total distance covered and 21.59 ± 23.66 m at high speeds over 21 km·h^{-1}, previous studies in female soccer players showed they covered distances over 10,000 m during a match (90 min duration), of which at least 600 m were at high speed running intensities [45]. The current results suggest that the effect provoked by field-hockey match-play did not decrease the isometric knee flexion strength in the dominant limb, which has been related to a higher hamstring injury risk in previous studies [46,47]. The absence of changes in the isometric knee flexion strength can be attributed to the lower distances covered at high running speeds. Extensive research [47] has shown that the hamstring muscles are most active in this phase, when their function is to increase stride frequency and produce a greater horizontal force as the contact time reduces. Given the distances covered in the different zones, it can be assumed that the hamstring muscles are not fatigued as in sports such as soccer, where a greater distance is covered at these intensities.

A reduced ROM during the straight leg raise test and dorsiflexion of the ankle ROM has been linked to the risk of hamstring injury [14,15]. Present results have shown an increase in the straight leg raise ROM levels for the non-dominant limb (+5.81%) and no significant differences in the dominant limb. No differences were seen for the ankle dorsiflexion ROM either for both legs after match-play. To the best of the authors' knowledge, only Wollin et al. (2017) analyzed the straight leg raise ROM after match-play on intermittent sports and report no significant differences after match-play [19]. Concerning ankle dorsiflexion ROM, some previous studies conducted in football [24], basketball [25] and Australian-rules football [44] showed different results. While Charlton et al. [44] reported reductions in ankle dorsiflexion ROM immediately after an Australian-rules football match, Wollin et al. [19] showed non-significant decrements in football players, and finally Moreno-Pérez et al. [24,25] in football and basketball observed increased ROM values post-match from pre-match in dominant and non-dominant limbs. While these sports involve the same multidirectional movements (e.g., accelerations, decelerations, changes of direction) during practice, they cannot be compared due to the differences in the characteristics of the sport; for example, the total duration is less in field hockey than other intermittent sports. Thus, increases in hip flexion ROM immediately post-match are likely due to the increase in tissue extensibility induced by temperature increment which leads to a reduction of the viscous resistance of muscle tissues and joints [48,49]. However, this is a speculation that needs to be elucidated in future studies.

As far as isometric hip ADD and ABD strength values and ADD/ABD strength ratio go, the results showed no significant differences between post-match from pre-match. While is difficult to establish comparisons, as no previous study has reported isometric hip ADD and ABD strength values in hockey players, the findings of the current study were similar to those of a previous study [50] conducted with 14 rugby players. However, the present results disagree with the results reported by a previous study in tennis players [23]. This lack of agreement between studies may be due to differences in physical demands and tactical aspects between the sports. In tennis, players are required to perform repetitive short high-intensity movements, which impose an elevated concentric and eccentric load on the ADD muscles, while hockey players perform multiple different movements at several intensities. One must also consider that hockey, like in rugby, permits rolling substitutions.

This study contains limitations that require acknowledgment. Firstly, as the current study was conducted with a specific sample of female field hockey players, conducted during a simulated match, the characteristics of the players do not permit a generalization of the results found. In addition, the selected measures in this study in response to hockey match-play were performed immediately post-match; future studies should evaluate these variables using several time-points to understand the recovery fatigue induces, for example registering data 48 h after the match. This study looked at the effect following a single simulated match; more data collected over a series of official matches, or even, over an entire season could provide more conclusive results about the effects of match-play on these potential risk factors.

5. Conclusions

A simulated field-hockey match increases hip isometric flexion strength and hip flexion ROM in the non-dominant limb and decreases jump height capacity. However, no differences were reported in isometric ADD and ABD strength and ADD/ABD strength ratio in the dominant and non-dominant limb, or in dominant hip flexion and ankle dorsiflexion ROM values between the pre- and post-match examinations. Finally, female hockey players who present a decrease in jump height capacity, may require additional rest between training and competitions.

Author Contributions: Conceptualization, V.S.-M., A.L.-S. and V.M.-P.; methodology, A.L.-S. and V.M.-P.; software, A.N.; validation, V.S.-M., A.L.-S., A.N. and V.M.-P.; formal analysis, V.S.-M., A.L.-S., A.N. and V.M.-P.; investigation, V.S.-M., A.L.-S., P.T.-M., E.M., V.F.-R., A.N. and V.M.-P.; resources, V.S.-M., A.L.-S., P.T.-M. and V.M.-P.; data curation, V.S.-M., A.L.-S., A.N. and V.M.-P.; writing-original draft preparation, V.S.-M., A.L.-S., A.N. and V.M.-P.; writing-review and editing, V.S.-M., A.L.-S., P.T.-M., E.M., V.F.-R., A.N. and V.M.-P.; visualization, A.L.-S. and V.M.-P.; supervision, A.L.-S. and V.M.-P.; project administration, A.L.-S. and V.M.-P.; funding acquisition, P.T.-M.; All authors have read and agreed to the published version of the manuscript.

Funding: This research received no external funding.

Acknowledgments: We would like to thank the participants for their uninterested participation in this project.

Conflicts of Interest: The authors declare no conflict of interest.

References

1. Reilly, T.; Borrie, A. Physiology applied to field hockey. *Sports Med.* **1992**, *14*, 10–26. [CrossRef] [PubMed]
2. Lythe, J.; Kilding, A.E. Physical demands and physiological responses during elite field hockey. *Int. J. Sports Med.* **2011**, *32*, 523–528. [CrossRef] [PubMed]

3.   White, A.D.; MacFarlane, N. Time-on-pitch or full-game GPS analysis procedures for elite field hockey? *Int. J. Sports Physiol. Perform.* **2013**, *8*, 549–555. [CrossRef] [PubMed]

4.   Rodas, G.; Pedret, C.; Yanguas, J.; Pruna, R.; Medina, D.; Hägglund, M.; Ekstrand, J. Male field hockey prospective injury study. Comparison with soccer. *Arch. Med. Deporte* **2009**, *129*, 22–30.

5.   Theilen, T.M.; Mueller-Eising, W.; Wefers Bettink, P.; Rolle, U. Injury data of major international field hockey tournaments. *Br. J. Sports Med.* **2016**, *50*, 657–660. [CrossRef] [PubMed]

6.   Barboza, S.D.; Joseph, C.; Nauta, J.; van Mechelen, W.; Verhagen, E. Injuries in Field Hockey Players: A Systematic Review. *Sports Med.* **2018**, *48*, 849–866. [CrossRef]

7.   Engebretsen, A.H.; Myklebust, G.; Holme, I.; Engebretsen, L.; Bahr, R. Intrinsic risk factors for groin injuries among male soccer players: A prospective cohort study. *Am. J. Sports Med.* **2010**, *38*, 2051–2057. [CrossRef]

8.   Moreno-Perez, V.; Lopez-Valenciano, A.; Barbado, D.; Moreside, J.; Elvira, J.L.L.; Vera-Garcia, F.J. Comparisons of hip strength and countermovement jump height in elite tennis players with and without acute history of groin injuries. *Musculoskelet Sci. Pract.* **2017**, *29*, 144–149. [CrossRef]

9.   Tyler, T.F.; Nicholas, S.J.; Campbell, R.J.; McHugh, M.P. The association of hip strength and flexibility with the incidence of adductor muscle strains in professional ice hockey players. *Am. J. Sports Med.* **2001**, *29*, 124–128. [CrossRef]

10.   Moreno-Perez, V.; Travassos, B.; Calado, A.; Gonzalo-Skok, O.; Del Coso, J.; Mendez-Villanueva, A. Adductor squeeze test and groin injuries in elite football players: A prospective study. *Phys. Ther. Sport Off. J. Assoc. Chart. Physiother. Sports Med.* **2019**, *37*, 54–59. [CrossRef]

11.   Lee, J.W.Y.; Mok, K.M.; Chan, H.C.K.; Yung, P.S.H.; Chan, K.M. Eccentric hamstring strength deficit and poor hamstring-to-quadriceps ratio are risk factors for hamstring strain injury in football: A prospective study of 146 professional players. *J. Sci. Med. Sport* **2018**, *21*, 789–793. [CrossRef]

12.   Ribeiro-Alvares, J.B.; Oliveira, G.D.S.; De Lima, E.S.F.X.; Baroni, B.M. Eccentric knee flexor strength of professional football players with and without hamstring injury in the prior season. *Eur. J. Sport Sci.* **2020**, 1–9. [CrossRef] [PubMed]

13.   Tol, J.L.; Hamilton, B.; Eirale, C.; Muxart, P.; Jacobsen, P.; Whiteley, R. At return to play following hamstring injury the majority of professional football players have residual isokinetic deficits. *Br. J. Sports Med.* **2014**, *48*, 1364–1369. [CrossRef] [PubMed]

14.   Witvrouw, E.; Danneels, L.; Asselman, P.; D'Have, T.; Cambier, D. Muscle flexibility as a risk factor for developing muscle injuries in male professional soccer players. A prospective study. *Am. J. Sports Med.* **2003**, *31*, 41–46. [CrossRef] [PubMed]

15.   Gabbe, B.J.; Finch, C.F.; Bennell, K.L.; Wajswelner, H. Risk factors for hamstring injuries in community level Australian football. *Br. J. Sports Med.* **2005**, *39*, 106–110. [CrossRef]

16.   Mosler, A.B.; Weir, A.; Serner, A.; Agricola, R.; Eirale, C.; Farooq, A.; Bakken, A.; Thorborg, K.; Whiteley, R.J.; Holmich, P.; et al. Musculoskeletal Screening Tests and Bony Hip Morphology Cannot Identify Male Professional Soccer Players at Risk of Groin Injuries: A 2-Year Prospective Cohort Study. *Am. J. Sports Med.* **2018**, *46*, 1294–1305. [CrossRef]

17.   van Dyk, N.; Bahr, R.; Burnett, A.F.; Whiteley, R.; Bakken, A.; Mosler, A.; Farooq, A.; Witvrouw, E. A comprehensive strength testing protocol offers no clinical value in predicting risk of hamstring injury: A prospective cohort study of 413 professional football players. *Br. J. Sports Med.* **2017**, *51*, 1695–1702. [CrossRef]

18.   van Dyk, N.; Farooq, A.; Bahr, R.; Witvrouw, E. Hamstring and Ankle Flexibility Deficits Are Weak Risk Factors for Hamstring Injury in Professional Soccer Players: A Prospective Cohort Study of 438 Players Including 78 Injuries. *Am. J. Sports Med.* **2018**, *46*, 2203–2210. [CrossRef]

19.   Wollin, M.; Thorborg, K.; Pizzari, T. The acute effect of match play on hamstring strength and lower limb flexibility in elite youth football players. *Scand. J. Med. Sci. Sports* **2017**, *27*, 282–288. [CrossRef] [PubMed]

20.   Morencos, E.; Romero-Moraleda, B.; Castagna, C.; Casamichana, D. Positional Comparisons in the Impact of Fatigue on Movement Patterns in Hockey. *Int. J. Sports Physiol. Perform.* **2018**, *13*, 1149–1157. [CrossRef]

21.   Rishiraj, N.; Taunton, J.E.; Niven, B. Injury profile of elite under-21 age female field hockey players. *J. Sports Med. Phys. Fit.* **2009**, *49*, 71–77.

22.   Small, K.; McNaughton, L.R.; Greig, M.; Lohkamp, M.; Lovell, R. Soccer fatigue, sprinting and hamstring injury risk. *Int. J. Sports Med.* **2009**, *30*, 573–578. [CrossRef] [PubMed]

23. Moreno-Perez, V.; Nakamura, F.Y.; Sanchez-Migallon, V.; Dominguez, R.; Fernandez-Elias, V.E.; Fernandez-Fernandez, J.; Perez-Lopez, A.; Lopez-Samanes, A. The acute effect of match-play on hip range of motion and isometric strength in elite tennis players. *PeerJ* **2019**, *7*, e7940. [CrossRef] [PubMed]

24. Moreno-Perez, V.; Soler, A.; Ansa, A.; Lopez-Samanes, A.; Madruga-Parera, M.; Beato, M.; Romero-Rodriguez, D. Acute and chronic effects of competition on ankle dorsiflexion ROM in professional football players. *Eur. J. Sport Sci.* **2020**, *20*, 51–60. [CrossRef]

25. Moreno-Perez, V.; Del Coso, J.; Raya-Gonzalez, J.; Nakamura, F.Y.; Castillo, D. Effects of basketball match-play on ankle dorsiflexion range of motion and vertical jump performance in semi-professional players. *J. Sports Med. Phys. Fit.* **2020**, *60*, 110–118. [CrossRef] [PubMed]

26. Navandar, A.; Veiga, S.; Torres, G.; Chorro, D.; Navarro, E. A previous hamstring injury affects kicking mechanics in soccer players. *J. Sports Med. Phys. Fit.* **2018**, *58*, 1815–1822. [CrossRef] [PubMed]

27. Mora-Rodriguez, R.; Garcia Pallares, J.; Lopez-Samanes, A.; Ortega, J.F.; Fernandez-Elias, V.E. Caffeine ingestion reverses the circadian rhythm effects on neuromuscular performance in highly resistance-trained men. *PLoS ONE* **2012**, *7*, e33807. [CrossRef]

28. Bastida-Castillo, A.; Gomez-Carmona, C.D.; De La Cruz Sanchez, E.; Pino-Ortega, J. Comparing accuracy between global positioning systems and ultra-wideband-based position tracking systems used for tactical analyses in soccer. *Eur. J. Sport Sci.* **2019**, *19*, 1157–1165. [CrossRef]

29. Foster, C.; Florhaug, J.A.; Franklin, J.; Gottschall, L.; Hrovatin, L.A.; Parker, S.; Doleshal, P.; Dodge, C. A new approach to monitoring exercise training. *J. Strength Cond. Res.* **2001**, *15*, 109–115.

30. Jodra, P.; Lago-Rodriguez, A.; Sanchez-Oliver, A.J.; Lopez-Samanes, A.; Perez-Lopez, A.; Veiga-Herreros, P.; San Juan, A.F.; Dominguez, R. Effects of caffeine supplementation on physical performance and mood dimensions in elite and trained-recreational athletes. *J. Int. Soc. Sports Nutr.* **2020**, *17*, 2. [CrossRef]

31. Thorborg, K.; Bandholm, T.; Petersen, J.; Weeke, K.M.; Weinold, C.; Andersen, B.; Serner, A.; Magnusson, S.P.; Holmich, P. Hip abduction strength training in the clinical setting: With or without external loading? *Scand. J. Med. Sci. Sports* **2010**, *20*, 70–77. [CrossRef] [PubMed]

32. Thorborg, K.; Petersen, J.; Magnusson, S.P.; Holmich, P. Clinical assessment of hip strength using a hand-held dynamometer is reliable. *Scand. J. Med. Sci. Sports* **2010**, *20*, 493–501. [CrossRef] [PubMed]

33. Calatayud, J.; Martin, F.; Gargallo, P.; Garcia-Redondo, J.; Colado, J.C.; Marin, P.J. The validity and reliability of a new instrumented device for measuring ankle dorsiflexion range of motion. *Int. J. Sports Phys. Ther.* **2015**, *10*, 197–202. [PubMed]

34. Cejudo, A.; Sainz de Baranda, P.; Ayala, F.; Santonja, F. Test-retest reliability of seven common clinical tests for assessing lower extremity muscle flexibility in futsal and handball players. *Phys. Ther. Sport Off. J. Assoc. Chart. Physiother. Sports Med.* **2015**, *16*, 107–113. [CrossRef] [PubMed]

35. Moreno-Perez, V.; Ayala, F.; Fernandez-Fernandez, J.; Vera-Garcia, F.J. Descriptive profile of hip range of motion in elite tennis players. *Phys. Ther. Sport Off. J. Assoc. Chart. Physiother. Sports Med.* **2016**, *19*, 43–48. [CrossRef]

36. Bosco, C.; Luhtanen, P.; Komi, P.V. A simple method for measurement of mechanical power in jumping. *Eur. J. Appl. Physiol. Occup. Physiol.* **1983**, *50*, 273–282. [CrossRef]

37. Cohen, J. A power primer. *Psychol. Bull.* **1992**, *112*, 155–159. [CrossRef]

38. Markovic, G.; Dizdar, D.; Jukic, I.; Cardinale, M. Reliability and factorial validity of squat and countermovement jump tests. *J. Strength Cond. Res.* **2004**, *18*, 551–555. [CrossRef]

39. Gathercole, R.J.; Sporer, B.C.; Stellingwerff, T.; Sleivert, G.G. Comparison of the Capacity of Different Jump and Sprint Field Tests to Detect Neuromuscular Fatigue. *J. Strength Cond. Res.* **2015**, *29*, 2522–2531. [CrossRef]

40. de Hoyo, M.; Cohen, D.D.; Sanudo, B.; Carrasco, L.; Alvarez-Mesa, A.; Del Ojo, J.J.; Dominguez-Cobo, S.; Manas, V.; Otero-Esquina, C. Influence of football match time-motion parameters on recovery time course of muscle damage and jump ability. *J. Sports Sci.* **2016**, *34*, 1363–1370. [CrossRef]

41. Andersson, H.; Raastad, T.; Nilsson, J.; Paulsen, G.; Garthe, I.; Kadi, F. Neuromuscular fatigue and recovery in elite female soccer: Effects of active recovery. *Med. Sci. Sports Exerc.* **2008**, *40*, 372–380. [CrossRef]

42. Kim, H.; Kipp, K. Number of Segments Within Musculoskeletal Foot Models Influences Ankle Kinematics and Strains of Ligaments and Muscles. *J. Orthop. Res. Off. Publ. Orthop. Res. Soc.* **2019**, *37*, 2231–2240. [CrossRef] [PubMed]

43. Magalhaes, J.; Rebelo, A.; Oliveira, E.; Silva, J.R.; Marques, F.; Ascensao, A. Impact of Loughborough Intermittent Shuttle Test versus soccer match on physiological, biochemical and neuromuscular parameters. *Eur. J. Appl. Physiol.* **2010**, *108*, 39–48. [CrossRef] [PubMed]

44. Charlton, P.C.; Raysmith, B.; Wollin, M.; Rice, S.; Purdam, C.; Clark, R.A.; Drew, M.K. Knee flexion strength is significantly reduced following competition in semi-professional Australian Rules football athletes: Implications for injury prevention programs. *Phys. Ther. Sport Off. J. Assoc. Chart. Physiother. Sports Med.* **2018**, *31*, 9–14. [CrossRef]

45. Datson, N.; Drust, B.; Weston, M.; Jarman, I.H.; Lisboa, P.J.; Gregson, W. Match Physical Performance of Elite Female Soccer Players During International Competition. *J. Strength Cond. Res.* **2017**, *31*, 2379–2387. [CrossRef] [PubMed]

46. Schache, A.G.; Dorn, T.W.; Williams, G.P.; Brown, N.A.; Pandy, M.G. Lower-limb muscular strategies for increasing running speed. *J. Orthop. Sports Phys. Ther.* **2014**, *44*, 813–824. [CrossRef]

47. Morin, J.B.; Gimenez, P.; Edouard, P.; Arnal, P.; Jimenez-Reyes, P.; Samozino, P.; Brughelli, M.; Mendiguchia, J. Sprint Acceleration Mechanics: The Major Role of Hamstrings in Horizontal Force Production. *Front. Physiol.* **2015**, *6*, 404. [CrossRef]

48. Bishop, D. Warm up I: Potential mechanisms and the effects of passive warm up on exercise performance. *Sports Med.* **2003**, *33*, 439–454. [CrossRef]

49. Opplert, J.; Babault, N. Acute Effects of Dynamic Stretching on Muscle Flexibility and Performance: An Analysis of the Current Literature. *Sports Med.* **2018**, *48*, 299–325. [CrossRef]

50. Roe, G.A.; Phibbs, P.J.; Till, K.; Jones, B.L.; Read, D.B.; Weakley, J.J.; Darrall-Jones, J.D. Changes in Adductor Strength After Competition in Academy Rugby Union Players. *J. Strength Cond. Res.* **2016**. [CrossRef]

Article

Jumping Side Volley in Soccer—A Biomechanical Preliminary Study on the Flying Kick and Its Coaching Know-How for Practitioners

Xiang Zhang [1,†], **Gongbing Shan** [1,2,3,*,†], **Feng Liu** [3] **and Yaguang Yu** [4]

1   Department of Physical Education, Xinzhou Teachers' University, Xinzhou 034000, China; xiangzhang@xztc.edu.cn
2   Biomechanics Lab, Faculty of Arts & Science, University of Lethbridge, Lethbridge, AB T1K 3M4, Canada
3   School of Physical Education, Shaanxi Normal University, Xi'an 710119, China; liufeng668@snnu.edu.cn
4   College of Martial Arts, Shandong Sport University, Rizhao 276800, China; yuyaguang@sdpei.edu.cn
*   Correspondence: g.shan@uleth.ca; Tel.: +1-403-329-2683
†   The authors contributed equally to this work.

Received: 4 May 2020; Accepted: 10 July 2020; Published: 12 July 2020

Abstract: The jumping side volley has created breathtaking moments and cherished memories for soccer fans. Regrettably, scientific studies on the skill cannot be found in the literature. Relying on the talent of athletes to improvise on the fly can hardly be considered a viable learning method. This study targeted to fill this gap by quantifying the factors of the jumping side volley and to contribute to the development of a coaching method for it. Using 3D motion capture (12 cameras, 200 Hz) and full-body biomechanical modeling, our study aimed to identify elements that govern the entrainment of skill execution. Given the rarity of players who have acquired this skill and the low success rate of the kick (even in professional games), we were able to achieve and review 23 successful trials from five college-level subjects and quantify them for the study. The results unveiled the following key elements: (1) the control of trunk rotation during jumping, (2) the angle between thighs upon take-off, (3) the whip-like control of the kicking leg while airborne, (4) timing between ball motion and limb coordination, and (5) damping mechanism during falling. An accurate kick can normally be achieved through repetitive training. This underlines the need for athletes to master a safe landing technique that minimizes risk of injury during practice. Therefore, training should begin with learning a safe falling technique.

Keywords: 3D motion capture; full-body biomechanical modeling; X-factor; hip flexibility; whip-like movement; dispersion of impact load during falling

---

1. Introduction

The great attraction of soccer for millions of fans may trace back to the basic idea of the game: the goal—an idea that never ceases to fascinate. Compared to many other sports, goals are relatively rare in soccer (on average <3 goals/game in the FIFA World Cup since the 1960s [1]). Because of the rarity, soccer goals are extremely exciting. The game can be thought of as an improvised drama, where emotional tension is built over long periods only to be fully released when a goal is achieved. In particular, the goals achieved by applying flying techniques such as the diving scorpion kick, bicycle kick, diving header, and jumping side volley are sources of rabid excitement. This uniqueness of soccer contributes to making the game the number one sport worldwide [2–5]. Among all the techniques, the jumping side volley is, no doubt, one of the infrequent scoring skills that fans invariably desire to see when attending games. Unfortunately, few players have performed this skill during national or international tournaments.

The jumping side volley is an acrobatic airborne technique (Figure 1). One can see its rarity from the European Championship 2016 (Euro 2016) where it was used for only one out of 108 goals [6]. However, this rare skill has created breathtaking moments and cherished memories for players and fans. A classic example is Wendell Lira's (Brazil) superb airborne side volley, which won the most prestigious FIFA Puskás Award 2015 [7]. In Euro 2016, Xherdan Shaqiri's (Swiss) jumping side volley was selected as the best goal of the tournament [8]. The novelty of the skill (just like the acrobatic bicycle kick) and its rarity are because these kicks are perceived as high-risk and low-return skills [9]. Regrettably, a scientific study on the skill cannot be found in the literature [4]. Relying on the talent of athletes to improvise on the fly can hardly be considered as a viable learning strategy. Therefore, in order to give efficient and effective information to practitioners, we launched a preliminary study [10] to give a scientific overview of the kick, to identify key features of the skill, and to formulate a scientific way of learning/training the technique. Specifically, this preliminary study had two objectives: (1) use of 3D motion capture technology to quantify the dominant factors contributing to the kick quality and (2) identify biomechanical elements that govern the entrainment of the jumping side volley in order to develop its coaching method.

Figure 1. Jumping side volley—a flying techniques for defeating goalies.

2. Materials and Methods

2.1. How to Establish a Lab Test Condition for Mimicking Reality in an Application-Oriented Investigation?

Since the study was an application-oriented investigation, the first challenge was how to gather realistic data. We went back to professional games to find a reasonable solution.

In reality, ball flying-in direction is arbitrary. Figure 2 supplies four successful examples selected from professional soccer games. Example 1 (Figure 2a) shows that the player used his head to set a flying-in ball vertically and then performed a jumping side volley. In examples 2 and 3 (Figure 2b,c), the player employed his chest and foot, respectively, to set a flying-in ball vertically for doing a jumping side volley. A vertically travelling ball can also be set by other players (Figure 2d, the goal keeper set a vertically travelling ball for a player). All four examples demonstrate that the vertically travelling ball is a frequent scenario for executing the jumping side volley. Therefore, the vertically travelling ball was used for each subject in our lab-based data collection.

Figure 2. Successful examples of the jumping side volley in professional games. The common feature of the examples is a vertically travelling ball. The ball condition can be created by a player using (**a**) his head; (**b**) his chest; (**c**) his foot; or (**d**) set by another player.

2.2. Motion Capture and Biomechanical Modeling

A 3D, twelve-camera VICON MX40 motion-capture system (VICON Motion Systems, Oxford Metrics Ltd., Oxford, England [11]) was used to measure the jumping side volley using 42 reflective, 9 mm markers on the body. The motion capture system tracked the markers at a rate of 200 frames/s. This capture rate has been widely applied in various analyses of complicated/elite sport-skills [12–16]. Figure 3 shows a 3D computer reconstruction of a single trial, including camera placement, capture volume, and a rendered stick figure. Markers were placed on subjects as follows: (1) four on the head, (2) trunk markers on the sternal end of the clavicle, xiphoid process of the sternum, C7 and T10 vertebrae, each scapula, left and right anterior superior iliac spine, posterior superior iliac spine, (3) upper-extremity markers on the right and left acromion, lateral side of upper arm, lateral epicondyle, lateral side of forearm, styloid processes of radius and ulna, and distal end of 3rd metacarpal bones, and (4) lower extremity markers on left and right lateral sides of thigh and shank, lateral tibial condyle, lateral malleolus, distal end of 5th metatarsal, calcaneus, and big toe. Please note that markers on the scapula, upper arms, forearms, thighs, and shanks, are referential, therefore, no accurate positions were required, i.e., as long as they are over the required bones, they will work. For reducing the risk of injury, soft markers were applied for the tests. These compress (and decompress) easily, therefore, their influence on skill performance is negligible. Calibration residuals were determined in accordance with VICON's guidelines [11] and yielded positional data accurate within 1 mm.

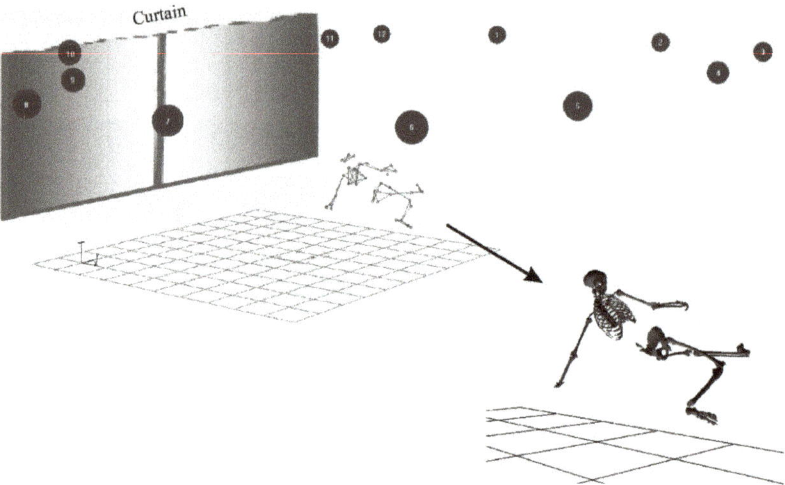

Figure 3. 3D motion capture reconstruction showing the 12 camera placements and a wire frame mesh reproduction of a jumping side volley (left) and a 15-segment biomechanical model built from the 3D data collected.

Additionally, three reflex markers (made from 3M reflective paper) were glued to the ball in order to quantify the soccer release speed.

VICON software triangulated the positions of each marker and rendered them in three-dimensional computer space. The raw data collected was processed by a five-point smoothing filter. The five-point filter was a premier filter in the time domain that reduced random noise while retaining a sharp step response [16]. It is widely applied to reduce noise from possible vibrations of the markers during 3D motion capture of sports and arts performances [14,17,18]. The resultant data supplied primary information, such as marker position, position changes, velocities, and accelerations. From the marker-position data, anatomical landmarks were established that allowed modeling of the skeletal structure for each participant. Using basic physics, simple positional data were translated into skeletal movement. VICON software provide tools for building a 15-segment biomechanical model of the soccer kick (Figure 3) [19–22]. Model segments were identified as follows: head, upper trunk, lower trunk, upper arms, lower arms, hands, thighs, shanks, and feet. The model calculated segment lengths, joint angles, and ranges of motion (ROMs) for the joints [23,24]. In such biomechanical modeling, inertial characteristics of the body are estimated using anthropometric norms found through statistical studies [25,26]. The modeling enables researchers to postulate motor control patterns. After model calculations, descriptive statistics (i.e., average and standard deviation) of body kinematic data (i.e., joint angles, joint ROMs, and coordination timing of joints) and correlation analyses among the body kinematic data with ball release speed were performed using EXCEL 2016. The ball release speed is commonly used to judge the kick quality [3–5], as such, the correlation analyses aimed to find the key/dominant factors among the kinematic data that could govern the entrainment of the jumping side volley.

Motion capture technology permits considerable freedom of movement for participants without negatively influencing their motor skill control. Taking advantage of this, we placed no restrictions on subjects' movements within the capture volume in an effort to preserve their normal "style". Given the rarity of individuals who have acquired this skill, we used the "self-identification" method in the search for subjects at three universities (all have the VICON system) and recruited 14 male soccer players. The subjects were informed of the testing procedures, signed consent forms, and voluntarily participated in the data collection. The universities' human-subject committees scrutinized and approved the test as meeting the criteria of ethical conduct for research involving human subjects. Through pretests, we found only five players who could actually perform this skill in the required test conditions, but not at a 100% success-rate. Two of them were Canadian and the others were Chinese. The anthropometrical characteristics and experience in soccer training were as follows: body height 1.74 ± 0.04 m, body weight 70.4 ± 3.8 kg, age 22.0 ± 1.6 years, and training 15.8 ± 1.5 years. After warm-up, each subject was asked to perform the skill six times. In total, 23 successful trials (i.e., the ball was accurately and powerfully kicked) were captured.

3. Results

Figure 4 shows that two events—take-off and ball contact—divide the jumping side volley into three phases: (1) the jumping phase, (2) the airborne phase, and (3) the landing phase. Our data reveal the following characteristics of the 1st phase: Before the volley, the athlete's trunk and pelvis rotate away from the goal. During the takeoff, the non-kick-side (NKS) leg is raised, at the same time, the trunk reverses rotational directions and twists toward the goal. In order to increase the range of motion (ROM) of trunk rotation, both arms abduct to near horizontal (over 80°, Table 1). Until the take-off, the ROM of the trunk twist (commonly known as the X-factor [27–29]) is about 40° (Table 1). At the end of the phase, the trunk-orientation approaches a more horizontal position (Figure 4). Correlation analyses confirm that the X-factor (α), angle between thighs (β) at take-off, and shoulder abduction during jumping are key/dominant factors, influencing the kick quality, i.e., the ball release speed (Figure 5a,b).

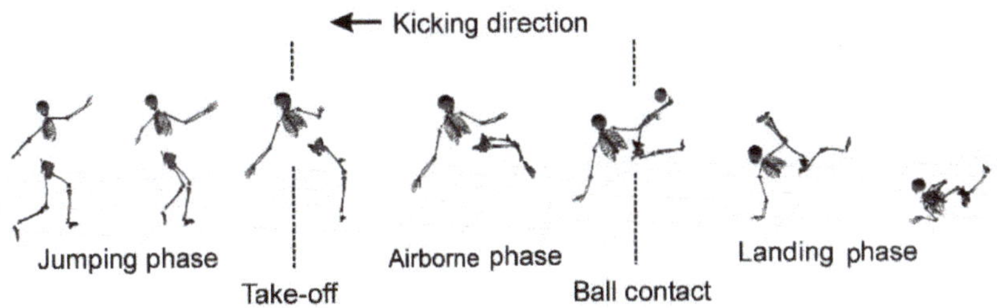

Figure 4. Phase identification based on 3D motion analysis data.

Table 1. Average and standard deviation of selected parameters, their confidence intervals (CI), coefficient of variation (CV), and their correlation with the ball release speed (r) ($p < 0.05$).

	Selected Parameters	Results	CI	CV	r
ROM (°)	Trunk Twist (X-factor α)	41.5 ± 5.1	40.5–42.6	0.12	0.86
	KS Hip (Flex/Ext)	51.6 ± 5.2	50.5–52.7	0.10	0.71
	KS Knee (Flex/Ext)	56.9 ± 6.1	55.6–58.1	0.11	0.76
	KS Ankle (Flex/Ext)	11.6 ± 2.2	11.1–12.1	0.19	0.52
Angle (°)	Max Shoulder Abduction *	83.7 ± 4.2	82.8–84.5	0.05	0.69
	Take-off Angle between Thighs (β)	75.2 ± 13.0	72.5–77.9	0.17	0.82
	Min KS Knee Angle	84.1 ± 9.6	82.1–86.1	0.11	−0.75
Coordination	Trunk → KS Hip	83.2 ± 4.1	82.3–84.1	0.05	0.57
Timing (%)	KS Hip → KS Knee	74.4 ± 15.4	71.2–77.6	0.21	0.89

X-factor: the angle between shoulder line (upper trunk) and hip line (lower trunk). Flex/Ext: Flexion/Extension, KS: Kick Side. *: the max value is the average of both shoulders.

Two notable characteristics of the 2nd phase (i.e., the airborne phase) are the flexed kick-side (KS) leg and the extended NKS leg (Figure 5c). Both legs form a scissor-movement, i.e., they move in opposite directions. The correlation analyses show that both the minimum angle and ROM of the KS knee are key factors affecting the kick quality (Table 1). From the timing perspective, the KS hip flexion starts when about 80% of the trunk twist toward the goal has finished. Similarly, the KS knee begins its extension after the KS hip finishes about 75% of its flexion (Table 1). Further, our data show that the explosive KS knee extension happens shortly before the ball contact, and is followed by an ankle flexion. The correlation analyses unveil that the sequential segment-coordination is also a key contributor to the kick quality.

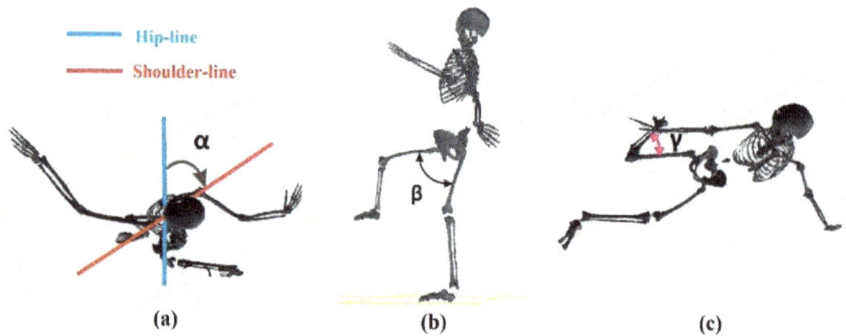

Figure 5. Key/dominate factors influencing kick quality: (**a**) X-factor α (top view); (**b**) Take-off angle between thighs β (side view); (**c**) Minimum KS knee angle γ (top-front view).

The main issue in the 3rd phase (i.e., the landing phase) is how to dissipate the impact load produced by falling in order to avoid potential injuries. Our data reveal that the well-trained athletes apply multiple landings to share the impact loads for reducing the risk of injury. The 1st landing is the flexed arm–hand chain (like a spring) for the 1st damping. The 2nd landing is the hip landing, followed by body rolling, sharing the rest load among multiple contact points.

4. Discussion

Since this study focuses on coaching practice, the results should be coach-friendly. A coach-friendly study, in our vision, should use a well-established scientific method to supply explanations on (1) scientifically-identified motor-control sequencing; (2) dominant factors (determined by correlation analyses) contributing to the control of the motor skill; and (3) instructions that can be understood by coaches and applied in their intervention in practice. Therefore, based on numerous analyses of our 3D data and their correlation analyses, the following key factors have been selected for illustrating the secrets of the control of the acrobatic skill in order to establish a practical way to learn and to train the jumping side volley:

- phase identification,
- trunk rotary control,
- hip flexibility,
- lower-limb control,
- upper-limb control, and
- coordination between upper- and lower-limbs.

A focused communication would help practitioners to understand the complex motor control in a timely and efficient way.

Regarding the determinants influencing the kick quality, the following key/dominant factors were revealed: There are two key factors in phase 1: the twisting control of the upper body (α) and the instance angle between the two thighs at take-off (β). The larger α and β are, the more powerful is the kick. Actually, these two factors play a crucial role in laying a foundation for performing the whip-like control of the kick in phase 2. In conjunction with the key factors in phase 2, the effects of α and β are elaborated below.

One notable feature of phase 2 is the asymmetric control of the legs, i.e., multi-segment control of the kick leg vs. quasi-single-segment control of the non-kick leg. This asymmetric control results in a difference between the moments of inertia of the two legs. The difference is vital for forwardly accelerating the kick leg; as such, it influences the quality of the whip-like movement. In the airborne phase, the human body follows the Law of Conservation of Angular Momentum. That means, the angular momentum of the forward action (the KS leg) equals the angular momentum of the backward reaction (the NKS leg). The flexed KS leg leads to a smaller moment of inertia (I), resulting in a faster forward motion in comparison to the extended NKS leg (a larger I creates lower backward motion) (Figure 6). In conjunction with phase 1, we found that the whip-system of the jumping side volley consists of four segments: trunk, KS-thigh, KS-shank, and kick foot. The whip-like kick is actually initiated in phase 1, beginning with the trunk twisting toward the goal, followed by hip flexion, knee extension, and ankle flexion, showing a sequential flow of energy and momentum transfer. It is well known that increasing the ROM of each segment will enhance the effect of the whip-like movement [14,19,28–31]. Therefore, flexibility of hip and knees should be emphasized during training, and the training should also pay attention to segment coordination (i.e., the whip-like control). Of course, timing is the most crucial element for coordination training.

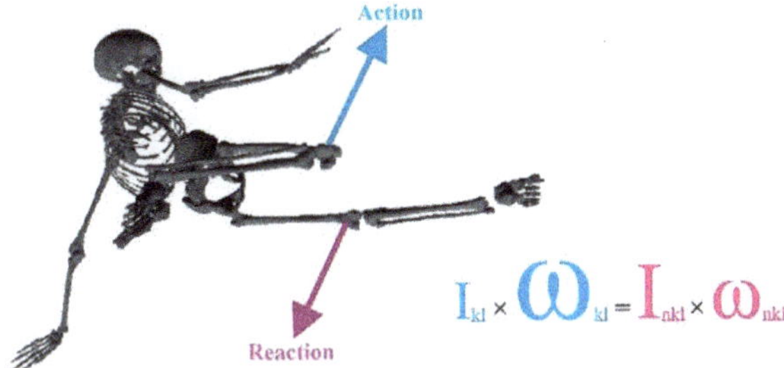

$$I_{kl} \times \omega_{kl} = I_{nkl} \times \omega_{nkl}$$

Figure 6. The asymmetric control of the legs during the airborne phase and their effects revealed by the Law of Conservation of Angular Momentum, i.e., the smaller the I, the faster the ω and vice versa.

For training of the jumping side volley, one should pay attention to the following three aspects: kick power, accuracy, and timing.

It is well known for practitioners that it is important to develop full-body whip-like control (from trunk to KS foot) for increasing kick power. For reaching this goal, jumping training should emphasize the upper-body twisting control, and the flexibility training should focus on the hip (one should always remember that the larger the β is, the more powerful is the kick). The airborne training should establish an asymmetrical control of legs as well as the ROM training of the KS-leg joints, i.e., KS leg performing a whip-like control and NKS leg remaining extended.

Repetitive training is a traditional way to increase the aspects of kick accuracy and timing. Those of us who are involved in the coaching practice know well that there are no short-cuts. The jumping side volley is considered high risk because of the inevitable fall to the ground, i.e., a realistic fear of injury to players/learners. We know that the feet and legs are naturally designed for absorbing landing impacts, but the arms and body are not. What does that mean? That means that bio-adaptation training for strengthening the arms and body against impact and safe-fall training to minimize injury risk during falling to the ground should first be considered in skill learning [32]. Hence, if we apply the repetitive-training approach in learning the jumping side volley, we should begin with phase 3, i.e., the learning and training of a safe falling technique, e.g., the multiple-landing technique revealed by the current study. Mastering a safe landing technique is the foundation that can ensure the repetitive training for improving kick accuracy and timing as well as the kick power.

Every study has limitations. As application-oriented research, the current study has two potential weaknesses. The obvious one is the low subject number and level. Future studies, if researchers cannot gain data from professional-level players, could first train advanced players based on the results of this study. As such, more subjects could be "produced" for future studies. The other one is the ecological approach to motor learning. For initiating the study, we simplified the ball movement. In reality, due to dynamic changes in the playing environment, a vertical travelling ball is only one of many possibilities. Therefore, identifying the learner–environment relationship as the basis for learning design in sport intentionality plays an important role in training. It will help to capture perception and action as intertwined processes underpinning individual differences in movement behavior [33,34]. The application of a nonlinear pedagogy, particularly in open and complex skills such as a jumping side kick, could reinforce the authors' intentions to help coaches to develop soccer-skills with their players. It would be a practitioner-desired topic for future studies.

5. Conclusions

The current study indicates one possibility to entrain the jumping side volley in soccer. Based on the results, the element training should focus on increasing the flexibility of the hip, the efficiency of the whip-like movement of the proximal-to-distal acceleration of the kicking leg, and the damping mechanism during falling. For skill training, one should focus on timing, because accurate timing is vital for a successful kick. One traditional way for improving timing is through repetitive training, which means repetitive falls to the ground during learning/training. Without safe fall protection, the skill learning/training presents a high risk of injury. This underlines the need for athletes to learn a safe landing technique that minimizes the risk of injury during practice. Therefore, if one applies the traditional approach in practice, training should begin with phase 3: mastering a safe fall.

Author Contributions: G.S., X.Z., F.L., and Y.Y. conceived and designed the experiments; G.S. and X.Z. performed the experiments; G.S., X.Z., F.L., and Y.Y. analyzed and discussed the data; G.S. and X.Z. contributed /materials/analysis tools; F.L. and Y.Y. prepared figures; G.S. and X.Z. wrote the paper; all authors contributed to the revisions and proof reading of the article. All authors have read and agreed to the published version of the manuscript.

Funding: The research project was supported by the National Sciences and Engineering Research Council of Canada (NSERC), grant number: RGPIN-2014-03648.

Conflicts of Interest: The authors declare no conflict of interest. The founding sponsors had no role in the collection, analyses, or interpretation of data; in the writing of the manuscript, and in the decision to publish the results.

References

1. FIFA. Average Number of Goals Scored per Game at the FIFA World Cup from 1930 to 2018. 2020. Available online: https://www.statista.com/statistics/269031/goals-scored-per-game-at-the-fifa-world-cup-since-1930/ (accessed on 26 January 2020).
2. Hyballa, P. The art of flying. *Success Soccer* **2002**, *5*, 19–26.
3. Reilly, T.; Williams, M. *Science and Soccer*, 2nd ed.; Routledge: London, UK, 2003.
4. Shan, G. Biomechanical Know-how of Fascinating Soccer-kicking Skills–3D, Full-body Demystification of Maximal Instep Kick, Bicycle Kick & Side Volley. In Proceedings of the 8th International Scientific Conference on Kinesiology, Zagreb, Opatija, Croatia, 10–14 May 2017.
5. Shan, G.; Zhang, X.; Wan, B.; Yu, D.; Wilde, B.; Visentin, P. Biomechanics of coaching maximal instep soccer kick for practitioners. *Interdiscip. Sci. Rev.* **2019**, *44*, 12–20. [CrossRef]
6. YouTube. All 108 UEFA EURO 2016 Goals: Watch Every One. 2016. Available online: https://www.youtube.com/watch?v=a4Qvh6VIoSs (accessed on 26 January 2020).
7. FIFA. FIFA Puskás Award. 2019. Available online: https://www.fifa.com/the-best-fifa-football-awards/puskas-award/ (accessed on 16 June 2019).

8.   UEFA. Top Ten Goals of UEFA EURO 2016 Revealed. 2016. Available online: https://www.uefa.com/uefaeuro-2020/news/0253-0d7e907eeaa5-b7d7554f6864-1000--top-ten-goals-of-uefa-euro-2016-revealed/ (accessed on 19 September 2018).

9.   Shan, G.; Visentin, P.; Zhang, X.; Hao, W.; Yu, D. Bicycle kick in soccer: Is the virtuosity systematically entrainable? *Sci. Bull.* **2015**, *60*, 819–821. [CrossRef]

10. Smith, P.G.; Morrow, R.H.; Ross, D.A. Preliminary studies and pilot testing. In *Field Trials of Health Interventions: A Toolbox*, 3rd ed.; Oxford University Press: Oxford, UK, 2015.

11. VICON. The Most Precise MoCap Ecosystem. 2020. Available online: https://www.vicon.com/applications/engineering/ (accessed on 26 January 2020).

12. Smith, S.L. Application of high-speed videography in sports analysis. In *Ultrahigh-and High-Speed Photography, Videography, and Photonics*; International Society for Optics and Photonics: Bellingham, WA, USA, 1993.

13. Chang, S.T.; Evans, J.; Crowe, S.; Zhang, X.; Shan, G. An innovative approach for Real Time Determination of Power and Reaction Time in a Martial Arts Quasi-Training Environment Using 3D Motion Capture and EMG Measurements. *Arch. Budo* **2011**, *7*, 185–196.

14. Wan, B.; Gao, Y.; Wang, Y.; Zhang, X.; Li, H.; Shan, G. Hammer Throw: A Pilot Study for a Novel Digital-route for Diagnosing and Improving Its Throw Quality. *Appl. Sci.* **2020**, *10*, 1922. [CrossRef]

15. Li, S.; Zhang, Z.; Wan, B.; Wilde, B.; Shan, G. The relevance of body positioning and its training effect on badminton smash. *J. Sports Sci.* **2017**, *35*, 310–316. [CrossRef] [PubMed]

16. Liu, Y.; Kong, J.; Wang, X.; Shan, G. Biomechanical analysis of Yang's spear turning-stab technique in Chinese martial arts. *Phys. Act. Rev.* **2020**, *8*, 16–22.

17. Michalowski, T. *Applications of MATLAB in Science and Engineering*; BoD–Books on Demand: Norderstedt, Germany, 2011.

18. Yu, D.; Yu, Y.; Wilde, B.; Shan, G. Biomechanical characteristics of the axe kick in Tae Kwon-Do. *Arch. Budo* **2012**, *8*, 213–218. [CrossRef]

19. Visentin, P.; Staples, T.; Wasiak, E.B.; Shan, G. A pilot study on the efficacy of line-of-sight gestural compensation while conducting music. *Percept. Mot. Skills* **2010**, *110*, 647–653. [CrossRef] [PubMed]

20. Shan, G.; Westerhoff, P. Full-body kinematic characteristics of the maximal instep soccer kick by male soccer players and parameters related to kick quality. *Sports Biomech.* **2005**, *4*, 59–72. [CrossRef] [PubMed]

21. Shan, G. Influences of Gender and Experience on the Maximal Instep Soccer Kick. *Eur. J. Sport Sci.* **2009**, *9*, 107–114. [CrossRef]

22. Shan, G.; Zhang, X. From 2D Leg Kinematics to 3D Full-body Biomechanics—The Past, Present and Future of Scientific Analysis of Maximal Instep Kick in Soccer. *Sports Med. Arthrosc. Rehabil. Ther. Technol.* **2011**, *3*, 23. [CrossRef] [PubMed]

23. Shan, G.; Yuan, J.; Hao, W.; Gu, M.; Zhang, X. Regression equations for estimating the quality of maximal instep kick by males and females in soccer. *Kinesiology* **2012**, *44*, 139–147.

24. Shan, G.; Daniels, D.; Wang, C.; Wutzke, C.; Lemire, G. Biomechanical analysis of maximal instep kick by female soccer players. *J. Hum. Mov. Stud.* **2005**, *49*, 149–168.

25. Wan, B.; Shan, G. Biomechanical modeling as a practical tool for predicting injury risk related to repetitive muscle lengthening during learning and training of human complex motor skills. *SpringerPlus* **2016**, *5*, 441. [CrossRef] [PubMed]

26. Shan, G.; Bohn, C. Anthropometrical data and coefficients of regression related to gender and race. *Appl. Ergon.* **2003**, *34*, 327–337. [CrossRef]

27. Winter, D.A. *Biomechanics and Motor Control of Human Movement*; John Wiley & Sons: Hoboken, NJ, USA, 2009.

28. Kearney, J.K.; Bhat, D.N.; Prasad, B.; Yuan, S.S. Efficient generation of whip-like throwing and striking motions. In *Models and Techniques in Computer Animation*; Magnenat-Thalmann, N., Thalmann, D., Eds.; Springer: Berlin, Germany, 1993; pp. 270–284.

29. Zhang, X.; Shan, G. Where do golf driver swings go wrong?—Factors Influencing Driver Swing Consistency. *Scand. J. Med. Sci. Sports* **2014**, *24*, 749–757. [CrossRef] [PubMed]

30. Zhang, Z.; Li, S.; Wan, B.; Visentin, P.; Jiang, Q.; Dyck, M.; Li, H.; Shan, G. The influence of X-factor (trunk rotation) and experience on the quality of the badminton forehand smash. *J. Hum. Kinet.* **2016**, *53*, 9–22. [CrossRef] [PubMed]

31. Wąsik, J.A.; Ortenburger, D.O.; Góra, T.O.; Shan, G.O.; Mosler, D.; Wodarski, P.; Michnik, R.A. The influence of gender, dominant lower limb and type of target on the velocity of taekwon-do front kick. *Acta Bioeng. Biomech.* **2018**, *20*, 133–138. [PubMed]
32. Ballreich, R.; Baumann, W. *Grundlagen der Biomechanik des Sports (The Basics of Biomechanics in Sports)*; Enke Verlag: Stuttgart, Germany, 1996.
33. Davids, K.; Araújo, D.; Hristovski, R.; Passos, P.; Chow, J.Y. Ecological dynamics and motor learning design in sport. In *Skill Acquisition in Sport: Research, Theory and Practice*, 2nd ed.; Hodges, N., Williams, M., Eds.; Routledge: London, UK, 2012; pp. 112–130.
34. Santos, S.D.; Memmert, D.; Sampaio, J.; Leite, N. The spawns of creative behavior in team sports: A creativity developmental framework. *Front. Psychol.* **2016**, *7*, 1282. [CrossRef] [PubMed]

Article

Effects of Plyometric Training on Surface Electromyographic Activity and Performance during Blocking Jumps in College Division I Men's Volleyball Athletes

Min-Hsien Wang [1], **Ke-Chou Chen** [2], **Min-Hao Hung** [3], **Chi-Yao Chang** [3], **Chin-Shan Ho** [3], **Chun-Hao Chang** [3] and **Kuo-Chuan Lin** [2,*]

[1] Department of Physical Education, Chinese Culture University, Taipei 11114, Taiwan; samwang6601@gmail.com
[2] Office of Physical Education, Chung Yuan Christian University, Taoyuan 320314, Taiwan; ivan@cycu.edu.tw
[3] Graduate Institute of Sports Science, National Taiwan Sport University, Taoyuan 333325, Taiwan; a2822180@gmail.com (M.-H.H.); anthonychang1124@gmail.com (C.-Y.C.); kilmur33@gmail.com (C.-S.H.); hao781106@gmail.com (C.-H.C.)
* Correspondence: misvb@cycu.edu.tw

Received: 29 April 2020; Accepted: 29 June 2020; Published: 30 June 2020

Abstract: In volleyball matches, there are three minute intervals between sets. Therefore, the improvement of the muscle output ratio is one of the most import foundational physical elements for the players. The purpose of this study was to investigate the effects of plyometric training on the changes in electrical signals in the lower limb muscles of male college volleyball players during continuous blocking and to examine the benefits of plyometric training on blocking agility and maximum vertical jump height. In this study, twenty elite male college volleyball players were recruited and divided into a plyometric training group (PTG) and a control group (CG). The wireless electromyography was used for data acquisition, and the electrodes were applied to the left and right rectus femoris, biceps femoris, tibialis anterior, and gastrocnemius. The median frequency was used as the measurement of the electromyographic signals during the jumping blocks. This study used covariate analysis methods, with previously measured results used as covariates to perform a two-way analysis of covariance for the independent samples. Based on the results of this study, after 6 weeks of training, the median frequency of the rectus femoris (2.13% to 4.75% improved) and that of the tibialis anterior muscles (4.14% to 7.71% improved) were significantly lower in the PTG than in the CG. Additionally, the blocking agility increased by 6.26% and the maximum vertical jump height increased by 3.33% in the PTG compared to the CG. The findings provide important insights on the neuromuscular status for volleyball players during continuous blocking jumps. Six weeks of appropriate plyometric training can facilitate the performance of volleyball players.

Keywords: muscular activity; blocking agility; maximum vertical jump height; median frequency

---

1. Introduction

Jumping ability is essential for performance in volleyball. Superior rebounding not only enables players to gain a competitive advantage on offense (increasing blocking height and attack angle) but also allows for a larger defensive range [1–3]. To increase the vertical jumping ability, weight-bearing jumping or plyometric jump-training methods can effectively improve leg muscles and explosive power, thereby improving overall strength and coordination in the legs. Vertical jumping is an important basic skill in many sports [4], but repetitive jumping is the primary cause of muscle fatigue. A study by Merletti and Parker [5] showed that the sites of neuromuscular fatigue can be divided into the following three major categories: central fatigue, fatigue of the neuromuscular junction, and muscle fatigue. Power-generating nerve signal transmission can be divided into central and peripheral nerve levels; thus, fatigue can be classified as central fatigue or peripheral fatigue based on the different sites of fatigue. Decreased muscle strength during fatigue is accompanied by central and peripheral effects. The former is associated with a decrease in the number of motor units involved in the action or a reduction in the frequency at which motor units are evoked.

It is well known that fatigue caused by vertical jumping can alter muscular characteristics, reduce muscle effectiveness, and change the maximum joint torque [6,7]. However, superior vertical jumping ability requires an excellent coordination of movements, that is, the ability to control and adjust musculoskeletal characteristics [8]. Therefore, fatigue may be a factor that causes changes in the central nervous system (CNS) and alters the coordination of the limbs during vertical jumps. The effects of post-jump fatigue on motor coordination have been discussed in previous studies [9–11], and the results showed that continuous jumping can reduce knee and ankle strength. Additionally, in the case of decreased muscle strength, it is not possible to improve strength even if the range of motion of the lower limb joints is increased.

Previous studies [12] have indicated that the rectus femoris, vastus lateralis, vastus medialis, and biceps femoris are the primary muscles used during squat jumps. After 50 repeated squat jumps, the maximum voluntary contraction of the knee extensors decreases by $25 \pm 11\%$, but there is no significant change in the knee flexors. In ve Dikey et al. [13], the correlation between maximum isokinetic strength, muscle activity, and jump height in 12 elite male volleyball players was investigated. The results of the study showed that hamstring/quadriceps ratios were greatest at an angular velocity of 240°/s. Additionally, the degree of biceps femoris activity was greater than that of the vastus lateralis and vastus medialis regardless of the angular velocity, and the degree of activity of the biceps femoris and the rectus femoris was consistent. The degree of muscular activity (MVC%) of the knee extensors and knee flexors increases with increasing fatigue. Therefore, some studies have recommended the evaluation of vertical jumping based on actual sports conditions [14,15]. In this method, individuals performed intermittent fixed-height jumps, such as intermittent jumps to 95% of the maximum vertical jump height, until the target jump height cannot be reached for three consecutive attempts. The degree of fatigue is estimated from the resting interval after each jump and the number of vertical jumps [14].

Plyometric training is a typical high-intensity exercise, and the generated peripheral fatigue can be significantly improved after plyometric training [16]. A main finding is that the maximum voluntary contraction and degree of activity generated by muscles are significantly improved after plyometric training [17]. Plyometric training can effectively enhance muscle activity and improve the muscle output ratio. This effect not only significantly increases the maximum jump height but also enhances the coordination of the lower limb muscles [18]. Plyometric training is a jumping training method that uses the physiological phenomenon of a stretch shortening cycle (SSC) to produce stronger contractions during the centripetal contraction phase [19]. Because volleyball players must perform movements such as repeatedly jumping, sprinting and changing direction, this training method is appropriate for targeting their physical training needs and is widely applicable [20,21].

According to previous studies, plyometric training has many positive effects on volleyball players. However, most of the prior studies focused on improvements in general physical abilities due to plyometric training. In this study, sport-specific tests were evaluated with game-like situations simulating volleyball blocking jumps. The tests required repetitions of blocking jumps, during which the fatigue state of the lower limb muscles was recorded. The measurement methods used in this study were modified from the study by Sheppard et al. [1], which used repetitive jumping blocks in competition-like conditions. Surface electromyography with electrodes attached to the lower limb muscle groups were used to elucidate the mechanisms and conditions caused by muscle fatigue. Plyometric training was used to strengthen muscle groups prone to fatigue so as to reduce the rate at which fatigue develops and enhance athletic performance. The initial hypothesis of this study was that plyometric training can induce positive changes in muscle activities in male college volleyball players.

2. Methods

2.1. Participants

This study was designed to be similar to a blocking action situation in actual competition. Players were divided into a plyometric training group (PTG) and a control group (CG) during the training period. Before entering the training cycle, the players in both groups underwent a pre-test. The PTG underwent six weeks of training after the pre-test, and the CG maintained the original training program. The two groups underwent a post-test after six weeks. In this study, 20 elite male college volleyball players were used as the study participants, and they were equally divided into the PTG (n = 10; mean age = 21.5 ± 1.2 years; mean height = 186.5 ± 5.1 cm; mean weight = 78.1 ± 4.7 kg) and the CG (n = 10; mean age = 22.1 ± 1.5 years; mean height = 176.5 ± 4.4 cm; mean weight = 77.4 ± 5.2 kg). All the participants in the study had participated in professional volleyball training for over five years and were registered in the men's Division I of the Republic of China University Volleyball League. The subjects were free of major musculoskeletal system disorders within the year preceding the study. This study conducted experiments using competition-like conditions during testing. Although slightly different from the actual competition conditions, verbal cues were used in the experiment, and the participants were required to make their best effort. All subjects gave their informed consent for inclusion before they participated in the study. The study was conducted in accordance with the Declaration of Helsinki, and the protocol was approved by the Ethics Committee of Fu Jen Catholic University Institutional Review Board (C103117).

2.2. Data Collection

In this study, 10 Vicon T40 motion capture systems (Vicon MX-Giganet, Oxford Metrics Ltd., UK) and Nexus software (Version 4.0.2.; Vicon Motion System Ltd, UK) were used with a frame rate of 250 Hz. One marker was attached on the seventh cervical vertebra (C7) to assess the jump height. In this study, wireless electromyography (BTS Free EMG; BTS Bioengineering Corp., USA) was used by attaching electrodes to muscle groups in the left and right legs, including the rectus femoris (RFM), biceps femoris (BFM), tibialis anterior (TAM), and gastrocnemius (GM) [22]. The sampling frequency was set to 1000 Hz. According to previous research [22], five muscles, each with their own specific anatomical function, have the major influences on jumping movements, as follows: RFM (hip joint flexion, knee joint extension), BFM (hip joint extension, knee joint flexion), vastus lateralis muscle (knee joint extension), GM (knee joint flexion, ankle plantar flexion), and TAM (ankle dorsiflexion, inversion). In this study, the athletes were required to perform blocking jumps in two directions for a total of 14 times. We aimed to observe the muscles that dominated in at least two functions during the blocking jumps. Therefore, the EMG measurements of the RFM, BFM, TAM, and GM were recorded. The blocking agility system (1000Hz sampling frequency) was used to record the total response time [23].

2.3. Experimental Design

Actual testing was performed as modified in the study by Sheppard et al. [1]. The blocking target was placed three meters away from the starting position of the movement, and the blocking height was set to 90% of the subject's individual jump height. The subjects performed 14 moving blocks. A total of three groups of tests (T1, T2, T3) were performed, with a three-minute rest between the groups. To give the subjects a fixed load, the blocking agility (BA) test from a study by Ho et al. [23] was used, with an eight-second interval between the blocks. All the data in this study were collected simultaneously using the Vicon server. The site layout is shown in Figure 1.

Figure 1. Diagram and explanation of the blocking agility test for volleyball. The athlete was required to perform a total of 14 blocks from two directions. The blocking agility test system and wireless electromyography (BTS Free EMG) were set in the front of the athlete.

2.4. Procedures

Before the start of the experiment, the participants were collectively informed about the purpose of the study and the experimental process, and any issues needing attention were addressed. Additionally, their rights during the study were disclosed, after which the subjects gave written informed consent. One week before testing, all the participants were provided with information about the experiment and administered two practice tests. During the course of the experiment, the subjects were instructed to not change their daily training program or training volume. The PTG underwent six weeks of plyometric training during the study. The primary purpose of their regimen was to train the muscle groups in the legs. The training program for this study was formulated as described by Makaruk and Sacewicz [18] (Table 1). During the six-week plyometric training program, the subjects participated in training sessions twice a week. The intensity and training volume of the regimen was based on the number of groups and repetitions proposed by Piper and Erdmann [24]. Two weeks of medium-intensity training and four weeks of high-intensity training were implemented in the program to improve the transmission of the CNS signals in the subjects and avoid excessive load or fatigue.

Table 1. Plyometric training program (PTG).

Plyometric Training Program Number of Sets × Number of Rebounds	Week 1–2 (Sets × Rebs)	Week 3–4 (Sets × Rebs)	Week 5–6 (Sets × Rebs)
Standing vertical hops	2 × 10		
Single foot hops	4 × 8		
Multiple two-foot hurdle jumps (hurdle height 0.55 m)	6 × 6		
Counter movement jumps	3 × 5		
Depth jumps (drop box height 0.20 m)	3 × 6		
Lateral two-foot jumps		2 × 10	
Two-foot jumps		4 × 8	
Counter movement jumps		3 × 5	
Multiple two-foot hurdle jumps (hurdle height 0.65 m)		6 × 6	
Depth jumps (drop box height 0.30 m)		3 × 6	
Two-foot jumps forward and backward:			2 × 10
Single foot jumps			2 × 8 on each foot
Counter movement jumps			3 × 5
Multiple two-foot hurdle jumps (hurdle height 0.76 m)			6 × 6
Depth jumps (drop box height 0.40 m)			3 × 6

2.5. Dependent Measures

The parameters investigated in this study were the myoelectric signals, blocking agility (BA), and the maximum vertical jump height. The collection and processing of each parameter are described as follows:

1. Collection and processing of the maximum vertical jump height

The participants were asked to stand on a force plate to perform a countermovement jump, which was measured using a three-dimensional motion capture system. The maximum vertical jump height test was performed three times. A program was written in Matlab software (Version R2008a; The MathWorks Inc., USA) to calculate the height of the highest point after the player takes off.

2. Blocking agility (BA) test

The subjects were required to warm up adequately for 10–15 min, with special attention focused on extending the joint ligaments in the legs. The subjects could complete the warm-up when they felt comfortable and ready. In the test, the blocking agility system was used to produce visual stimuli and record the total response time during blocking. The examiner gave a voiced signal to the athlete and triggered the blocking agility test system at the same time. The light cue was activated as stimulation 8 s after the system was triggered. During the BA test, the subjects were instructed to land on both feet. The actions were performed seven times in each direction (left and right) for a total of 14 blocks, and three repeated sets of records were analyzed. The subjects were directed to stand in the preparation area, which was three meters away from the blocking point, and to wait for a light to turn on. When the light turned on, the subject completed the blocking action as quickly as possible by touching the target.

3. Collection and processing of the muscle median frequency

Free EMG was used to wirelessly transmit data to a computer, and the collected EMG data were stored and Matlab 7.0.1 was used to read and write a program. The EMG signals were band-pass filtered (10–500 Hz) with a fourth order Butterworth filter. The median frequency (MDF) was derived from the EMG data; MDF is the frequency value that divides the power spectrum into two equal regions and is considered a reliable method for assessing muscle fatigue during exercise [25]. Studies [26,27] have shown that when muscles become fatigued, high-frequency motor units are evoked less frequently, and low-frequency motor units are evoked more frequently and recruited in larger numbers. This increases the slow contraction effects and slows muscle fiber conduction velocity, such that the MDF of the power spectrum is biased towards low frequencies (Figure 2); this characteristic can be used to assess muscle fatigue.

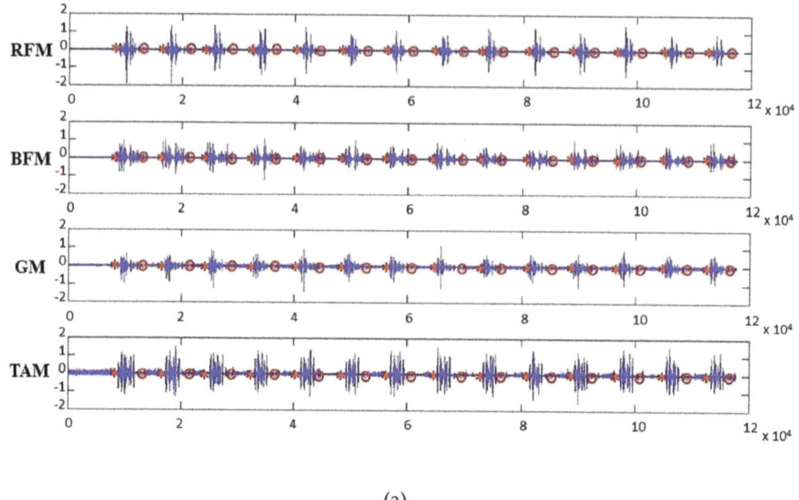

(a)

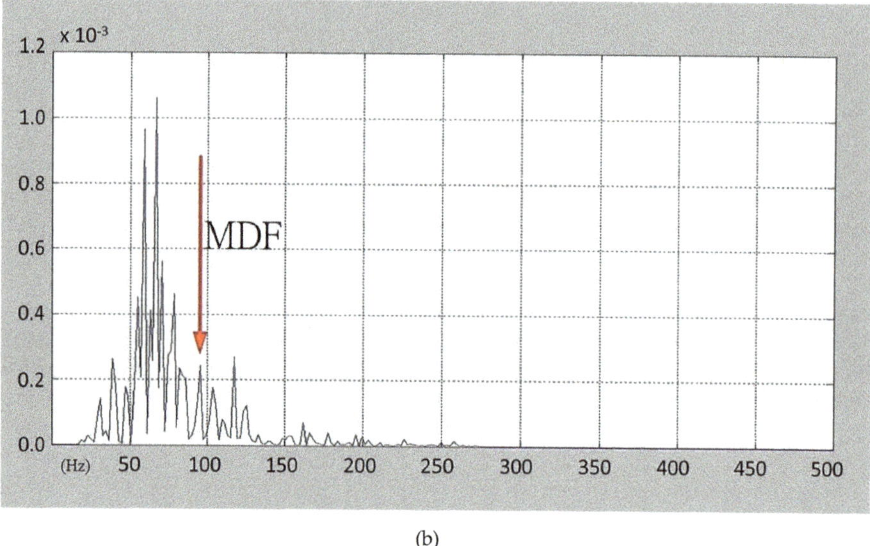

(b)

Figure 2. The signals of surface electromyographic activity and the results of median frequency (MDF) on the lower extremity muscles. (**a**) The EMG signals for the rectus femoris (RFM), biceps femoris (BFM), tibialis anterior (TAM), and the gastrocnemius (GM) during the blocking jumps. The time series of the EMGs were captured and analyzed between the subject contact "*" with the force-plate and when it leaves "o". (**b**)The median frequency of the signal spectrum.

2.6. Statistical Analysis

The Statistical Package for the Social Sciences (SPSS) 20.0 software (version 20.0; SPSS Inc., Chicago, IL, USA) was used for the statistics and data analysis. First, the reliability of the measured data was tested using the intra-class correlation coefficient (ICC). This study primarily investigated the effects of plyometric training and continuous blocking on the MDFs of the RFM, TAM, GM, and BFM of the study participants. Descriptive statistics methods were used to describe the median frequency of the participants tested before and after the blocking rounds. As some interfering factors in actual experimental situations can affect experimental results, a two-way analysis of covariance (ANCOVA) with pre-test performance as the covariate was used to correct for sources of error and increase the accuracy. If the effect of the two factors reached the level of significance, a simple effects test (Bonferroni post-hoc test) was performed. If the interaction did not reach the level of significance, a main effects test was performed. The significance level was set at alpha ≤ 0.05.

3. Results

The pre- and post-test ICCs for the MDF (RFM, TAM, GM, and BFM) and jump performances are shown in Table 2. All the variables showed moderate to excellent results. ANCOVA, which was used in this study, requires that the variables in the data are homogeneous. The results of Levene's test showed that all the variables were homogeneous ($p > 0.05$) and that there were no interactions between the covariate and the independent variables that met the basic assumptions for ANCOVA.

Table 2. The pre- and post-test intra-class correlation coefficient (ICC) for the MDF (RFM, TAM, GM, and BFM) and jump performances (blocking agility (BA) and maximum vertical jump height).

| | MDF | | | | Jump Performances | |
	RFM	TAM	GM	BFM	BA	Maximum Vertical Jump Height
Pre-test	0.991	0.976	0.614	0.660	0.973	0.964
Post-test	0.978	0.936	0.560	0.690	0.988	0.983

3.1. Muscle Activities of Blocking Jumps

The results of pre- and post-testing in the PTG and CG (Table 3) show that as the number of rounds increased, the MDF of both the RFM and TAM decreased, but the MDFs of the GM and BFM did not. The frequency was highest in the BFM followed by the TAM; the frequencies of the RFM and GM were similar. The differences in the RFM reached the level of significance with respect to the group and group × round but not with respect to the round. The comparison of the marginal means showed that the PTG > the CG. The differences in the TAM reached the level of significance with respect to the group and group × round but not with respect to the round. The comparison of the marginal means showed that the PTG > the CG. The differences in the GM did not reach the level of significance with respect to the group, round, or group × round. The differences in the BFM did not reach the level of significance with respect to the group, round, or group × round. As the interaction between the RFM and the TAM reached the level of significance with respect to the group × round, a simple main effects test was performed. The results for the RFM in the CG showed that the MDF decreased by 11.55% and 20.61% in the second and third rounds of the pre-test, respectively, and by 11.04% and 21.52% in the second and third rounds of the post-test, respectively, exhibiting similar declines. In the PTG, the MDF decreased by 13.49% and 22.42% in the second and third rounds of the pre-test, respectively, and by 11.36% and 17.67% in the second and third rounds of the post-test, respectively; the decrease in the MDF in the PTG was significantly slower than that in the CG. The results of the MDF for the TAM in the CG showed that the MDF decreased by 23.96% and 31.79% in the second and third rounds of the pre-test, respectively, and by 22.20% and 30.99% in the second and third rounds of the post-test, respectively; the degrees of decline in the MDF were nearly the same in both groups. In the PTG, the MDF decreased by 25.08% and 35.02% in the second and third rounds of the pre-test, respectively, and by 20.94% and 27.31% in the second and third rounds of the post-test, respectively.

Table 3. Values are the mean ± standard deviation. Data are reported for the MDF of the RFM, BFM, TAM and the GM during the blocking jumps. The unit of MDF is hertz (Hz).

		PTG			CG			Interaction Effect (F-value)	Simple Effects (F-value)	Post-hoc Test
		T1	T2	T3	T1	T2	T3			
RFM	Pre-test	109.7 ± 7.2	94.9 ± 3.7	85.1 ± 3.1	108.2 ± 6.4	95.7 ± 4.4	85.9 ± 3.9	23.624 **	T1 = 0.301	PTG > CG
	Post-test	109.2 ± 6.6	96.8 ± 3.6	89.9 ± 3.5	107.8 ± 6.4	95.9 ± 3.7	84.6 ± 4.0		T2 = 8.136 * T3 = 40.690 **	PTG > CG
BFM	Pre-test	200.4 ± 9.5	172.2 ± 32.8	172.4 ± 27.3	188.1 ± 19.4	174.8 ± 35.8	178.7 ± 30.1	1.639		
	Post-test	193.6 ± 5.6	174.3 ± 27.7	170.9 ± 27.1	187.4 ± 18.6	188.1 ± 18.4	183.1 ± 27.8			
TAM	Pre-test	182.2 ± 7.7	136.5 ± 4.3	118.4 ± 7.2	174.9 ± 15.9	133.0 ± 2.9	119.3 ± 6.7	11.086 **	T1 = 1.284	PTG > CG
	Post-test	180.5 ± 6.2	142.7 ± 6.4	131.2 ± 2.7	170.7 ± 17.7	132.8 ± 3.0	117.8 ± 7.6		T2 = 95.390 ** T3 = 109.149 **	PTG > CG
GM	Pre-test	101.1 ± 28.9	94.4 ± 11.2	100.9 ± 14.2	95.3 ± 9.3	95.5 ± 8.1	100.9 ± 97.2	1.093		
	Post-test	96.1 ± 29.3	87.5 ± 14.0	92.9 ± 10.2	92.0 ± 21.4	93.4 ± 9.7	99.1 ± 9.1			

PTG: plyometric training group; CG: control group. The table indicates significant change (post–pre) when using the pre-test score as a covariate: * $p < 0.05$; ** $p < 0.01$. RFM: rectus femoris muscle; BFM: biceps femoris muscle; TAM: tibialis anterior muscle; GM: gastrocnemius muscle.

3.2. Jump Performances

Table 4 shows the BA of the two groups measured before and after training. The average BA test score over the three rounds in the PTG and CG was 2.08 ± 0.14 s and 2.09 ± 0.14 s in the pre-test, respectively, and 1.95 ± 0.11 s and 2.08 ± 0.17 s in the post-test, respectively. This shows that the magnitude of decrease in the post-test was larger in the PTG than in the CG. The ANCOVA results showed that the BA test reached the level of significance with respect to the group but not with respect to the round or group × round ($p > 0.05$). Since there were only two levels of group factors, comparing the marginal means showed that the PTG > the CG; that is, the BA of the PTG was significantly higher than that of the CG. However, the results of the descriptive statistics showed that in the CG, the BA increased by 0.48% and 1.93% in the second and third rounds of the pre-test, respectively, and by 0.97% and 1.46% in the second and third rounds of the post-test, respectively, exhibiting similar increases. In the PTG, the BA increased by 1.96% and 3.43% in the second and third rounds of the pre-test, respectively, and by 1.04% and 1.55% in the second and third rounds of the post-test, respectively.

Table 4. Values are the mean ± standard deviation. Data are reported for the blocking agility (BA) test. The unit of BA is second (s).

	PTG			CG			Interaction Effect (F-value)	Main Effects (F-value)	Post-hoc Test
	T1	T2	T3	T1	T2	T3			
Pre-test	2.04 ± 0.16	2.08 ± 0.13	2.11 ± 0.13	2.07 ± 0.15	2.08 ± 0.16	2.11 ± 0.13	0.618	58.287 **	CG > PTG
Post-test	1.93 ± 0.12	1.95 ± 0.12	1.96 ± 0.11	2.06 ± 0.16	2.08 ± 0.16	2.09 ± 0.18			

PTG: plyometric training group; CG: control group. The table indicates significant change (post–pre) when using the pre-test score as a covariate: ** $p < 0.01$.

Table 5 shows the maximum vertical jump height of the two groups measured before and after training. The average jump height in the PTG and CG were 67.04 ± 3.83 cm and 66.86 ± 4.06 cm in the pre-test, respectively, and 69.27 ± 3.87 cm and 66.79 ± 3.76 cm in the post-test, respectively. This shows that the PTG had a large magnitude of improvement after six weeks of plyometric training. There were significant differences between the groups. Comparing the marginal means showed that the PTG > the CG, that is, the plyometric training could effectively improve the maximum vertical jumping ability of the study participants.

Table 5. Values are the mean ± standard deviation. Data are reported for the maximum vertical jump height. Units in centimeters (cm).

	PTG	CG	F-value	Post-hoc Test
Pre-test	67.04 ± 3.83	66.86 ± 4.06	26.862 **	PTG > CG
Post-test	69.27 ± 3.87	66.79 ± 3.76		

PTG: plyometric training group; CG: control group. The table indicates significant change (post–pre) with the pre-test score as a covariate: ** $p < 0.01$.

4. Discussion

Based on the results of this study, the decrease in the MDF of the RFM and TAM was significantly lesser in the PTG after six weeks of training compared to that of the CG, and there was significant improvement in both the BA performance and maximum vertical jump height in the PTG.

From the results, the MDF increased in both the RFM and TAM as the number of rounds increased. In other words, under the effects of the exercise load in this study, the contraction ability of the muscles was reduced, and the characteristics of the power spectrum were altered. Among the MDF data for the four muscles, the TAM showed the greatest improvement. However, the MDF of these four muscles still decreased as the number of rounds in the blocking test increased. The amount of muscle recruitment in various parts of the legs when jumping was studied [22], and the results showed that the principal muscle groups recruited when performing squat jumps are the TAM > RFM > GM >

BFM; that is, the TAM and RFM are the principal muscle groups used when jumping. In the results of the continuous jump blocking test conducted on the study participants in the present study, the decrease in MDF in the legs improved in the PTG after a six-week interventional training; this result is consistent with that of previous studies [16]. The report [16] indicated that the plyometric training (PT) increased central fatigue significantly by about 15–20%, but significantly decreased peripheral (muscle) fatigue during the 2-min MVC by about 10% in the quadriceps femoris. In this study, the MDF of the RMF and TAM of the PTG and CG decreased by similar percentages in the pre-test. However, the MDF of the PTG was significantly higher than that of the CG in the second round (by 2.13% and 4.14% for the PTG) and the third round (by 4.75% and 7.71% for the PTG). In other words, plyometric training can delay the decrease in MDF, but it cannot alter the development of fatigue.

The ANCOVA results show that the BA of the PTG improved significantly after six weeks of training. Although not reflected as statistically significant, the BA increased as the number of rounds increased. These results show that the change occurred more gradually in the PTG. With respect to the average percentage of improvement, the PTG improved by 6.26% after six weeks of interventional training, whereas the CG improved by only 0.48%. In volleyball, blocking is not only a defensive skill but also a skill essential for scoring. In team sports, the defensive ability is considered to be a display of individual agility [28,29]. Agility is an important index in team sports and therefore must be improved through functional training. Improving the ability of the neuromuscular system to adapt and control may be factors that improve BA. Plyometric training can stimulate the CNS signal transmission, which can improve the stretch-shortening cycle ability in the leg muscles. Through the training process, exercise patterns involving the ability to change directions in a short period of time can produce the effect of muscle stretching–contraction–circulation, yielding the benefits of reflexive muscle strength and stretch reflex characteristics and improving the instant reaction ability.

The ANCOVA results showed that the standing jump height in the PTG after six weeks of training was significantly higher than that of the CG, increasing by 2.23 cm (3.33%) in the PTG and slightly decreasing by 0.07 cm (–0.10%) in the CG. Although the improvement in the PTG was only approximately 2 cm, a difference in height of 2 cm is a sign of approaching the ability to break through the maximum limits of the body in players performing at high levels. The results of the previous studies have shown that undergoing plyometric training two to three times a week can effectively improve the maximum output power and the rebounding ability of muscles [30], and the results of the present study are consistent with those of previous studies.

The results from this study show that repeated jumping causes fatigue-related declines in the RFM and TAM. In particular, blocking jumps require fast SSC muscle activity. Because of the lower leg muscle activity in the braking phase, the muscle stiffness decreases and simultaneously diminishes the efficacy of SSC actions [31,32]. To achieve the maximum rebound in muscle elastic strain energy during the push-off phase, the muscle must maintain a high stiffness during the braking phase [33]. Therefore, the EMG activities change during the jump, depending on the stiffness of the muscle. In this study, the power spectrum of MDFs (RFM and TAM) was biased towards low frequencies. The blocking agility test in this experiment comprised 14 blocking jumps, which required the athletes to perform repeated SSC movements, which in turn led to declines in muscle activities. In this research, a six-week plyometric training period was applied to provide additional loads on the athletes and thereby stimulate adaptation. At the same time, the training program included sufficient rest time, which is important for the recovery of muscle functions, thereby delaying the occurrence of muscle fatigue and improving the CNS function after the training period. In addition, enhancements in the acquired abilities through plyometric training primarily result from the stimulation of the muscular and nervous systems, and the process of adaptation.

The athletes recruited in this study were sufficiently proficient and physically fit to fulfill the requirements of volleyball matches. While we found effective improvements in these volleyball players after plyometric training, we produced no conclusive results regarding the settings that should be used for optimal effects. Further studies should help determine the appropriate intensity of plyometric training. The results of this study may not be generalizable to populations differing from the skill level, sex and age. In this study, a potential weakness of the data collection method was that the BA test program included a fixed exercise intensity. However, the exercise intensity in real games is variable; therefore, the time intervals could be varied in future studies.

5. Conclusions

Based on the results of this study, six weeks of appropriate plyometric training can delay the decrease in MDF in the RFM and TAM, improve vertical jump height, and significantly shorten BA in volleyball players when movement and jumping are combined. Increasing the muscle strength and coordination of movements in the legs facilitates rapid and complete movements during a moving block; that is, the total power output during the nerve conduction and muscle contraction processes is improved. For volleyball athletes, jumping high and fast and maintaining a high level of performance on the court is the goal of training. Based on fundamental training principles (e.g., variation, periodicity, individualization), recommendations to coaches include the use of different jumping loads to stimulate adaptation in the athletes. Neuromuscular fatigue during continuous jumping can be evaluated using the MDF after EMG spectrum analysis. Such markers can therefore be used to assess the fatigue resistance of athletes in competition-like situations.

Author Contributions: Conceptualization, M.-H.W. and K.-C.C.; methodology, M.-H.W.; software, C.-S.H.; validation, M.-H.H., C.-Y.C. and K.-C.L.; formal analysis, M.-H.H.; investigation, M.-H.W.; resources, K.-C.C.; data curation, C.-H.C.; writing—original draft preparation, C.-Y.C.; writing—review and editing, K.-C.L.; visualization, K.-C.L.; supervision, M.-H.W.; project administration, C.-S.H. All authors have read and agreed to the published version of the manuscript.

Funding: This research received no external funding.

Conflicts of Interest: The authors declare no conflict of interest.

References

1. Sheppard, J.M.; Gabbett, T.; Taylor, K.L.; Dorman, J.; Lebedew, A.J.; Borgeaud, R. Development of a repeated-effort test for elite men's volleyball. *Int. J. Sport Physiol.* **2007**, *2*, 292–304. [CrossRef]

2. Sheppard, J.; Borgeaud, R.; Strugnel, A. Influence of stature on movement speed and repeated efforts in elite volleyball players. *J. Aust. Strength Cond.* **2008**, *16*, 12–14.

3. Sheppard, J.M.; Cronin, J.B.; Gabbett, T.J.; McGuigan, M.R.; Etxebarria, N.; Newton, R.U. Relative importance of strength, power, and anthropometric measures to jump performance of elite volleyball players. *J. Strength Cond. Res.* **2008**, *22*, 758–765. [CrossRef]

4. Menzel, H.J.; Chagas, M.H.; Szmuchrowski, L.A.; Araujo, S.R.; Campos, C.E.; Giannetti, M.R. Usefulness of the jump-and-reach test in assessment of vertical jump performance. *Percept. Mot. Skills.* **2010**, *110*, 150–158. [CrossRef]

5. Merletti, R.; Parker, P. *Electromyography—Physiology, Engineering and Noninvasive Applications*, 1st ed.; John Wiley & Sons Inc.: Hoboken, NJ, USA, 2004.

6. Bojsen-Møller, J.; Magnusson, S.P.; Rasmussen, L.R.; Kjaer, M.; Aagaard, P. Muscle performance during maximal isometric and dynamic contractions is influenced by the stiffness of the tendinous structures. *J. Appl. Physiol.* **2005**, *99*, 986–994. [CrossRef]

7. Ishikawa, M.; Komi, P.V.; Finni, T.; Kuitunen, S. Contribution of the tendinous tissue to force enhancement during stretch–shortening cycle exercise depends on the prestretch and concentric phase intensities. *J. Electromyogr. Kines.* **2006**, *16*, 423–431. [CrossRef]

8. Bobbert, M.F.; Van Der Krogt, M.M.; Van Doorn, W.H.; de Ruiter, C.J. Effects of Fatigue of Plantarflexors on Control and Performance in Vertical Jumping. *Med. Sci. Sports Exerc. Off. J. Am. Coll. Sports Med.* **2011**, *43*, 673–684. [CrossRef]

9. Horita, T.; Komi, P.V.; Nicol, C.; Kyrolainen, H. Effect of exhausting stretch-shortening cycle exerciseon the time course of mechanical behaviour in the drop jump: Possible role of muscle damage. *Eur. J. Appl. Physiol.* **1999**, *79*, 160–167. [CrossRef]

10. Horita, T.; Komi, P.; Hämäläinen, I.; Avela, J. Exhausting stretch-shortening cycle (SSC) exercise causes greater impairment in SSC performance than in pure concentric performance. *Eur. J. Appl. Physiol.* **2003**, *88*, 527–534. [CrossRef]

11. Kuitunen, S.; Avela, J.; Kyrolainen, H.; Nicol, C.; Komi, P.V. Acute and prolonged reduction in joint stiffness in humans after exhausting stretch-shortening cycle exercise. *Eur. J. Appl. Physiol.* **2002**, *88*, 107–116.

12. Neyroud, D.; Samararatne, J.; Kayser, B.; Place, N. Neuromuscular Fatigue After Repeated Jumping With Concomitant Electrical Stimulation. *Int. J. Sport Physiol.* **2017**, *12*, 1335–1340. [CrossRef] [PubMed]

13. ve Dikey, A.; Şimşek, D.; Kırkaya, İ.; Güngör, E.O.; Soylu, A.R. Relationships among Vertical Jumping Performance, EMG Activation, and Knee Extensor and Flexor Muscle Strength in Turkish Elite Male Volleyball Players. *Turk. Klin. J. Sports Sci.* **2016**, *8*, 46–56.

14. Pereira, G.; de Freitas, P.B.; Rodacki, A.; Ugrinowitsch, C.; Fowler, N.; Kokubun, E. Evaluation of an innovative critical power model in intermittent vertical jump. *Int. J. Sports Med.* **2009**, *30*, 802–807. [CrossRef] [PubMed]

15. Pereira, G.; Correia, R.; Ugrinowitsch, C.; Nakamura, F.; Rodacki, A.; Fowler, N.; Kokubun, E. The rating of perceived exertion predicts intermittent vertical jump demand and performance. *J. Sports Sci.* **2011**, *29*, 927–932. [CrossRef]

16. Skurvydas, A.; Brazaitis, M.; Streckis, V.; Rudas, E. The effect of plyometric training on central and peripheral fatigue in boys. *Int. J. Sports Med.* **2010**, *31*, 451–457. [CrossRef]

17. Nordlund, M.M.; Thorstensson, A.; Cresswell, A.G. Central and peripheral contributions to fatigue in relation to level of activation during repeated maximal voluntary isometric plantar flexions. *J. Appl. Physiol.* **2004**, *96*, 218–225. [CrossRef]

18. Makaruk, H.; Sacewicz, T. Effects of plyometric training on maximal power output and jumping ability. *Hum. Mov.* **2010**, *11*, 17–22. [CrossRef]

19. Markovic, G.; Mikulic, P. Neuro-musculoskeletal and performance adaptations to lower-extremity plyometric training. *Sports Med.* **2010**, *40*, 859–895. [CrossRef]

20. Kim, Y.Y.; Park, S.E. Comparison of whole-body vibration exercise and plyometric exercise to improve isokinetic muscular strength, jumping performance and balance of female volleyball players. *J. Phys. Ther. Sci.* **2016**, *28*, 3140–3144. [CrossRef]

21. Pereira, A.M.; Costa, A.; Santos, P.; Figueiredo, T.; Vicente João, P. Training strategy of explosive strength in young female volleyball players. *Medicina* **2015**, *51*, 126–131. [CrossRef]

22. Wulf, G.; Dufek, J.S.; Lozano, L.; Pettigrew, C. Increased jump height and reduced EMG activity with an external focus. *Hum. Mov. Sci.* **2010**, *29*, 440–448. [CrossRef] [PubMed]

23. Ho, C.S.; Lin, K.C.; Chen, K.C.; Chiu, P.K.; Chen, H.J. System design and application for evaluation of blocking agility in volleyball. *Proc. Inst. Mech. Eng. Part P J. Sports Eng. Technol.* **2016**, *230*, 195–202. [CrossRef]

24. Piper, T.J.; Erdmann, L.D. A 4 step plyometric program. *Strength Cond. J.* **1998**, *20*, 72–73. [CrossRef]

25. Hayashibe, M.; Zhang, Q.; Guiraud, D.; Fattal, C. Evoked EMG-based torque prediction under muscle fatigue in implanted neural stimulation. *J. Neural Eng.* **2011**, *8*, 064001. [CrossRef] [PubMed]

26. Stulen, F.B.; De Luca, C.J. Frequency parameters of the myoelectric signal as a measure of muscle conduction velocity. *IEEE Trans. Biomed. Eng.* **1981**, *7*, 515–523. [CrossRef]

27. Merletti, R.; Conte, L.R.L. Surface EMG signal processing during isometric contractions. *J. Electromyogr. Kines.* **1997**, *7*, 241–250. [CrossRef]

28. Sheppard, J.M.; Young, W.B.; Doyle, T.L.; Sheppard, T.A.; Newton, R.U. An evaluation of a new test of reactive agility and its relationship to sprint speed and change of direction speed. *J. Sci. Med. Sport.* **2006**, *9*, 342–349. [CrossRef]

29. Farrow, D.; Young, W.; Bruce, L. The development of a test of reactive agility for netball: A new methodology. *J. Sci. Med. Sport.* **2005**, *8*, 52–60. [CrossRef]

30. Potteiger, J.A.; Lockwood, R.H.; Haub, M.D.; Dolezal, B.A.; Alumzaini, K.S.; Schroeder, J.M.; Zebas, C.J. Muscle power and fiber characteristic following 8 weeks of plyometric training. *J. Strength Cond. Res.* **1999**, *13*, 275–279.

Appl. Sci. **2020**, *10*, 4535

31. Arampatzis, A.; Stafilidis, S.; Morey-Klapsing, G.; Brüggemann, G.P. Interaction of the human body and surfaces of different stiffness during drop jumps. *Med. Sci. Sports Exerc.* **2004**, *36*, 451–459. [CrossRef]
32. Lesinski, M.; Prieske, O.; Demps, M.; Granacher, U. Effects of fatigue and surface instability on neuromuscular performance during jumping. *Scand J. Med Sci Spor.* **2016**, *26*, 1140–1150. [CrossRef] [PubMed]
33. Avela, J.; Komi, P.V.; Santos, P.M. Effects of differently induced stretch loads on neuromuscular control in drop jump exercise. *Eur. J. Appl. Physiol. Occup. Physiol.* **1996**, *72*, 553–562. [CrossRef] [PubMed]

Article

Can We Rely on Flight Time to Measure Jumping Performance or Neuromuscular Fatigue-Overload in Professional Female Soccer Players?

Estrella Armada-Cortés [1,2], Javier Peláez Barrajón [1], José Antonio Benítez-Muñoz [3], Enrique Navarro [1] and Alejandro F. San Juan [1,*]

[1] Department of Health and Human Performance, Sport Biomechanics Laboratory, Facultad de Ciencias Actividad Física y Deporte—INEF, Universidad Politécnica de Madrid, 28040 Madrid, Spain; cortesarmadaestrella@gmail.com (E.A.-C.); javi.pelaezb@gmail.com (J.P.B.); enrique.navarro@upm.es (E.N.)

[2] Department of Physical Activity and Sports Science, Faculty of Sport Sciences, University of Castilla-La Mancha, 45071 Toledo, Spain

[3] Department of Health and Human Performance, LFE Research Group, Facultad de Ciencias Actividad Física y Deporte—INEF, Universidad Politécnica de Madrid, 28040 Madrid, Spain; jbenitez020@gmail.com

* Correspondence: alejandro.sanjuan@upm.es

Received: 27 May 2020; Accepted: 24 June 2020; Published: 27 June 2020

Featured Application: It is recommended that high-level sportswomen and men should be assessed with the force platform through the take-off velocity method in a vertical jump as gold standard technology to ensure correct performance and/or fatigue-overload control during the sport season.

Abstract: The main purpose of this study was to compare the validity of the take-off velocity method (TOV) measured with a force platform (FP) (gold standard) versus the flight time method (FT) in a vertical jump to measure jumping performance or neuromuscular fatigue-overload in professional female football players. For this purpose, we used a FP and a validated smartphone application (APP). A total of eight healthy professional female football players (aged 27.25 ± 6.48 years) participated in this study. All performed three valid trials of a countermovement jump and squat jump and were measured at the same time with the APP and the FP. The results show that there is a lack of validity and reliability between jump height (JH) calculated through the TOV method with the FP and the FT method with the FP ($r = 0.028$, $p > 0.84$, intraclass correlation coefficient (ICC) = -0.026) and between the JH measured with the FP through the TOV method and the APP with the FT method ($r = 0.116$, $p > 0.43$, ICC = -0.094 (-0.314–0.157)). A significant difference between the JH measured through the TOV with the FP versus the APP ($p < 0.05$), and a trend between the JH obtained with the FP through the TOV and the FT ($p = 0.052$) is also shown. Finally, the JH with the FP through the FT and the APP did not differ ($p > 0.05$). The eta-squared of the one-way ANOVA was $\eta 2 = 0.085$. It seems that only the TOV measured with a FP could guarantee the accuracy of the jump test in SJ+CMJ and SJ, so it is recommended that high-level sportswomen and men should be assessed with the FP through TOV as gold standard technology to ensure correct performance and/or fatigue-overload control during the sport season.

Keywords: vertical jump; flight time; take-off velocity; fatigue; muscle overload; performance; force platform; smartphone application; APP; female soccer

---

1. Introduction

The vertical jump (VJ) is one of the most widely used performance tests. Its popularity is due both to its simplicity and effectiveness [1], and to the similarity between its movement patterns and sports, explosiveness, speed, and intensity [2,3]. Moreover, it is a useful tool to assess physical fitness in a healthy population (i.e., children, adults, and elderly people) [4–6], and to control an adequate musculoskeletal injury recovery [7]. There are different jump protocols; two of the most frequently used are: (1) The squat jump (SJ), which provides a basic leg strength parameter and is described by some authors as pure concentric [3]; (2) the counter movement jump (CMJ) adds an elastic-reactive component of the subject. It is estimated that the CMJ value can be up to 25% higher than the SJ in athletes [8]. Their performance has been used to: (1) monitor the positive effects of strength, plyometric, resistance, and speed training; and (2) control mechanical and neuromuscular fatigue status in individual and team sports [9]. Several researchers have found that CMJ performance is an interesting objective marker of fatigue and overcompensation for athlete performance [9], being one of the factors related to the high incidence of injuries (e.g., muscle overload) in the lower limb muscles [10,11]. Thus, a relationship has been observed between height loss in the CMJ and metabolic markers such as lactate or ammonium in the sprint [12], and after a Wingate test [13]. This suggests that through decreases in the CMJ's mechanical variables such as jump height (JH), it should be possible to estimate the metabolic stress, neuromuscular fatigue, and overload of the subject [5,12,14].

Specific performance factors for both the CMJ and SJ include a number of different kinematic and kinetic variables. Some of these variables are more sensitive than others to determine an athlete's neuromuscular status and may depends of the age and level of the subjects: JH, power, velocity and force (peak, relative peak, and mean of the last three), rate of force development (RFD), and mean impulse [9]. In addition, it has been proven that it is better to average the number of jumps executed than to choose the one that obtained the maximum height [9].

To obtain the height value of the jump we can use two variables: the take-off velocity (TOV) or the flight time (FT) [15,16]:

1. The TOV is the vertical velocity of the center of mass at takeoff. The height value of the jump will be the result of velocity at takeoff squared divided by twice the gravity:

$$\left(h = \frac{V^2}{2g} \right) \tag{1}$$

2. The FT is the time period between takeoff and land contact (i.e., the subject is on the air, without land contact). The height value of the jump will be the result of the gravity value multiplied by FT squared, divided by eight:

$$\left(h = \frac{gt^2}{8} \right) \tag{2}$$

There are many tools used to evaluate the performance of the jump (e.g., force platforms (FP), contact mats, video analysis through a smartphone applications (APPs), accelerometers, infrared systems, *Vertec* Vertical Jump Measuring Device). Most devices use the FT to calculate the JH [17]. However, to measure it in an accurate manner, the height of the center of mass at takeoff and landing has to be the same. Any difference between the athlete's takeoff and landing center of mass position (e.g., different joint position of ankle, knee, or hip) could increase or decrease the FT, and then change the result of the JH estimation [15,17]. Considering these technical implications, TOV has been suggested as the more suitable method to evaluate the JH [18]. If the TOV can be measured easily by a FP, then it is considered the gold standard for measuring the height of the VJ [19,20]. A FP can measure and use either the FT or the TOV methodology, the latter being the most accurate method for determining the height of the VJ [19].

The need to quantitatively evaluate the performance of the athlete's JH has promoted the emergence of APPs for this purpose. Some authors have reported that these APPs have a high reliability and accuracy in their measures [21,22], while others qualify their results as equivocal, so it is not yet clear what is the ideal methodology to use [23].

Although the VJ and its performance can be extrapolated to any sport level, the greater the level the subjects have, the more correlation exists in its variables, so it is in high performance sport where it has more utility [24]. Moreover, the accuracy of the results is essential from a competitive point of view in high-level sports, for example in elite-cyclists improvements of around 0.6% are sufficient to make a difference [25].

Therefore, the main purpose of this study was to compare the validity of the TOV method using a FP versus the FT method using a FP and a smartphone application to measure jumping performance or neuromuscular fatigue-overload in professional female football players.

2. Materials and Methods

2.1. Participant Selection: Inclusion and Exclusion Criteria

A total of eight healthy professional female football players of the First and Second Spanish Football Divisions (Liga Iberdrola and Reto Iberbrola, respectively) (aged 27.25 ± 6.48 years; body mass 56.73 ± 4.86 kg; height 1.61 ± 0.06 m. Values expressed as mean ± standard deviation) participated in this study.

Inclusion criteria were: (1) playing in a female professional soccer league; (2) not having suffered a musculoskeletal injury one year prior to the date of the protocol (i.e., checked through a previous exclusion questionnaire); (3) not presenting any cardiovascular, musculoskeletal, and/or neurological disease, nor previous ones that could affect participation in the study. Exclusion criteria were: (1) aged younger than 18 years; (2) having consumed any narcotic and/or psychotropic agents or drugs during the test. We have selected this specific study population to homogenize the level of the sample and to deepen the knowledge of women's professional soccer.

At the study outset, participants were informed of the study protocol, schedule, and nature of the exercises and tests to be performed before signing an informed consent form. The study protocol adhered to the tenets of the Declaration of Helsinki and was approved by the Ethics Committee of the Technical University of Madrid (Madrid, Spain).

2.2. Experimental Design

This study consisted of a single evaluation session (see Figure 1) in a laboratory in the same time frame to avoid the detrimental performance effects associated with circadian rhythm [26]. Subjects were required to avoid physically demanding activity in the 24 h prior to the session.

Three valid trials of a CMJ and SJ were conducted with each subject, with 60 s recovery between trials, and three minutes between each jump test (see Figure 1). All jumps were measured at the same time with the mobile application and on the FP.

The sessions began with a 10-min general warm up consisting of continuous running, specific running, joint mobility, and ballistic stretching exercises, followed by a specific pre-test warm-up where participants performed five SJs and CMJs with 30 s between each jump. After three minutes of rest, the participants started the jumping test.

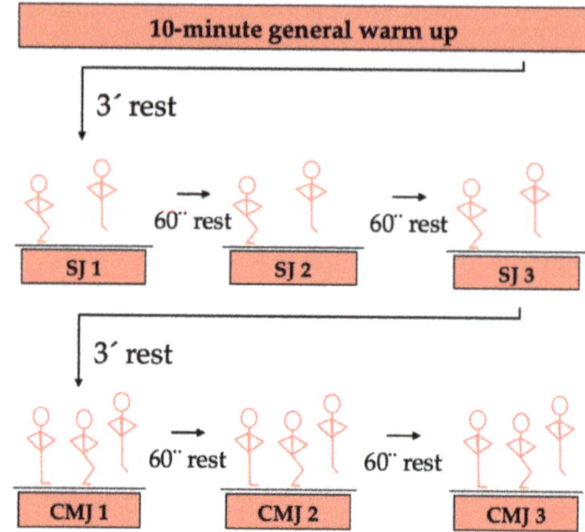

Figure 1. Experimental design. CMJ = countermovement jump test; SJ = squat jump.

2.3. Vertical Jump Test

The VJs included in the present study were the CMJ and SJ, which have been previously demonstrated to be reliable measures (intraclass correlation coefficient (ICC) = 0.97 and 0.96, respectively) [27]. All participants were completely familiarized with both techniques (i.e., they realized these tests during the soccer season). Moreover, they practice before the test during the specific warm-up (see warm-up description above). Participants were always instructed to jump as high as possible, keeping their hands on their hips and that their legs should remain straight during flight, making contact with the ground with the tips of their feet and knees extended [28]. In both modalities, the aim was to reach the maximum height by means of the jump; nevertheless, each modality of jump consists of different characteristics [29].

The CMJ (Figure 2a) initial position consists of a static standing position with hands on the hips. From this position, a continuous and fast triple hip, knee, and ankle flexion movement is executed until reaching ≈ 90° of knee flexion, followed by the triple extension of the same joints in a fluid, fast, and continuous way [29]. In this type of VJ, there is a stretching-shortening cycle (SSC), which takes place during the consecutive eccentric, isometric, and concentric phases [29].

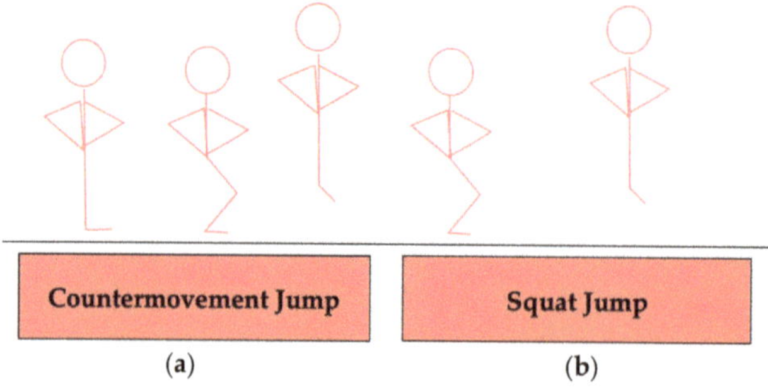

Figure 2. (a) Countermovement Jump; (b) Squat Jump.

The SJ (Figure 2b) only presents the concentric component. The initial position consists of maintaining the hips and knees flexed at ≈ 90° for approximately four seconds to avoid countermovement and the elastic component. From this position, the concentric phase of the jump with an explosive extension of the lower limb joints is performed. All participants were asked to jump as high as possible and without performing a SSC [29]. The SJ forces before takeoff were checked on the FP to ensure that the participant did not perform countermovement.

For both jumps, participants were asked to take off and land at the same place to avoid lateral or horizontal displacement. Only successful trials were considered. Participants were asked to repeat the trial if a jump was incorrectly performed.

2.4. Instruments

2.4.1. Force Platform

Two 600 mm × 400 mm piezoelectric platforms (Type 9286AA; Kistler Instruments AG, Winterthur, Switzerland) mounted together on the floor according to the manufacturer's instructions, were used in this experiment. The force sensors of these platforms were constituted by piezoelectric transducers 5 kN, total max 20 kN. Data were recorded at a sampling frequency of 1000 Hz. The FPs were connected to a portable computer with specific software, the Measurement, Analysis and Reporting Software "Kistler MARS" (Kistler Instruments AG, Winterthur, Switzerland).

2.4.2. Smartphone Application

The validated smartphone application used to measure the performance of the jump was My Jump2 (APP) [30] installed on a mobile phone (iPhone 6s Apple, Cupertino, CA, USA) at a sampling rate of 240 Hz. My Jump was designed for analyzing VJ measuring the time (in ms) between two frames selected by the user and subsequently to calculate the JH using the equation based on FT. The position of the camera was always at ~1.5 m from the middle line of the subject so that it was constantly aligned with the reference joint points.

After the recording, one researcher manually selected the frames in which the subject performed exactly the moments of take-off and landing. A second person was consulted in case of doubt when analyzing the jumps with the APP. The JH was calculated based on the FT between the selected take-off and landing frame, as previously described [30].

2.5. Statistical Analysis

The normality and homogeneity of the data were analyzed using the Shapiro–Wilk and Levene tests, respectively. The normality and homogeneity of the dependent variables was confirmed ($p > 0.05$); all of the data are provided as their means and standard deviations.

Various statistical analyses were used to prove the APP validity and reliability in comparison with the FP using FT and TOV to calculate JH. The Pearson correlation coefficient (r) was used to calculate the concurrent validity between the APP and the FP using different measurement methods.

To expound the magnitude of the relationship between height measurements by the APP and the FP, Cohen's convention was used. To measure reliability between the APP and the FP, the coefficient of variation (CV), an absolute agreement Intraclass Correlation Coefficient (ICC), and Cronbach's Alpha were used. The CV was used to observe the uniformity of the values with respect to the mean. A Bland–Altman plot was created to graphically represent the agreement between the APP's measured heights and FP FT measured heights.

The mean differences between the measurements for each jump (SJ and CMJ) and for both jumps together (CMJ + SJ) were calculated through a one-way ANOVA, and Tukey's post-hoc test was used to analyze pairwise comparisons between means. The standard error of estimate (SEE) was used to show the typical error in measurement. Significance was set at $p < 0.05$. All statistical analyses were performed using the software package SPSS® version 25 (IBM Co., USA).

3. Results

The coefficients of variation values were very small for the CMJs with the FP through the TOV, the FP through the FT and the APP (2.82%, 3.56%, and 3.55%, respectively), and for the SJ (12.44%, 5.24%, and 4.88%, respectively).

The Pearson's correlation showed a poor relationship and a small reliability between the JH measured through the FT and the TOV, both measured with the FP ($r = 0.028$, $p > 0.84$, ICC $= -0.026$, 95% CI $= -0.276$–0.24) (Figure 3), and between the JH measured with the FP through the TOV and the APP ($r = -0.116$, $p > 0.43$, ICC $= -0.094$, 95% CI $= -0.314$–0.157) (Figure 4).

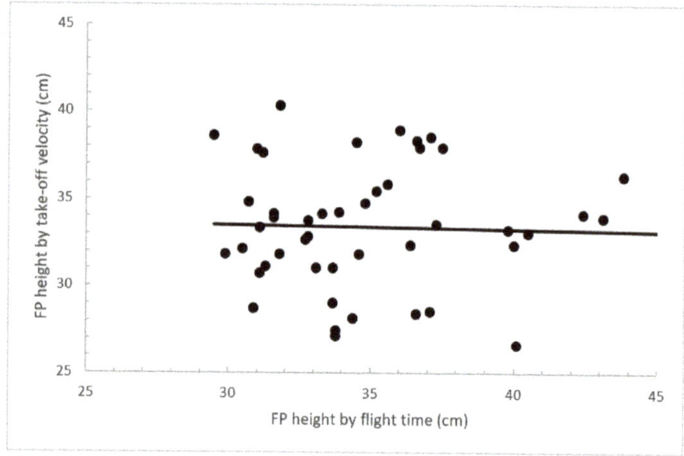

Figure 3. Pearson's correlation between jump heights measured by the FP using take-off velocity and the FP using flight time; FP, force platform; graph made by the authors.

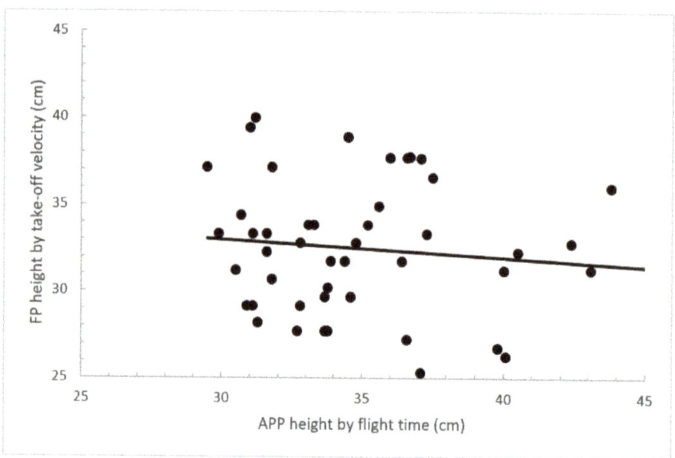

Figure 4. Pearson's correlation between jump heights measured by the FP using take-off velocity and the APP using flight time; FP, force platform; APP, My Jump App; graph made by the authors.

On the other hand, a high and significant relationship between the JH measured with the FP through the FT and the APP ($r = 0.872$, $p < 0.01$) was found (Figure 5). Moreover, there was a very high agreement between these two variables as revealed by the ICC $= 0.843$ (95% CI $= 0.681$–0.919) and Bland–Altman plots (Figure 6).

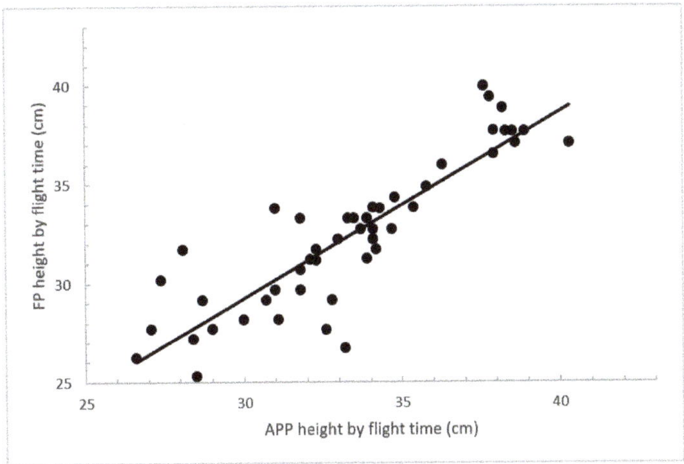

Figure 5. Pearson's correlation between jump heights measured by the FP using flight time and the APP using flight time; FP, force platform; APP, My Jump App; graph made by the authors.

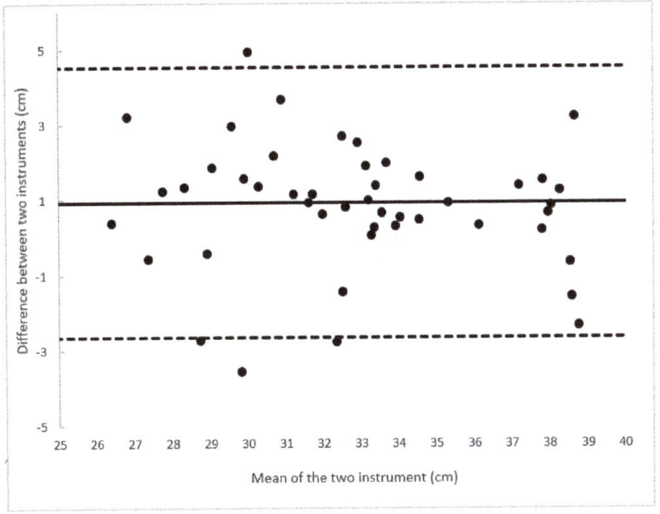

Figure 6. Bland–Altman plot between the FP using flight time and the APP using flight time to measure jump height. The central line represents the systematic bias between instruments, while the upper and the lower lines represent ± 1.96 SD; FP, force platform; APP, My Jump App; graph made by the authors.

The one-way ANOVA shows a significant difference between the JH measured with the FP through the TOV versus the APP ($p = 0.002$), and a trend versus the FP through the FT ($p = 0.052$). Finally, the JH analyzed with the FP through the FT and the APP did not differ ($p = 0.457$) (Table 1). The eta squared of the one-way ANOVA was η2 = 0.085.

Specifically, in SJ, the one-way ANOVA shows a significant difference in JH measured with the FP through the TOV versus the FP and the APP both through the FT ($p < 0.01$). Finally, the JH analyzed with the FP through FT and the APP did not differ ($p = 0.826$) (Table 2). The eta squared of the one-way ANOVA used to compare SJ was η2 = 0.374.

Table 1. Values of the different methods for measuring jump height combining CMJ and SJ.

	Mean ± SD		Mean ± SD	*p* Value
FP–TOV	35.21 ± 4.22	FP–FT	33.36 ± 3.48	0.052
FP–TOV	35.21 ± 4.22	APP	32.43 ± 3.77	0.002 *
FP–FT	33.36 ± 3.48	APP	32.43 ± 3.77	0.457

Data are presented as mean ± standard deviation. FP–TOV: Force Platform–Take-Off Velocity; FP–FT: Force Platform–Flight Time; APP: My Jump Application (flight time). * Significant difference.

Table 2. Values of the different methods for measuring SJ jump height.

	Mean ± SD		Mean ± SD	*p* Value
FP–TOV	37.23 ± 4.68	FP–FT	31.53 ± 3.09	0.000 *
FP–TOV	37.23 ± 4.68	APP	30.89 ± 3.33	0.000 *
FP–FT	31.53 ± 3.09	APP	30.89 ± 3.33	0.826

Data are presented as mean ± standard deviation. FP–TOV: Force Platform–Take-Off Velocity; FP–FT: Force Platform–Flight Time; APP: My Jump Application (flight time); SJ: squat jump. * Significant difference.

Lastly, in CMJ, the one-way ANOVA did not show significant differences between JH measured by the different methods ($p = 0.07$, η2 = 0.073) (Table 3).

Table 3. Values of the different methods for measuring CMJ jump height.

	Mean ± SD		Mean ± SD	*p* Value
FP–TOV	33.18 ± 2.44	FP–FT	35.19 ± 2.87	0.060
FP–TOV	33.18 ± 2.44	APP	33.96 ± 3.60	0.644
FP–FT	35.19 ± 2.87	APP	33.96 ± 3.60	0.336

Data are presented as mean ± standard deviation. FP–TOV: Force Platform–Take-Off Velocity; FP–FT: Force Platform–Flight Time; APP: My Jump Application (flight time); CMJ; countermovement jump.

4. Discussion

The main purpose of this study was to compare the validity of the TOV method using a FP versus the FT method using a FP and a smartphone application to measure jumping performance or neuromuscular fatigue-overload in professional female soccer players.

The major findings were the lack of reliability and validity between JH measured with the FP through the TOV and JH measured through the FT either by the FP ($r = 0.028$, $p > 0.84$, ICC = −0.026 (−0.276–0.24)) or the APP ($r = 0.116$, $p > 0.43$, ICC = −0.094 (−0.314–0.157)). To our knowledge, this is the first study that has researched the JH of SJ and CMJ together (SJ + CMJ) with these technologies: FP trough the TOV and trough FT, and APP through FT. Two previous studies [20,31] have analyzed the SJ, CMJ and CMJ with arms (CMJA), together comparing JH with FP vs. contact mat (both through FT method) [20], and with FP (TOV and FT methods) vs. contact mat (FT method) [31]. This latter study [31] concluded that the TOV must be used instead of the FT as a more valid and accurate way to calculate the VJ performance. It provides a more sensitive approach to scientifically establishing jump determinants, greater sensitivity in analyzing changes in performance factors after training (i.e., overcompensation, overload), and greater accuracy in monitoring performance within the session (i.e., fatigue) [18,19]. Further, the mean comparison shows a significant difference between the JH measured with the FP through the TOV versus the APP ($p = 0.002$), and a trend versus the FP through the FT ($p = 0.052$). To our knowledge, there is only one previous study that researched the differences between the FP through the TOV versus a smartphone application (FT method) [19], and it showed near perfect correlations (ICC = 0.996; $p < 0.001$). These differences with our results may be explain by some reasons: 1) In our study the subjects performed three different CMJs and SJs, while in the previous study the subjects performed five CMJs [19]; (2) The distance between the researcher and the subject was of 1 m in the previous study [19], while in our study it was of 1.5 m; (3) In addition, rest between jumps was not specified in the previous study [19], so it is not possible to know if fatigue could have affected their results through the FT method (i.e., Different joint position between take-off and landing), whilst our subjects rested 60 s between trials and three minutes between each jump test. Despite these methodological differences, their results comparing the FP through the TOV versus a smartphone application (FT method) regarding only CMJ did not differ with our CMJ results. Based on the present study and the literature [19,20], it seems that the flight time method, independent of the technology used (e.g., FP, contact mat, smartphone application), is an imprecise means of assessment.

On the other hand, there is a high correlation and good reliability between the JH SJ+CMJ measured through the FT with the FP and the APP ($r = 0.872$, $p < 0.01$, ICC = 0.843, CV = 4.88%). These results are in accordance with other studies that compare the SJ+CMJ+CMJA height using a FP and a contact mat (both through FT) ($r = 0.99$, $p < 0.001$, ICC = 0.985, CV% = 2.7) [21]. Additionally, the mean comparison of the present study shows that the JH analyzed with the FP through the FT and the APP did not differ ($p = 0.457$). Therefore, the APP and the contact mat seem to equal the accuracy of the FP with the FT method. However, it is important to highlight the fact that some studies compare the My Jump APP based on FT, with other instruments also based on FT (i.e., contact mat or FP through the FT) [21,30,31], instead of comparing it with the FP through the TOV (gold standard) [16]. Therefore, all these technologies that use the FT method are not as precise as the TOV method. Bearing in mind that the most accurate and considered the gold standard method to measure the height of the jump is the FP based on the TOV [19,20], these are the results that should be trusted.

Analyzing specifically the different jumps, it was found in SJ significant differences between the JH measured with the FP through the TOV and the FP and the APP through the FT ($p < 0.01$ in both cases). However, we found no significant difference comparing the JH measured with the FP and the APP (both through the FT) ($p = 0.826$). Further, CMJ data showed no significant differences among the three methods (i.e., FP through the TOV versus the FP and the APP through the FT) ($p > 0.05$ in both cases). Our results are in line with the only one previous study that researched the differences between the FP through the TOV versus a smartphone application (FT method) [19], and it showed near perfect correlations (ICC = 0.996; $p < 0.001$). Results of the reliability and validity obtained in the present study comparing the CMJ height through the FT with the FP and the APP ($r = 0.872$, $p < 0.01$, ICC = 0.843 and CV = 4.88%) are in accordance with the first validation study of My Jump App [30] that compare the CMJ height using a FP through FT ($r = 0.995$, $p < 0.001$, ICC = 0.997, CV% = 3.5), and other study that compare the CMJ height using a contact mat and the My Jump App (both through FT) ($r = 0.999$, $p < 0.001$, ICC = 0.948, CV% = 10.096) [21]. Despite the differences between the sampling frequency of the devices (i.e., FP $\approx$1000 Hz and the smartphone camera $\approx$ 240 Hz), the height values obtained were very similar between the two methods, as reflected by the ICC.

The literature that evaluates the smartphone applications in this area has analyzed the CMJ as the only vertical jump option discarding the SJ [19,21]. However, with this study it has been demonstrated that the results of the CMJ cannot be extrapolated to the SJ. The height of the SJ and CMJ is defended by several authors as a method to detect fatigue and individualize load prescriptions, so the reduction of the height of the VJ can be interpreted as evidence of the deterioration of neuromuscular function (i.e., fatigue, neuromuscular overload) [12]. Its measurement could provide a relatively simple and accessible tool for quantifying the degree of neuromuscular, mechanical, and metabolic fatigue [5,12,32]. This fatigue can remain for up to 48–72 h after a high-intensity session [33]. It could therefore provide useful feedback to coaches to determine the impact of training loads on player recovery and performance [12]. The neuromuscular factor is considered the most influential factor in training and is also an intrinsic risk factor of an athlete's injury [34]. The lack of strength and poor coordination capacity are part of it [35]. Several studies have observed that after a fatiguing exercise bout and during the landing phase of different jump types, there is a change in the neuromuscular control strategies used by these subjects [36,37]. These motor control alterations are associated with different injury risk factors, such as reduced knee and hip flexion, increased knee valgus, increased ground reaction force, and greater stabilization time [37]. In high-level sports, measurement accuracy is critical in the assessment of performance and in the control and assimilation of training loads to avoid muscle–tendon overloads [25]. In addition, some authors recommend the evaluation of jumping performance, especially at high and elite levels [24]. In this sense and referring to the exclusive sample of this study, as they are professional female soccer players, these small variations described above should be considered and limited as far as possible. In our study, the differences in JH between the FP through the TOV versus through the FT was 5.23% (1.84 cm) and the differences between the FP through the TOV versus the APP was 7.90% (2.78 cm). The only previous study that researched the differences between the FP (TOV method) versus a smartphone application (FT method) [19], found a small overestimated JH obtained from the APP of 0.78% compared to the FP based on TOV [19]. This small difference (0.78%) is more than ten times lower than the differences reported by us (7.90%), more than six times the difference between FP–TOV vs. FT methods (5.23%), and more than three times lower than the differences between the FP (FT method) and the APP (2.8%). This latter difference reported in the present study between the FP (FT method) and the APP agree with other studies comparing the My Jump application and a FP (FT method) [30], and an infrared contact mat and a FP (both through the FT method) [19]. More research regarding the issue is needed to clarify if smartphone applications may replace a FP (TOV method) in professional environments where accuracy is a key factor (i.e., high-level sport, medical clinic).

Furthermore, we observed that the difference of JH between the FP and the APP both through the FT was smaller, only 2.8% (0.94 cm). These latter results are in accordance with previous studies that observed differences in JH between an infrared contact mat and a FP, both through the FT of $\approx$ 2.5% (1.06 cm) [19], and between the My Jump application and a FP both through the FT (1.1 ± 0.5 cm) [30]. These concordances are produced because they use the FT method (i.e., less accurate) to compare different technologies. When the FT method is used with any technology, even the FP, there are many possibilities to produce measurement errors because the takeoff and the landing positions are different (e.g., different degree of lower limbs joint/s flexion) [38]. Then, it seems recommendable that high-level sportswomen and men should be assessed with this gold standard technology and method (i.e., FP and TOV method) to ensure correct performance and/or overload control during the sport season.

5. Conclusions

The results of the present study showed that in female professional soccer players, there is a lack of validity and reliability between JH in SJ + CMJ and SJ calculated with the TOV method versus the FT method. It seems that only the TOV measured with a FP could guarantee the accuracy of the jump test.

In order to be able to generalize our conclusions to different athlete populations, future research is needed, aiming at comparing the validity and reliability between JH of different kind of jumps calculated with the TOV method versus the FT method on high-level and recreationally trained athletes from different sport modalities and both sexes.

Author Contributions: Conceptualization, E.A.-C., J.A.B.-M. and A.F.S.J.; methodology, E.A.-C., J.P.B., J.A.B.-M., E.N. and A.F.S.J.; validation, E.A.-C. and A.F.S.J.; formal analysis, E.A.-C., J.P.B. and A.F.S.J.; investigation, E.A.-C., J.P.B., J.A.B.-M. and A.F.S.J.; resources, E.A.-C., E.N. and A.F.S.J.; supervision, E.N. and A.F.S.J.; project administration, E.N. and A.F.S.J. All authors have read and agreed to the published version of the manuscript.

Funding: This research received no external funding.

Conflicts of Interest: The authors declare no conflicts of interest.

References

1. Cronin, J.; Gill, N.D. Fatigue Monitoring in High Performance Sport: A Survey of Current Trends. *J. Aust. Strength Cond.* **2012**, *20*, 12–23.
2. Yamauchi, J.; Ishii, N. Relations Between Force-Velocity Characteristics of the Knee-Hip Extension Movement and Vertical Jump Performance. *J. Strength Cond. Res.* **2007**, *21*, 703–709. [PubMed]

3. Nagahara, R.; Naito, H.; Miyashiro, K.; Morin, J.B.; Zushi, K. Traditional and ankle-specific vertical jumps as strength-power indicators for maximal sprint acceleration. *J. Sports Med. Phys. Fitness* **2014**, *54*, 691–699. [PubMed]

4. Ramírez-Campillo, R.; Castillo, A.; de la Fuente, C.I.; Campos-Jara, C.; Andrade, D.C.; Álvarez, C.; Martinez, C.; Castro-Sepulveda, M.; Pereira, A.; Marques, M.C.; et al. High-speed resistance training is more effective than low-speed resistance training to increase functional capacity and muscle performance in older women. *Exp. Gerontol.* **2014**, *58*, 51–57. [CrossRef] [PubMed]

5. Maté-Muñoz, J.L.; Lougedo, J.H.; Garnacho-Castaño, M.V.; Veiga-Herreros, P.; Lozano-Estevan, M.D.C.; García-Fernández, P.; de Jesus, F.; Guodemar-Perez, J.; San Juan, A.F.; Dominguez, R. Effects of β-alanine supplementation during a 5-week strength training program: A randomized, controlled study. *J. Int. Soc. Sports Nutr.* **2018**, *15*, 19. [CrossRef]

6. Fernandez-Santos, J.R.; Ruiz, J.R.; Cohen, D.D.; Gonzalez-Montesinos, J.L.; Castro-Piñero, J. Reliability and Validity of Test to Assess Lower-Body Muscular Power in Children. *J. Strength Cond. Res.* **2015**, *29*, 2277–2285. [CrossRef]

7. Setuain, I.; Millor, N.; Alfaro, J.; Gorostiaga, E.; Izquierdo, M. Jumping performance differences among elite professional handball players with or without previous ACL reconstruction. *J. Sports Med. Phys. Fitness* **2015**, *55*, 1184–1192.

8. Sánchez, W.G.; Gómez, D.A.; Quiceno, H.B.; Alzate, S.J. Análisis comparativo intrasujeto en salto vertical 2d: Squat jump y counter-movement jump Resumen Introducción. *VIREF Rev. Educ. Fís.* **2016**, *5*, 1–17.

9. Gustavo, J.; Cronin, J.; Mezêncio, B.; Travis, D.; Mcguigan, M.; Tricoli, V.; Carlos Amadio, A.; Cerca Serrao, J. The countermovement jump to monitor neuromuscular status: A meta-analysis. *J. Sci. Med. Sport* **2017**, *20*, 397–402.

10. Greig, M.; Siegler, J.C. Soccer-specific fatigue and eccentric hamstrings muscle strength. *J. Athl. Train.* **2009**, *44*, 180–184. [CrossRef]

11. Science, E.; Kingdom, U.; Activity, P.; Kingdom, U. Effect of Timing of Eccentric Hamstring Strengthening Exercises During Soccer Training: Implications for Muscle Fatigability. *J. Strength Cond. Res.* **2009**, *23*, 1077–1083.

12. Jiménez-Reyes, P.; Pareja-Blanco, F.; Cuadrado-Peñafiel, V.; Ortega-Becerra, M.; Párraga, J.; González-Badillo, J.J. Jump height loss as an indicator of fatigue during sprint training. *J. Sports Sci.* **2018**, *37*, 414. [CrossRef] [PubMed]

13. Reeve, T.C.; Tyler, C.J. The validity of the smartjump contact mat. *J. Strength Cond. Res.* **2013**, *27*, 1597–1601. [CrossRef] [PubMed]

14. Aragón-Vargas, L. Evaluation of Four Vertical Jump Tests: Methodology, Reliability, Validity and accuracy. *Meas. Phys. Educ. Exerc. Sci.* **2000**, *4*, 215–228. [CrossRef]

15. Buckthorpe, M.; Morris, J.; Folland, J.P. Validity of vertical jump measurement devices. *J. Sports Sci.* **2012**, *30*, 63–69. [CrossRef] [PubMed]

16. Moir, G.L. Three different methods of calculating vertical jump height from force platform data in men and women. *Meas. Phys. Educ. Exerc. Sci.* **2008**, *12*, 207–218. [CrossRef]

17. Ferro, A.; Floría, P.; Villacieros, J.; Muñoz-lópez, A. Maximum velocity during loaded countermovement jumps obtained with an accelerometer, linear encoder and force platform: A comparison of technologies. *J. Biomech.* **2019**, *95*, 109281. [CrossRef]

18. Jiménez-Reyes, P.; Pareja-Blanco, F.; Rodríguez-Rosell, D.; Marques, M.C.; González-Badillo, J.J. Maximal velocity as a discriminating factor in the performance of loaded squat jumps. *Int. J. Sports Physiol. Perform.* **2016**, *11*, 227–234. [CrossRef]

19. Carlos-Vivas, J.; Martin-Martinez, J.P.; Hernandez-Mocholi, M.A.; Perez-Gomez, J. Validation of the iphone app using the force platform to estimate vertical jump height. *J. Sports Med. Phys. Fitness* **2018**, *58*, 227–232.

20. Glatthorn, J.F.; Gouge, S.; Nussbaumer, S.; Stauffacher, S.; Impellizzeri, F.M.; Maffiuletti, N.A. Validity and Reliability of Optojump Photoelectric Cells for Estimating Vertical Jump Height. *J. Strength Cond. Res.* **2011**, *25*, 556–560. [CrossRef]

21. Cruvinel-Cabral, R.M.; Oliveira-Silva, I.; Medeiros, A.R.; Claudino, J.G.; Jiménez-Reyes, P.; Boullosa, D.A. The validity and reliability of the "my Jump App" for measuring jump height of the elderly. *PeerJ* **2018**, *6*, e5804. [CrossRef] [PubMed]

22. Brooks, E.R.; Benson, A.C.; Bruce, L.M. Novel technologies found to be valid and reliable for the measurement of vertical jump height with jump-and-reach testing. *J. Strength Cond. Res.* **2018**, *32*, 2038–2845. [CrossRef] [PubMed]

23. Leard, J.S.; Cirillo, M.A.; Katsnelson, E.; Kimiatek, D.A.; Miller, T.W.; Trebincevic, K.; Garbalosa, J.C. Validity of Two Alternative Systems for Measuring Vertical Jump Height. *J. Strength Cond. Res.* **2007**, *21*, 1296–1299. [PubMed]

24. Jiménez-Reyes, P.; Samozino, P.; García-Ramos, A.; Cuadrado-Peñafiel, V.; Brughelli, M.; Morin, J.-B. Relationship between vertical and horizontal force-velocity-power profiles in various sports and levels of practice. *PeerJ* **2018**, *6*, e5937. [CrossRef]

25. Paton, C.D.; Hopkins, W.G. Variation in performance of elite cyclists from race to race. *Eur. J. Sport Sci.* **2006**, *6*, 25–31. [CrossRef]

26. Drust, B.; Waterhouse, J.; Atkinson, G.; Edwards, B.; Reilly, T. *Circadian Rhythms in Sports Performance—An Update*; Chronobiology International; Taylor and Francis Inc.: Abingdon, UK, 2005; Volume 22, pp. 21–44.

27. Markovic, G.; Dizdar, D.; Jukic, I.; Cardinale, M. Reliability and Factorial Validity of Squat and Countermovement Jump Tests. *J. Strength Cond. Res.* **2004**, *18*, 551–555.

28. Jiménez-Reyes, P.; Cuadrado-Peñafiel, V.; González-Badillo, J.; Alfonso El Sabio, U.X.; Jiménez Reyes Universidad Alfonso el Sabio, P.X. Analysis of Variables Measured in Vertical Jump Related to Athletic Performance and its Application to Training. *Cult. Cienc. Deport.* **2011**, *6*, 113–119.

29. Bosco, C. *La Fuerza Muscular. Aspectos Metodológicos*, 1st ed.; INDE, Ed.; INO Reproducciones, S.A.: Barcelona, Spain, 2000; pp. 177–187.

30. Balsalobre-Fernández, C.; Glaister, M.; Lockey, R.A. The validity and reliability of an iPhone app for measuring vertical jump performance. *J. Sports Sci.* **2015**, *33*, 1574–1579. [CrossRef]

31. Gallardo-Fuentes, F.; Gallardo-Fuentes, J.; Ramírez-Campillo, R.; Balsalobre-Fernández, C.; Martínez, C.; Caniuqueo, A.; Canas, R.; Banzer, W.; Loturco, I.; Nakamura, F.; et al. Intersession and intrasession reliability and validity of the my jump app for measuring different jump actions in trained male and female athletes. *J. Strength Cond. Res.* **2016**, *30*, 2049–2056. [CrossRef]

32. San Juan, A.F.; López-Samanes, Á.; Jodra, P.; Valenzuela, P.L.; Rueda, J.; Veiga-Herreros, P.; Perez-Lopez, A.; Dominguez, R. Caffeine supplementation improves anaerobic performance and neuromuscular effciency and fatigue in Olympic-level boxers. *Nutrients* **2019**, *11*, 2120. [CrossRef]

33. Hellsten-westing, Y.; Norman, B.; Balsom, P.D.; Sjodin, B. Decreased resting levels of adenine nucleotides in human skeletal muscle after high-intensity training. *J. Appl. Phys.* **1993**, *74*, 2523–2528. [CrossRef] [PubMed]

34. Griffin, L.Y.; Albohm, M.J.; Arendt, E.A.; Bahr, R.; Beynnon, B.D.; DeMaio, M.; Dick, R.W.; Engebretsen, L.; Garrett, W.E.; Hannafin, J.A.; et al. Understanding and preventing noncontact anterior cruciate ligament injuries: A review of the Hunt Valley II Meeting, January 2005. *Am. J. Sports Med.* **2006**, *34*, 1512–1532. [CrossRef] [PubMed]

35. Fort Vanmeerhaeghe, A.; Romero Rodriguez, D. Neuromuscular risk factors of sports injury. In *Apunts Medicina de l'Esport*; Elsevier España, S.L.: Cataluña, Spain, 2013; Volume 48, pp. 109–120.

36. Borotikar, B.S.; Newcomer, R.; Koppes, R.; McLean, S.G. Combined effects of fatigue and decision making on female lower limb landing postures: Central and peripheral contributions to ACL injury risk. *Clin. Biomech.* **2008**, *23*, 81–92. [CrossRef] [PubMed]

37. McLean, S.G.; Felin, R.E.; Suedekum, N.; Calabrese, G.; Passerallo, A.; Joy, S. Impact of fatigue on gender-based high-risk landing strategies. *Med. Sci. Sports Exerc.* **2007**, *39*, 502–514. [CrossRef] [PubMed]

38. Kibele, A. Possibilities and Limitations in the Biomechanical Analysis of Countermovement Jumps: A Methodological Study. *J. Appl. Biomech.* **1998**, *14*, 105–117. [CrossRef]

MDPI

St. Alban-Anlage 66

4052 Basel

Switzerland

Tel. +41 61 683 77 34

Fax +41 61 302 89 18

www.mdpi.com

Applied Sciences Editorial Office

E-mail: applsci@mdpi.com

www.mdpi.com/journal/applsci